SCREEN TOURISM AND AFFECTIVE LANDSCAPES

This book explores ways in which screen-based storyworlds transfix, transform, and transport us imaginatively, physically, and virtually to the places they depict or film. Topics include fantasy quests in computer games, celebrity walking tours, dark tourism sites, Hobbiton as theme park, surf movies, and social gangs of Disneyland.

How physical, virtual, and imagined locations create a sense of place through their immediate experience or visitation is undergoing a revolution in technology, travel modes, and tourism behaviour. This edited collection explores the rapidly evolving field of screen tourism and the affective impact of landscape, with provocative questions and investigations of social groups, fan culture, new technology, and the wider changing trends in screen tourism. We provide critical examples of affective landscapes across a wide range of mediums (from the big screen to the small screen) and locations.

This book will appeal to students and scholars in film and tourism, as well as geography, design, media and communication studies, game studies, and digital humanities.

Erik Champion is Enterprise Fellow at University of South Australia, Emeritus Professor at Curtin University, Australia, and formerly UNESCO Chair and Visualisation Theme Leader of the Curtin Institute of Computation. His recent books are *Rethinking Virtual Places* (2021), *Organic Design in Twentieth-Century Nordic Architecture* (Routledge, 2019), *Critical Gaming: Interactive History and Virtual Heritage* (Routledge, 2016), and *Playing With the Past* (2011). He is the editor of *Virtual Heritage: A Guide* (2021), *The Phenomenology of Real and Virtual Places* (Routledge, 2019), and *Game Mods: Design, Theory and Criticism* (2012), and the co-editor of *Cultural Heritage Infrastructures in Digital Humanities* (Routledge, 2018).

Christina Lee is Senior Lecturer in English and Cultural Studies at Curtin University, Australia. She is the author of *Screening Generation X: The Politics and Popular Memory of Youth in Contemporary Cinema* (2010), and the editor of *Violating Time: History, Memory, and Nostalgia in Cinema* (2012) and *Spectral Spaces and Hauntings: The Affects of Absence* (Routledge, 2017). Her research interests include spaces of imagination and spectrality (including screen tourism sites), popular culture, and fandom.

Jane Stadler is Honorary Professor of Film and Media Studies at The University of Queensland in Australia. She led a collaborative Australian Research Council project on landscape and location in Australian cinema, literature, and theatre (2011–2014) and co-authored *Imagined Landscapes: Geovisualizing Australian Spatial Narratives* (2016). She is the author of *Pulling Focus: Intersubjective Experience, Narrative Film and Ethics* (2008) and the co-author of *Screen Media* (2009) and *Media and Society* (2016). Her recent research focuses on the audience's affective responses to cinema.

Robert Moses Peaslee is Associate Professor in the Department of Journalism and Creative Media Industries at Texas Tech University, USA. He is the co-editor of *The Joker: A Serious Study of the Clown Prince of Crime* (2015), *The Supervillain Reader* (2020), and *Web-Spinning Heroics: Critical Essays on the History and Meaning of Spider-Man* (2012). He also co-edited *Marvel Comics Into Film: Essays on Adaptations Since the 1940s* (2016). Robert studies relationships between media and place, as well as popular culture, adaptation, and fandom.

Routledge Cultural Heritage and Tourism Series

Series editor: Dallen J. Timothy
Arizona State University, USA

The Routledge Cultural Heritage and Tourism Series offers an interdisciplinary social science forum for original, innovative and cutting-edge research about all aspects of cultural heritage-based tourism. This series encourages new and theoretical perspectives and showcases ground-breaking work that reflects the dynamism and vibrancy of heritage, tourism and cultural studies. It aims to foster discussions about both tangible and intangible heritages, and all of their management, conservation, interpretation, political, conflict, consumption and identity challenges, opportunities and implications. This series interprets heritage broadly and caters to the needs of upper-level students, academic researchers, and policy makers.

Tourism and Development in the Himalaya
Social, Environmental and Economic Forces
Edited by Gyan Nyaupane and Dallen J. Timothy

Cultural Tourism and Cantonese Opera
Jian Ming Luo

Screen Tourism and Affective Landscapes
The Real, the Virtual, and the Cinematic
Edited by Erik Champion, Christina Lee, Jane Stadler and Robert Moses Peaslee

Cultural Heritage and Tourism in Africa
Edited by Dallen J. Timothy

For more information about this series, please visit: www.routledge.com/Routledge-Cultural-Heritage-and-Tourism-Series/book-series/RCHT

SCREEN TOURISM AND AFFECTIVE LANDSCAPES

The Real, the Virtual, and the Cinematic

Edited by Erik Champion, Christina Lee, Jane Stadler, and Robert Moses Peaslee

Routledge
Taylor & Francis Group
LONDON AND NEW YORK

Cover image: © Getty Images

First published 2023
by Routledge
4 Park Square, Milton Park, Abingdon, Oxon OX14 4RN

and by Routledge
605 Third Avenue, New York, NY 10158

Routledge is an imprint of the Taylor & Francis Group, an informa business

British Library Cataloguing-in-Publication Data
A catalogue record for this book is available from the British Library

ISBN: 978-1-032-35595-5 (hbk)
ISBN: 978-1-032-35596-2 (pbk)
ISBN: 978-1-003-32758-5 (ebk)

DOI: 10.4324/9781003327585

Typeset in Bembo
by Apex CoVantage, LLC

CONTENTS

LIST OF FIGURES

LIST OF CONTRIBUTORS

Erin Ashenhurst is a faculty member at the Wilson School of Design at Kwantlen Polytechnic University, Canada, teaching design history, popular culture, and contemporary issues. She has published in the edited books *The Fantasy of Reality: Critical Essays on The Real Housewives* (2015) and *Imaginations: A Journal of Cross-Cultural Image Studies.*

David Bailey assisted the UCF Blackstone LaunchPad and Center for Innovation and Entrepreneurship and directed research with the UCF Film Department and Themed Experience program. He currently studies digital media towards a PhD degree in texts and technology. He has theme park experience as a performer, creative director, and leader within Orlando-based theme parks.

Carissa Baker, is Assistant Professor of Theme Park and Attraction Management at the University of Central Florida. Her scholarship focuses primarily on narratives in the theme park space. Dr. Baker has researched and taught in China in addition to being a visiting scholar at Breda University of Applied Sciences in the Netherlands.

Gemma Blackwood is Media Lecturer at the University of Tasmania and a researcher of visual culture, focusing on screen media. She edited *Motion Pictures: Travel Ideals in Film* (2016), and is currently researching visual and filmic representations of ecological disaster and how this influences environmental communication strategies.

Kornelia Boczkowska is Assistant Professor in the Department of Studies in Culture, Faculty of English, Adam Mickiewicz University, Poznan, Poland. She is the author of two books and circa 40 other publications on independent, experimental, and documentary film. Her current research is funded by a National Science Center post-doctoral grant.

Ian Brodie, MNZM, is a film tourism consultant, award-winning photographer, and internationally successful author who has published 22 books to date. These include *The Lord of the Rings Location Guidebook* (2011) and *Cameras in Narnia* (2005). Ian was the former Media and Communications Manager of Hobbiton Movie Set Tours.

Erik Champion is Enterprise Fellow at University of South Australia, Emeritus Professor at Curtin University, Australia, Honorary Research Professor at ANU, and Honorary Research Fellow at UWA. His recent books include *Rethinking Virtual Places* (2021) and *Critical Gaming: Interactive History and Virtual Heritage* (Routledge, 2016).

Charlotte Crofts is Associate Professor of Filmmaking at the University of the West of England (UWE Bristol) and Founder and Director of the biennial Cary Grant Comes Home for the Weekend Festival. She teaches on the BA (Hons.) in filmmaking, and she researches and develops locative smartphone apps (including the award-winning Curzon Memories App [2012]).

Ray Eddy is Lecturer in the College of Business at the University of Central Florida. Prior to his teaching career, he was an actor and stunt performer who played Indiana Jones in the "Indiana Jones Epic Stunt Spectacular" at Walt Disney World and numerous roles in theme parks, commercials, television, and film.

Richard Ek is Professor of Human Geography at the Department of Geography, Media and Communication, Karlstad University, Sweden. His research spans regional development, planning, social media technologies, hedonistic tourism, and place marketing. Nowadays, his focus is mainly on introducing political-philosophical ideas in tourism studies.

Kenneth Fox is Principal Lecturer and School Director of Learning & Teaching in the School of Creative Arts & Industries at Canterbury Christ Church University, specializing in film landscapes. He recently published: "Paddy the Cope, Michael Powell and the story of the unmade film." Donegal Annual: Journal of the Donegal Historical Society. 2019, Vol. 71.

Lesley Hawkes is Associate Professor of Professional Communication in the School of Communication at Queensland University of Technology. Her areas of research interest are cultural tourism and spatial understanding in relation to literary texts and cultural studies. She co-edited *Popular Appeal: Books and Films in Contemporary Youth Culture* (2013).

Mia Larson is Professor of Business Administration at the Center for Service Research at Karlstad University, Sweden. Her publications cover subjects such as network collaboration, service innovation, event management, destination marketing and development, pop-culture tourism, social media, digital marketing, and sharing economy. Her current research focuses on the platform economy.

Christina Lee is Senior Lecturer in English and Cultural Studies at Curtin University, Australia. She wrote *Screening Generation X: The Politics and Popular Memory of Youth in Contemporary Cinema* (2010) and edited *Violating Time: History, Memory, and Nostalgia in Cinema* (2012) and *Spectral Spaces and Hauntings: The Affects of Absence* (Routledge, 2017).

Jane Lovell worked at the Royal Opera House Covent Garden and at Canterbury City Council. At Canterbury Christ Church University, Jane specializes in teaching heritage and creative Industry management. Her research interests and publications focus on tourism, heritage, authenticity, and places (magical spaces, film locations, and light installations).

Can-Seng Ooi is Professor of Cultural and Heritage Tourism at the University of Tasmania. As a sociologist, his academic career started at the National University of Singapore and obtained his PhD degree from Copenhagen Business School. He studies and theorizes tourist experiences, destination branding, and cultural tourism. His personal website is www.cansengooi.com.

Robert Moses Peaslee is Associate Professor at the Department of Journalism and Creative Media Industries, Texas Tech University, USA. He co-edited *The Joker: A Serious Study of the Clown Prince of Crime* (2015) and *The Supervillain Reader* (2020). Robert researches media, place, and fandom.

Bobby Schweizer is Assistant Professor of Journalism and Creative Media Industries at Texas Tech University, USA. He co-edited *Meet Me at the Fair: A World's Fair Reader* (2014). Bobby's research focuses on how the theme park as a medium has historically expressed stories through its attractions.

Jane Stadler is Honorary Professor of Film and Media Studies at The University of Queensland in Australia. She led a collaborative Australian Research Council project on landscape and location in Australian cinema, literature, and theatre (2011–2014) and co-authored *Imagined Landscapes: Geovisualizing Australian Spatial Narratives* (2016).

Stephanie Williams-Turkowski is Lecturer at Texas Tech University in the College of Media and Communication. Her research focuses on how fan communities influence media companies. She also studies the aesthetic aspect of television and films and how they impact the audience's perception of characters, settings, and reality.

Alberto Zambenedetti is Assistant Professor at the Department of Italian Studies and the Cinema Studies Institute, University of Toronto. His latest book is *Acting Across Borders: Mobility and Identity in Italian Cinema* (2021). He edited two recent books and co-edited *Federico Fellini. Riprese, riletture, (re)visioni* (2016).

ACKNOWLEDGEMENTS

We would like to thank the authors in this volume for their patience and good humour, and those prospective authors with excellent ideas that we could not quite fit into this book.

A special thanks to the anonymous academic reviewer and to the staff at Routledge.

1

INTRODUCTION

Erik Champion, Christina Lee, Jane Stadler, and Robert Moses Peaslee

From the deepest oceans[1] to space above us,[2] screen texts, screen production, and the traces they leave behind[3] surround and reshape both us and the environment. The stories and landscapes we view onscreen energize, depress, attract, or repel us, drawing us into a narrative or, in the case of screen tourism, enticing us to visit locations featured in or associated with a production. Consider the bucolic Shire in *The Lord of the Rings* (Matamata, Aotearoa/New Zealand), the gritty cityscape of *Deadpool* (Vancouver, Canada), the bustling streets of Kings Landing in *The Game of Thrones* (Dubrovnik, Croatia), and the interactive, out-of-this-world land of *Star Wars: Galaxy's Edge* located within Disneyland Park (Anaheim and Orlando, USA). The popularity of such locations shows the affective pull of places that are of narrative significance to people, and the rich tourism opportunities that arise from capitalizing on the interplay between story and place.

How do screen cultures impact on our understandings of and engagements with "place"? What are the shared qualities of real-world places and their onscreen counterparts? How do viewers-as-tourists negotiate landscapes and the slippage between fact and fiction? And how might locations featured onscreen be filmed or framed to elicit an affective response? Are there creative opportunities still to be explored in blending interactive media with tourism?

In *Screen Tourism and Affective Landscapes: The Real, the Virtual, and the Cinematic*, we have cast our net wide with a range of contemporary case studies that cross geographic boundaries. They range from *The Hobbit* to Fellini, from Nordic noir to surf documentaries set in Bali, from augmented reality walking tours that trace the journeys of screen legends to immersive role-playing historical games. Acknowledging the changes in viewing practices, digital technologies and portable computing devices, expanded markets, and the increasingly transmediated nature of the entertainment industry, this interdisciplinary book reflects the hybrid

DOI: 10.4324/9781003327585-1

developments within tourism and screen media (including film, television, videogames, and online videos). The authors come from a variety of disciplines and professions: the film industry, screen tourism, tourism, theatre, geography, design and art, communication studies, film studies, digital media, and digital humanities.

Virtual, augmented, and mixed reality technology, along with wearable computing, biosensors, portable supercomputers (our smartphones), and improved machine intelligence suggest that hybrid cinematic, virtual, and real-world landscapes are just around the corner. According to Computer World: "It's official: Smartphones are powerful enough to serve as laptop or desktop computers."[4] We are now taking our movies (and other media) with us.

With powerful smartphones, tablets, and the sophisticated entertainment systems now built into automobiles (Apple and Sony are designing cars[5] while Google has released a new, totally reconfigurable Android Auto car interface app),[6] the road trip is also becoming more "cinematic." The Xpeng P5 is an electric Chinese car prototype that includes a blow-up mattress, small drinks fridge, projector, and projector screen so that you can lie down and watch films from the rear seat in comfort. Meanwhile, Audi has partnered with Disney to explore virtual reality in car options along with a new platform called Holoride.[7] It is little surprise that journalist Jasper Jolly reports that Apple, Spotify, and WeChat also view the car as a future entertainment lounge.[8] While novel and not currently mainstream, these developments hold promise for the more subtle undercurrents of atmosphere and nuance in screen media to be experienced, filtered, and re-lived in various new modes of tourism and travel.

We are now also immersed in a streaming era where there has been a marked shift away from the traditional cinema-going experience towards viewing on smaller screens and consoles.[9] In 2019, before the onset of the global pandemic, Netflix's revenue exceeded the revenue of Disney and Warner Brothers films combined.[10] In recent decades, the mainstreaming of subscription streaming services has forced the entertainment industry to adapt its business model to cater for a public that now expects to be able to access any content, anywhere, on demand. At the same time, this may well help build greater desire for unique travel experiences. This opportunity was recognized in a White Paper produced by the United Nations World Tourism Organization in partnership with Netflix in 2021, which advocated for greater integration between internet entertainment services and the tourism sector to build "cultural affinity" that would benefit both industries.[11]

During the peak of the Covid-19 pandemic, when severe restrictions and lockdowns were in place across continents, the screen interface (mobiles, tablets, and television) became windows to other places and experiences. From the relative safety of our homes and screens, we tuned into media that provided comfort, escapism, and distraction. This included stories revolving around the unusual, the suspenseful, the frightful, and the uneasy. These stories, perhaps conversely, reminded us that we were still living and feeling in a time when our lives had collectively been put on halt. Given that dark tourism is political, complex, and contested[12] (explored by several chapters in this book), could new, more personal and portable forms of media and range of devices change the way in which atmosphere, suspense, and unease are conveyed? Following an extended period of social and physical isolation, and increased time in front of the screen, do we now seek to transfer onscreen thrills to the real, physical world of screen tourism?

To answer these questions, the book is divided into three major parts. *Part I* focuses on the attraction of screen locations, and our movements within them, both physically and virtually. These affective environments are accessed via increasingly powerful yet also shrinking portable devices, and the transformation of basic modes of transportation into sophisticated entertainment platforms. Does technology more effectively immerse us, or separate us, from the immediate cinematic and narrative experience of a road and walking tour? In this first section, the authors explore screen location tours, road trips with virtual windshields, film pilgrimages, surf tourism, and augmented reality walking tours that follow in the footsteps of screen legends.

In *Chapter 2*, screen tourism consultant Ian Brodie provides an account of his extensive experience within the screen tourism industry. He was the former Media and Communications Manager of Hobbiton Movie Set Tours in Aotearoa/New Zealand, and is the author of books such as *The Lord of the Rings Location Guidebook* and *Cameras in Narnia*. Brodie shares his expertise and insights gleaned from decades of working on film sets and in tourism, specifically those factors that are critical to establishing and sustaining a thriving screen tourism industry and the role augmented reality can play.

In *Chapter 3*, Kornelia Boczkowska explores the relationship between the North American fascination with automobiles and road trips, national parks, and (small) screen culture. Boczkowska discusses the niche area of windshield tourism, whether it be via physically driving through a natural landscape or observing such an experience through a screen interface. The chapter looks at discrepancies between different cinematic landscapes and the affinities between languorous scenic cinematography and scenic drive videos with their "slow" movement and the meditative state of windshield tourists. Boczkowska's study of YouTube's most watched scenic drive videos shows that they intensify travellers' interest in visiting national parks, revealing the role of digital media in assisting, mediating, and boosting self-drive tourism.

In *Chapter 4*, Gemma Blackwood focuses on tropes of paradise and freedom in the Australian production *Morning of the Earth*, and the power of cinematic texts to virtualize affective (travel) experiences. As a poetic sports documentary and ethnographic travelogue, *Morning of the Earth* features compelling sequences of surfing and travel that were filmed in Australia, Hawaii, and Indonesia. Blackwood analyses how the film's loose narrative, disjunctive editing, folk-rock music soundtrack, and sublime seascapes convey affective spaces of freedom and possibility that were opened up by this film for travel. However, as the author writes, this also problematically elides the real impact of tourism in Bali and the troubling aspects of neocolonialism in *Morning of the Earth*.

The cinema as a metaphoric portal to nostalgic travel and a bygone era is of course not restricted to Asia and the Pacific. In *Chapter 5*, Alberto Zambenedetti articulates the cinematic affects of pioneering Italian director Federico Fellini that extend beyond film. The chapter recounts the theme of independence and mobility in relation to *La dolce vita* and *Amarcord*, and then discusses pilgrimage and *saudade* (melancholic feelings of longing) in terms of Cinecittà, the famous film studio founded in 1937 by Mussolini. Zambenedetti explores how Fellini's felt presence and endurance in the culture articulates a complex relationship between cinematic memories, media, and public discourse. His analysis of the advertising campaign for the retro model Fiat 500 La Prima electric car is a striking example of the ways in which the atmosphere of Fellini's films infiltrates objects and places to become palpable in the present.

The cinematic pilgrimage can also be in unexpected places. Not all know that the debonair icon of Hollywood, Cary Grant, was born Archie Leach in England. *Chapter 6* presents Charlotte Croft's discussion of fandom communities and embodied experiences arising from the *Looking for Archie* walking tour of Grant's working-class hometown of Bristol. The "star-induced" tourism experience is physical, online, communal, a crisscross of celebrity and the less known granules of history; be they paparazzi encounters, rebuilt Cathedrals, or World War II bomb sites, interspersed with "magic moments." For many, this tour changed their perception not only of the actor but also of Bristol. The author concludes with a discussion on a potential extended reality (XR) phone app that will enhance the walking tour as embodied experience.

In *Part II*, the authors examine landscapes that are suffused with criminality, darkness, fear, and unknowability. The various chapters explore how unease is conveyed, and the ways in which specific places are framed or staged to produce affective atmospheres.

Sometimes the most atmospheric, alien, and uncanny settings are filmed in the familiar. In *Chapter 7*, Erin Ashenhurst takes us on a photographic journey through the "unsettled landscapes" of Vancouver – a city that has come to stand in for "Anywhere, USA" on screen – and other locales in British Columbia that have featured as anywhere-but-there. Using an autoethnographic approach, the author recounts scenes where fictional narratives seep into and entangle with the personal, lived experience of a place. The result is the illusiveness of Vancouver, which is constantly reimagined as a site teeming with horror, haunting, and mystery. The chapter also looks at the loosely defined movement termed "photoconceptualism," particularly its relation to film locations in Vancouver. Ashenhurst conveys what it is like to live in a city replete with false identities, and cinematic moments where there is slippage between real and unreal, and fact and fiction.

The 16 Emmy Award-winning television series *Breaking Bad* is now over a decade old, and yet it continues to live on in the imaginations of fans-as-tourists. Revolving around the methamphetamine trade, and the relationship between the two protagonists (Walter White and Jesse Pinkman) and others in their social orbit, the series was immortalized by powerful visual storytelling and the eternal, mythical desert setting that is still a major drawcard for tourism to Albuquerque, New Mexico. The local tourism authority referred to Albuquerque itself as a "character" in the television show, illustrating how place was not simply the background but played a critical role in the series.[13] Jane Lovell and Kenneth Fox's *Chapter 8* examines five "desert tropes" exemplified by *Breaking Bad*, and Southwestern imagery on screen.

Even after moving to cooler climes, such as Scandinavia, onscreen landscapes that convey unease and languid lachrymosity can captivate maudlin-magnetized tourists. Can-Seng Ooi, Richard Ek, and Mia Larson's *Chapter 9* investigates how the genre of Nordic noir – characterized by bleak landscapes, an atmosphere of melancholia, biting cold, and violent crime – have become tourism hotspots. The authors discuss television series such as *The Killing*, *Borgen*, *Midnattssol*, and *Bron/Broen* that blend the mundane with the exotic (locations, language), the familiar with the unfamiliar.[14] Importantly, these small screen series often represent the landscape in a way that is epic and cinematic in scale while simultaneously accentuating a sense of desolation. But the sun doesn't always hide, the rain doesn't always weep. On bright and warm Nordic days, imagination and foresight are needed of the Nordic noir tourist. How do the sparse, gloomy visual clues of the Nordic noir landscape mesh with personal experiences?

In *Chapter 10*, Jane Stadler and Lesley Hawkes investigate the ethics of the dark tourism trade and analyse the role that genre plays in shaping the affective qualities of screen experiences and tourists' visits to sites of terror and trauma. The chapter examines *Wolf Creek* and *Snowtown*, which are films based on the infamous backpacker murders and "bodies in barrels" crimes that took place in Australia. Where thriller films offer a visceral emotional charge combining horror with disturbing closeness to the killer and their victims, social realism and true crime provide sociocultural context and invite audience alignment with an investigative or evaluative role (detective or juror). The authors demonstrate that the macabre fascination with visiting film and television locations to collect "murderabilia" souvenirs distinguishes dark tourists from fans of horror and true crime and echoes the killer's compulsion to revisit crime scenes and keep mementos.

In *Part III*, we examine the fans and the audience. Despite the range, power, and suitability of technology like 5G, in recent years, there have been many anxieties and restrictions impeding participation in touristic activities and cultural events that were previously taken for granted. During extended lockdown periods, when many public spaces closed and large gatherings were prohibited, virtual spaces and armchair tourism were sought out as opportunities for education, entertainment, and to feel a sense of connection with others and the world beyond the home. These dynamic changes, too, were affecting museums and galleries worldwide:

> *In 2020, museums were closed for an average of 155 days, and since the beginning of 2021, many of them have again had to shut their doors, resulting, on average, in a 70% drop in attendance and a 40–60% decline in revenue compared to 2019.*[15]
>
> (UNESCO 2021)

To entice visitors back to museums and galleries, these institutions are investing in advanced 3D technology, immersive exhibits, and playful interfaces that leverage their performative areas and facilities. Heritage sites are hosting after-hours themed adventures. Theme parks are mutating into rich, diverse experiences, with holograms and augmented reality. Videogames are influencing tourism and creating e-sports events and escape rooms. Single-viewer experiences are being transformed into social interactive and immersive experiences. How are natural and artificial places used or fashioned to engender these social encounters? How are myth and magic spun to create these enchanted places and magical circles for groups and families?

In *Chapter 11*, Stephanie Williams-Turkowski discusses the adult fandom practices and identity formations of Disneyland Social Clubs that are known as "Disneyland Gangs." These groups have developed a sense of community based on shared interest in and affection for the Disney franchise (screen productions and theme park attractions). Disneyland Gangs illustrate how fans negotiate and play within transmediated landscapes. Williams-Turkowski's chapter provides an understanding of affective landscapes through the lens of fandom culture, and how Disneyland's restrictions and regulations due to the pandemic have impacted members' sense of magic, freedom, and expression of individuality within the space.

The complex nature of theme parks is explored in *Chapter 12* by Carissa Baker, Ray Eddy, and David Bailey. From Disneyland's *Indiana Jones* Adventure and *Pirates of the Caribbean* ride to Universal Studio's *Jaws* attraction, the authors explain how amusement park adventures and behind-the-scenes tours of legendary television and film studios have generated their own form of screen tourism. This type of tourism allows tourists to not only experience sites and scenes from fictional narratives, but the nature of theme parks also invites participation and physical engagement with the surrounds. The authors explore how audiences are brought into the film(making) world and how the shows and tours have morphed into more immersive experiences.

Theme parks can also be smaller in scope and size, and still attract millions of visitors. One of the most famous of these is the Hobbiton Movie Set Tour (on the site of a working sheep farm), which is located in the farming town of Matamata in Aotearoa/New Zealand, and features in the country's top ten tourist destinations. The Hobbiton site doubled as the Shire in *The Lord of the Rings* and *The Hobbit* film franchises, and now operates as a permanent theme park. Robert Moses Peaslee and Bobby Schweizer's *Chapter 13* reflects on two decades of screen tourism success with Hobbiton as both landscape and "authentic" relic of film production. A film location, landscape, theme park, and museum – the authors investigate the various appeals of this fictional Shire.

Moving to the digital adventure world, Erik Champion's *Chapter 14* does battle with misconceptions on the appeal and market value of videogame-induced tourism. Given the videogame market is one of the most lucrative segments in the entertainment industry, and videogames have become sophisticated in their modes of storytelling and world-building (also evident from the numerous screen adaptations of computer games into film and television shows), the author asks: why has the prospect and potential of videogame-induced tourism only recently been discussed in academic publications? While *S.T.A.L.K.E.R.: Shadow of Chernobyl* uncovered a new type of videogame tourist, one who adventured into places game designers explicitly did not intend them to go, videogame tourism has also resulted in festivals and events held in heritage sites.[16] Focusing on *Assassin's Creed: Odyssey*, Champion suggests new developments that may accelerate the impact of videogames on tourism, and the related experiencing of affective landscapes.

Notes

1 D. Hardawar, "James Cameron's New Nat Geo Show Will Explore the Mysteries of the Ocean," *Engadget*, July 15, 2021, www.engadget.com/james-cameron-oceanxplorers-national-geographic-180020914.html.
2 "Russian Space Movie Crew Set for Blast-Off," *Reuters*, updated September 17, 2021, www.reuters.com/lifestyle/science/russian-space-movie-crew-set-blast-off-2021-09-16/.
3 Alice Gorman, "From Poo Politics to Rubbish Disposal: 5 Big Questions About the International Space Station Becoming a Movie Set," *The Conversation*, September 23, 2021, https://theconversation.com/from-poo-politics-to-rubbish-disposal-5-big-questions-about-the-international-space-station-becoming-a-movie-set-168282.
4 Mike Elgan, "With Smartphones Like These, Why Do We Need Laptops?," *ComputerWorld Australia, ComputerWorld Australia*, October 6, 2017, www.computerworld.com/article/3241233/with-smartphones-like-these-why-do-we-need-laptops.html.
5 Jasper Jolly, "A Living Room on a Skateboard: How Electric Vehicles Are Redefining the Car," *The Guardian*, July 16, 2022, www.theguardian.com/business/2022/jun/11/living-room-on-a-skateboard-how-electric-vehicles-are-redefining-the-car.
6 Ron Amadeo, "Google's New Android Auto Interface Works With Any Screen Size," *Ars Technica*, July 17, 2022, https://arstechnica.com/gadgets/2022/05/googles-new-android-auto-interface-works-with-any-screen-size/.
7 "Audi Partners With Disney to Entertain Car Passengers With VR," *Mashable*, updated January 7, 2019, https://mashable.com/article/audi-disney-vr-holoride-ces-2019.
8 Katie Alexander, "Waking the Sleeping Electric Vehicle Dragons," *Evision Electric Vehicles*, July 17, 2021, www.evisionevs.co.uk/2021/11/18/waking-the-sleeping-electric-vehicle-dragons/.
9 Mustafa Yarımbaş, "Is Netflix Reshaping the Film Industry?," *Illumination*, June 22, 2020, https://medium.com/illumination/is-netflix-reshaping-the-film-industry-18c60d70058b.
10 Brandon Katz, "Every Movie Theater vs. Streaming Release is Riddled With Pros and Cons," *The Observer*, August 7, 2021, https://observer.com/2021/07/hollywood-movie-theaters-vs-streaming-pros-cons/.
11 UNWTO and Netflix, "Cultural Affinity and Screen Tourism – The Case of Internet Entertainment Services," *UNWTO eLibrary*, September 2021, www.e-unwto.org/doi/book/10.18111/9789284422838.
12 John Lennon, "Dark Tourism," *Oxford Research Encyclopedia of Criminology and Criminal Justice* (2017), https://doi.org/10.1093/acrefore/9780190264079.013.212.
13 "*Breaking Bad* in Albuquerque: Set and Filmed in Albuquerque," *Visit Albuquerque 2022*, accessed July 15, 2022, www.visitalbuquerque.org/about-abq/film-tourism/breaking-bad/.
14 Anne Waade, "'Just Follow the Trail of Blood': Nordic Noir Tourism and Screened Landscapes," in *Danish Television Drama: Global Lessons from a Small Nation*, ed. Anne Marit Waade, Eva Novrup Redvall, and Pia Majbritt Jensen (Cham: Springer, 2020), 231–52.
15 "Supporting Museums: UNESCO Report Points to Options for the Future," *UNESCO*, updated April 13, 2021, https://en.unesco.org/news/supporting-museums-unesco-report-points-options-future.
16 Ismail Shaheer, "The Nexus of Video Games and Heritage Attractions," *Current Issues in Tourism* 25, no. 9 (May 3, 2022): 1356–60. https://doi.org/10.1080/13683500.2021.1915254.

Bibliography

Alexander, Katie. "Waking the Sleeping Electric Vehicle Dragons." *Envision Electric Vehicles*, July 17, 2021. www.evisionevs.co.uk/2021/11/18/waking-the-sleeping-electric-vehicle-dragons/.

Amadeo, Ron. "Google's New Android Auto Interface Works With Any Screen Size." *Ars Technica*, July 17, 2022. https://arstechnica.com/gadgets/2022/05/googles-new-android-auto-interface-works-with-any-screen-size/.

"Audi Partners With Disney to Entertain Car Passengers With VR." *Mashable*, January 7, 2019. https://mashable.com/article/audi-disney-vr-holoride-ces-2019.

"*Breaking Bad* in Albuquerque: Set and Filmed in Albuquerque." *Visit Albuquerque 2022*, accessed July 15, 2022. www.visitalbuquerque.org/about-abq/film-tourism/breaking-bad/.

Elgan, Mike. "With Smartphones Like These, Why Do We Need Laptops?" *ComputerWorld Australia*, October 6, 2017. www.computerworld.com/article/3241233/with-smartphones-like-these-why-do-we-need-laptops.html.

Gorman, Alice. "From Poo Politics to Rubbish Disposal: 5 Big Questions About the International Space Station Becoming a Movie Set." *The Conversation*, September 23, 2021. https://theconversation.com/from-poo-politics-to-rubbish-disposal-5-big-questions-about-the-international-space-station-becoming-a-movie-set-168282.

Hardawar, D. "James Cameron's New Nat Geo Show Will Explore the Mysteries of the Ocean." *Engadget*, July 15, 2021. www.engadget.com/james-cameron-oceanxplorers-national-geographic-180020914.html.

Jolly, Jasper. "A Living Room on a Skateboard: How Electric Vehicles Are Redefining the Car." *The Guardian*, July 16, 2022. www.theguardian.com/business/2022/jun/11/living-room-on-a-skateboard-how-electric-vehicles-are-redefining-the-car.

Katz, Brandon. "Every Movie Theater Vs. Streaming Release is Riddled With Pros and Cons." *The Observer*, August 7, 2021. https://observer.com/2021/07/hollywood-movie-theaters-vs-streaming-pros-cons/.

Lennon, John. "Dark Tourism." *Oxford Research Encyclopedia of Criminology and Criminal Justice* (2017). https://doi.org/10.1093/acrefore/9780190264079.013.212.

"Russian Space Movie Crew Set for Blast-Off." *Reuters*, September 17, 2021. www.reuters.com/lifestyle/science/russian-space-movie-crew-set-blast-off-2021-09-16/.

Shaheer, Ismail. "The Nexus of Video Games and Heritage Attractions." *Current Issues in Tourism* 25, no. 9 (May 3, 2022): 1356–60.

"Supporting Museums: UNESCO Report Points to Options for the Future." *UNESCO*, April 13, 2021. https://en.unesco.org/news/supporting-museums-unesco-report-points-options-future.

Waade, Anne. "'Just Follow the Trail of Blood': Nordic Noir Tourism and Screened Landscapes." In *Danish Television Drama: Global Lessons from a Small Nation*, edited by Anne Marit Waade, Eva Novrup Redvall, and Pia Majbritt Jensen, 231–52. Cham: Springer, 2020.

Yarımbaş, Mustafa. "Is Netflix Reshaping the Film Industry?" *Illumination*, June 22, 2020. https://medium.com/illumination/is-netflix-reshaping-the-film-industry-18c60d70058b.

2

SCREEN TOURISM

Marketing the Moods and Myths of Magic Places

Ian Brodie

Introduction: The Power of Cinematic Landscape

Tourism as a result of watching scenes created in film or television has been evident as a motivator for travel for more than 50 years.[1] Arguably one of the most famous screen tourism locations is Salzburg, Austria, where many visitors have flocked to since the release of *The Sound of Music* in 1965. These visitors wanted to see Schloss Leopoldskron, which doubled as the Villa Von Trapp's backyard. In a normal year (i.e. pre-Covid), there are more than 300,000[2] people visiting Salzburg because of the Von Trapp family.

Despite being the birthplace of the famous composer Wolfgang Amadeus Mozart, for many, such sites as Mirabell Palace and Gardens, Hellbrunn Palace, and Frohnburg Palace hold much greater significance and emotional meaning because of their appearance in the film. So, while people have been making pilgrimages to locations seen on-screen for many decades, it was only in the early 1990s that any efforts were made to properly study and provide evidence of the relationship between film and tourism.

The medium of film leaves a lasting impression on the viewer. Scenes of passion, action, or landscapes spark an emotion that lasts much longer than the film itself. To quote one of the most imaginative directors of the last century, Stanley Kubrick: "The screen is a magic medium. It has such power that it can retain interest as it conveys emotions and moods that no other art form can hope to tackle." The inter-relationship between the imagination of film and the reality of landscape is an interesting mix. Although a director can call on any form of computer generation to create a world, the basis of the creation is nearly always the tangibility of a real landscape.

DOI: 10.4324/9781003327585-2

FIGURE 2.1 Evocative scenes create memories. The mountains of Glencoe have appeared in film and television for over 70 years.

The impact of landscape in film is a result of the subject and how the landscape is revealed (Figure 2.1). However, it is present in every film to some degree.

- The viaduct and train (Scotland) from *Harry Potter*
- Death Valley (Eastern California) in *Zabriskie Point*
- The Devils Tower (Wyoming) from Close Encounters of the Third Kind
- The Swiss Alps in *The Sound of Music*
- The Highlands of Scotland in *Braveheart*

The mention of these places creates an image of the film in your mind, even if you have never been there. Your psyche is unconsciously programmed to expect a location to appear in reality the same as the celluloid image from film.

The question is: why? Why do people travel to a film location, and what do they expect when they get there? The location can be seen as the only "real" link for the visitor to experience. This is just part of the answer.

One of the most significant arguments may be the diversity of the visitor experience. Many people may arrive with ill-defined reasons, but the actual visit invokes and awakens a range of reactions that can include a sense of nostalgia or of longing for the particular kind of world they associate with the film. Generalizations are valid but each visitor has some individual form of chemistry with the place, its presentation, and its associated characters and events, real-life or fictional. They want their memories to be validated by the introduction of reality – in the form of a landscape.

The Magic of Film Locations

I was first drawn to the relation of film location and touristic landscape in the mid-1970s. As a teenage film enthusiast, I was enamored with the social revolution that had been taking place in the USA, often featuring an escapist landscape of freedom from traditional teenage norms where longer hair, drug use, and ideals differed greatly from WWII-era parents. These ideals were shaping a number of films.[3] *Zabriskie Point,*[4] *Easy Rider, Vanishing Point,* and *Two Lane Blacktop* are famous examples of this genre where, invariably, the main protagonists shun expected norms of life in 1960s America and escape to the road, taking the viewer with them on a journey of escape that features fantastical landscapes (such as Zabriskie Point in Death Valley) mixed with the popular music of the time. The finale in Zabriskie Point is an architecturally designed desert house (owned by one of the "normal" protagonists) exploding to a soundtrack from the popular rock band Pink Floyd.

Zabriskie Point impacted me in several ways. The ideals that were presented intrigued me, probably because of my own drug experimentation and thoughts of rebellion that most teenagers seem to embrace at some time during those formative years. The landscape and music combined to provide a strong desire to visit Zabriskie Point and experience some of the emotions that I felt while watching the film. The sweeping desert of Nevada was something that (as a New Zealander) I had never experienced. It showed lonesome roads, diners often filled with disapproving adults, clear blue skies, and heat that seemed to penetrate the soul. Especially though, the open road was a means of escape from the norm – it was adventure. It is this mixture that is a key motivator[5] in the film tourism traveller.

The Road to Mordor

In the Central Otago region of New Zealand, there is a dusty track climbing and winding its way from the Ida Valley to a dam high in the surrounding hills. Built during The Great Depression, it supplied water to the parched paddocks of the surrounding valleys. For over 70 years, the track was only known as a means of access to a few small cabins that had been built around the water for local fisherman to spend a day fly-fishing for the introduced trout and revelling in a night away from civilization.

In December 2001, much to the chagrin of those fishermen, everything changed. This remote location was just one of over 150 places in New Zealand that filmmaker Peter Jackson, and a literal army of filmmakers had transformed into Middle-earth and presented to the world with the release of *The Fellowship of The Ring*. As book fans and a group of screen neophytes were introduced to a three-dimensional live adaptation of Tolkien's world, there was growing interest from a practical and theoretical perspective in the practice of screen tourism.

The impact of screen tourism in New Zealand after the release of *The Lord of the Rings* trilogy took the country by surprise.

This perfect storm mixture needs further analysis, because it is these key features that together combined to provide a 50 percent increase in visitor numbers

FIGURE 2.2 Where literary description meets film adaptation. The Green Dragon Inn at Hobbiton Movie Set near Matamata, New Zealand.

after the release of *The Fellowship of the Ring*.[6] The monetary impact of "Lord of the Rings" tourism continues to this day, and there are many statistics that point to the success of the two film trilogies in New Zealand. Tourism increased from 1.7 million in 2000 to 2.4 million per annum in 2010, which is a 40 percent increase (or 700,000 visitors). Figures from Tourism New Zealand indicate that at least 20 percent of visitors came to NZ because of their exposure to the country on screen (Figure 2.2). If a conservative 5 percent of that number is a film tourist and increasing in number, it means that an extra 35,000 people visited because of the films. If this is multiplied by the average spend of $2798.00, there is an additional income of NZ$98 million.[7]

The results have become a famous example of the positive effects of film tourism and what a powerful series of films can do for local tourism and the economy of a country. New Zealand's process of promoting the films and the attractions it has to offer has been an illuminating lesson in what joint consultation and cooperation between film and tourism can achieve. The combination of mythical blockbuster and tranquil natural sublime has provided the perfect mix for New Zealand. There was a large existing Tolkien fan base that identified with both the production values and the landscapes used in the film trilogy. The many diverse geographical descriptions that Tolkien provided of Middle-earth allowed the location scouts to explore the full gamut of the landscapes of New Zealand. The diversity and relative compactness of the country meant that a unit base could be set up in a central situation that allowed multiple locations to be filmed.

A good example of this is the Queenstown region. From this base, over 20 locations were filmed covering all three films. These included Paradise (doubling as Lothlorien Forest), Deer Park Heights (the road from Edoras to Minith Tiras, among other locations), and Twelve Mile Delta (Ithilien). Importantly, this region is also an existing tourism hub for visitors to New Zealand with experienced tourism operators and a broad supply of accommodation.

During the year following the release of *The Fellowship of the Ring*, travel operators in the region noticed a new genre of tourists arriving. These tourists were asking specifically to visit the places that had featured in the film. They did not know where they were in the Queenstown Lakes, but they did know their geographical location in Middle-earth.

This leads to a very important aspect of screen tourism. As the operators started to develop tours to these locations, there emerged distinct divisions. Those who knew the works of Tolkien (as well as those who had seen the films) and those who had not read either the books or seen any of the movies. Anecdotally, there was only one operator (Nomad Safaris in Queenstown) whose founders had read the books and had created tours into the Skippers Canyon (before the release of the films), titled "The Road to Mordor."

The operators unfamiliar with Tolkien's books soon fell afoul of the actual tourist they were trying to entice. A misplaced word, an inaccurate description of a location, or any type of Middle-earth "faux pas" alerted the tourist immediately that here was somebody that is not a fan and is just trying to collect my money. Other operators who were fans that started operating tours received the opposite response. Tourism numbers increased and word soon spread in the community about the preferred operators.

A third type of operator was also motivated to expand their existing business by encouraging the screen tourist. These operators identified the traveller and understood the importance of authenticity when dealing with the emotive needs of these tourists. They consulted with people who knew the world of Tolkien before releasing the product.

The resultant discovery here is important in the development of screen tourism products. The operator needs to identify the visitor and understand their specific and varied needs. These can be quite detailed, and in the case of *The Lord of the Rings* locations some visitors wanted to know the exact position of trees or rocky viewpoints.[8]

Some of the actual locations used in these films also created issues. Locations are mostly chosen by scouts to suit the production with scant regard for what comes after. The crew move in, film, and then the land is returned to its original state. In the case of *The Lord of the Rings*, some places were rather remote and while ideal for a vision of Middle-earth they were impractical for a large number of fan visitations.

An example of this is Mount Sunday, in the Rangitata Valley of the South Island. The crew spent six months building a set of the Great Hall of Rohan at Edoras. This location was a conundrum for those working on site. The area is subject to very strong foehn winds, winter snow, and summer heat, so working conditions were quite severe. There is only one access road (51 km and unsealed) into the point which did allow for security and privacy. There were no facilities within 75 km of the location and access was achieved by fording a stream, which can be very difficult after rain.

Edoras was a pivotal scene in *The Two Towers* and the location became an important place to visit by fans. Despite the removal of the actual buildings, this matters little to the screen tourist. Imagination is imperative to deepening our understanding of the film tourism experience[9] *Lord of the Rings* tourists in New Zealand are seen as "eager to test the connection between imagination and geographical places by physically travelling to the film location(s)" (Buchmann 2010: 79) and demonstrate "a playful attempt to touch something untouchable" (Peaslee 2011: 42). Edoras has been an imagined place to readers of The Lord of the Rings or a place in a film that looks fantastical and can't possibly exist. By visiting and affirming the "reality" of Mount Sunday, the visitor receives affirmation of a non-existent place.

As tourist numbers grew to the location, the local council became aware of several issues. Cars were breaking down (with no mobile coverage available) and some visitors were suffering dehydration and other medical issues with no support. The response by the council was to produce small leaflets that explained the remoteness of the location and encouraged visitors to be prepared.

The remoteness also played its part in the adventure of the film tourist. Crossing a stream in the inaccessible Southern Alps created a sense of belonging and allowed the tourist to take on the personality of the cast character, thus adding another dimension to the experience. This was particularly apparent with those members of Red Carpet Tours.[10] Arriving at Mount Sunday, the group would cross the stream, helping each other (and in some instances carrying prop replicas) and revelling in something that many people from overseas had never experienced before.

This aspect of satisfying the participant's experience is an important consideration by the tour operator in the development of the product. It provides another link to the senses of the tourist and their wish to feel that they are part of the film and are fulfilling some of the physical challenges faced by the character.

The Real Middle-earth

The subsequent filming and production of *The Hobbit* film trilogy in New Zealand from 2011 to 2012 provided another opportunity to promote the country as "The real Middle-earth." In the negotiations that preceded the final agreement to film in New Zealand, Tourism NZ played their part and required specific deliverables from Warner Bros related to tourism. These actions are also an important aspect of negotiation, ensuring that copyright or naming rights are not broken during the promotion of the film locations to the tourist. This was a particular issue that had to be addressed during the production of these six films. The Rights holders (a combination of The Tolkien Estate, HarperCollins, Universal, New Line, Warner Bros, and Saul E. Zaentz) all created several issues relating to the promotion.

Liaison between local councils, tourism entities, and production companies before filming is a key aspect of satisfying future tourism activities. Often the regions themselves contribute revenue (either directly or with taxation easements) to the film production, and it is imperative at this time that ongoing promotion of the locations (for tourism operations) is considered. It is my experience that some councils within Australia are adding specific film tourism clauses into film production contracts. This adds a requirement for production to undertake specific EPK videos and quotations providing background information on the location.

The Breaking Bad Experience

While *The Lord of the Rings* filmed in New Zealand used the fantastical landscapes as a means of promotion, coupled with a best-selling book and large fan base, other types of television or film can also be very successful in encouraging tourists. This is particularly apparent with the *Breaking Bad* series.[11] *Breaking Bad* deals with a schoolteacher who develops cancer and enters the underworld of crystal meth production – an unlikely combination to inspire tourism.[12] This mixture does some have parallels with the New Zealand experience, however, and this is the link between the heroes/antagonists and the desert locations. The viewer's relationship with these people becomes very strong as the series develops and the appeal of something illegal and desperate provides a what-if to the viewer. The desert landscapes where deals are completed, the seedy motels of meth users or local diners visited all become places of pilgrimage for the tourist who seek a means of feeling and emotion related to the show itself.

This has led to a unique tour operated by two fans who shifted to Albuquerque because of their passion for both the landscapes and the series. Purchasing two similar recreational vehicles to that used in the series, they undertake daily tours of Breaking Bad locations in the city which include a vehicle car wash, industrial laundry, derelict motel, and fast-food outlets. The interior of the RV is fitted out to look like the vehicle used in filming. The success of this operation can be seen as a means of tourism revenue (this was a secondary consideration initially), but, importantly, it is operated by fans who know the series intimately. Just like those tour operators in NZ that created the right ambience and knowledge of the locations in *The Lord of the Rings*, this company identifies with and acknowledges the fandom base.

The downsides of film tourism also need to be addressed here. The owners of the fictional Walter White's house in Albuquerque have been forced to fence the property and place no-parking signs outside. Fans visit the home daily to photograph themselves in the environment and relive some of the scenes from the series.[13] The owners never considered this issue when they leased their home to the production company and now have an ongoing problem with gawking tourists.

This is also apparent in Matamata, New Zealand, where the number of tourists travelling along Buckland Road to the Hobbiton Movie Set required major roadworks to ease congestion. The i-Site Visitor Centre in Matamata also required road redirection and additional parking had to be added.

What Makes a Good Production for Film Tourism?

One of the questions often asked of me is "*how do we know what is a good production for film tourism*?" Several subjects need to be considered to answer this succinctly, and by looking at some particular productions some conclusions can be drawn. Often the pure location can provide an opportunity for tourism promotion. This can be seen in the 2000 release of *The Beach*. The film itself was a moderate box office success but received mixed to negative reviews from critics and has a score of 20 percent on the Rotten Tomatoes website. But the location of Ko Phi Phi Le became an immediate success specifically because of its beauty. The number of tourists visiting this location (up to 5,000 per day[14]) forced its closure to visitors in 2016 because of the damage to the reef caused by visitors. It was reopened in 2022 after rehabilitation and a downturn in visitor numbers due to Covid-19.

In this instance, it was the sheer beauty of the location that encouraged the tourist. Another successful example is the joint New Zealand/Scottish production of *The Waterhorse*. The film was a success with the critics and the locations of Lake Wakatipu in New Zealand's southern lakes (Figure 2.3) and Ardkinglas Estate in the highlands of Scotland combined to provide an opportunity for both countries to promote the landscape beauty. In this instance, the Scottish locations were heavily promoted, but the New Zealand locations were not. Many people who watched this film were unaware that Lake Wakatipu provided much of the Highland loch landscape. New Zealand possibly considered that the juxtaposition of the two countries was a reason not to promote the landscape. In this instance, it was a successful film combined with landscapes that provided an opportunity, but the misplaced geographical locations created (unfounded) issues for New Zealand promotion.

This conundrum in NZ reversed itself in 2022 with the release of *The Power of the Dog*. Filmed in Central Otago but portraying the wilds of Montana provided a promotional opportunity for New Zealand but was not mentioned in Montana.

FIGURE 2.3 Mount Nicholas Station, on the banks of Lake Wakatipu in New Zealand, becomes a Scottish loch for The Water Horse.

The 2008 film *Australia* was considered a good opportunity for film tourism by several Australian tourism entities. Buoyed by the achievement of Crocodile Dundee, it was considered the combination of epic landscapes, a well-known cast and a dramatic plotline provided a good recipe for success. In this instance,[15] the film itself failed to achieve international box office success but importantly the remoteness and inaccessibility of many of the locations became an issue.

I believe that every production can be important for film tourism, but the investment needs to be commensurate with the expenditure. Tourism infrastructure and accessibility are important factors contributing to the success of film tourism. Close cooperation between production, tourism, consultants, and researchers before the production starts can work a long way towards devising if the key elements are in place.

Another vital piece that needs to happen during production is the cast, and crew need to LOVE where they are. This can be used in tourism promotion as they talk about it, extolling its virtues, and they are enthusiastic, and this enthusiasm is recorded. This was particularly effective within NZ and filming of the Middle-earth series where a cast and crew driven narrative became an integral part of Tourism NZ marketing. Several case studies undertaken can provide support for this argument.

The 2019 production *Rams* was filmed exclusively in the Mount Barker region of Western Australia. The author was employed as a Unit Stills photographer by the production which allowed access to the locations, the cast, and the production. In addition to obtaining the required production stills, he was also able to capture additional landscape imagery and research locations for tourism promotion. The subsequent online guide was produced with the full cooperation of the production, the region's Business Development office, and the state's film commission (ScreenWest). The cost to produce this guide was less than AU$20,000, but the result was a tool that was used by all parties and became another implement in the tourism and film promotion of the region and state. Research showed a 150 percent increase in visitor numbers to the Mount Barker Visitor Centre and a 200 percent increase in visitor numbers immediately after the release of the film and guide.

Concurrently an online location guide was created by me (for the same stakeholders) for the film *H is for Happiness*.[16] In this instance, the author was embedded on set as an additional photographer to capture stills for use in the guide (but additionally available for production). This resulted in an uptake in domestic tourism and the creation of a local bus tour that operates on-demand to some of the locations in the city of Albany. This guide also incorporated some additional popular tourist locations unrelated to the film but allowed for the guide to be used as an adjunct to traditional tourism material.

Imitation Landscapes

When locations stand in for other places (or countries), there is an opportunity for both places to promote this fact. They encourage visitors to explore and understand how and why the locations were used. The use of stand-in locations and geographical mismatch can be a common occurrence in production. The multi-season production of *Vikings* is filmed in Northern Ireland with the use of plate shots of Norway as required to "set the scene." Despite this geographic discombobulation, there are opportunities for both Norway and Northern Ireland to promote the series. One prospect is a joint promotion between the two countries' tourism agencies.

Another example is the recent production *The Power of the Dog*. All the landscapes were filmed in the Central Otago area of New Zealand, but the entire film was based on events in rural Montana. At the time of writing, this film had been nominated for (and received) multiple accolades and award nominations. The NZ Film Commission[17] has promoted this fact as a film location but there is a further opportunity that has not been fully explored or exploited.

On the other hand, sometimes this can lead to interesting new myths, where inaccuracy becomes local home truth and legend becomes reality. I offer two examples of this concerning *The Lord of the Rings* film locations in New Zealand.

The first example is related to Mount Doom and the use of Mount Ngauruhoe. This volcanic peak in the centre of the North Island does share a remarkable similarity to Mount Doom, and even director Peter Jackson has made this comparison in some media conferences. But the mountain was never used in the film; it is held sacred by the Maori and was not filmed in any way. The closest filming was undertaken around the nearby Mount Ruapehu but even long shots where Mount Ngauruhoe (Figure 2.4) could be seen were digitally altered in post-production to remove the peak. In my location guide for the films, I advise on this fact (after receiving confirmation from production). Despite this, tourism organizations and operators (including Tourism New Zealand) advertise Mount Ngauruhoe as Mount Doom – the famed Tongariro Crossing walking track features a number of Mount Doom references. In this case, convenience and marketing have taken over from reality.

Another example of this is a small waterfall that can be reached via a short walk while driving to the Chateau Tongariro. I visited this location in 1974 and was

FIGURE 2.4 Mount Ngauruhoe creates its own mythology. Despite a likeness to the Tolkien Mount Doom, the mountain does not appear in the film trilogy.

immediately drawn to the likeness of the falls and landscape to Henneth Annûn, as described by Tolkien in *The Lord of the Rings*. I returned here many times, and in my location guide, I wrote about this as a place to visit while travelling to the real locations at nearby Mount Ruapehu. I specifically noted in the book that this location does not appear in the films, but it is reminiscent of Tolkien's descriptive text. Despite this, fandom now chronicles this as a real location.

Building Successful Film Tourism Destinations

Screen tourism occurs because the feature captures the imagination and creates a desire for people to visit. This does not necessarily mean the film or location need to be traditional tourism visitor points. It just has to capture an emotion and a feeling with the viewer. This, of course, differs in every person, and thus we have a rich mix of landscapes and places.

Tourism and Film-Making

Cooperation between production, local stakeholders, and the local film commission prior to filming creates a key leverage point to produce film tourism material. The relevant film tourism organization should be part of the production sequence with the development of the film. This symbiotic relationship not only allows for cross-pollination of views and ideas that will provide more information on a location for production but also allows production to showcase a location that might not have been ever considered by the tourism group.

FIGURE 2.5 A local hotel in Mount Barker, Western Australia, used in the film Rams.

I used this premise through my Lord of the Rings *Location Guidebook*. While researching the locations, I obtained information from the Unit Publicist regarding cafes, accommodation, and means of transport. These became benchmarks included within the book.

Another example of this is in my work with ScreenWest in Western Australia. An online location guidebook (Brodie, 2022) based on the *Rams* feature film (Figure 2.5) included a tour of the local wineries in the film location with text by one of the main actors, New Zealander Sam Neill, who owns a famous vineyard in New Zealand.

Consultation with the community and engagement of local stakeholders to gain ownership is crucial. This allows councils and local governmental organizations to be involved in the production process, resulting in both groups gaining from the experience and understanding each other's needs. I suggest, for example, to have a "town hall meeting" to explain the making of the film. Showcase some of the talents of the filmmakers to encourage future interns and future employment within the industry.

Behind the Scenes Authenticity

Film locations are not the only property that should be included or promoted to the screen tourist. It is important that accommodation, cafes/restaurants, and other locations that the cast may have used or visited are included. This provides further information for the tourist about the film, but it also appeals to their senses by providing a glimpse behind-the-scenes[18] into the cast and crew. It provides a further direct link that the film tourist can have to the (hero) cast. A good example of this is the Powderhorn Hotel in Ohakune in NZ. The major cast of *The Lord of the Rings* stayed here during filming and the hotel now promotes certain rooms such as the "Orlando Bloom Room" and the "Elijah Wood Room." The tourist explores the locations from the film and then spends the night in a room that was used in the production.

However, authenticity and engaging experiences are also highly subjective. Many of the websites and applications developed to share film locations are created by fans and concentrate on a particular series, pet theory, or favourite location. Like much of the information on the internet, accuracy can be a problem, although peer reviews do take place with enthusiasts weighing in on discussions to ensure geographical correctness.

Empathy

Film tourism is not just about visiting sites – it is the complete package. We are not building a theme park, but creating a place of pilgrimage and memories. We are dealing with emotion, often deep-rooted, that requires empathy. Role-playing, knowledgeable guides that can identify with and provide "insider" information for the tourist are essential. Any front of house people involved with the screen tourist needs to have a basic understanding of the film and the reasons why someone may visit, for empathy is so important.

Companions

When the needs of the film tourist are considered by tourism operators (or individuals creating guides), the requirements of the accompanying traveller should be taken into account.

It is the passion for the subject and the need to identify, and visit, the location that pushes the film tourist. One scholar, Niki Macioni, has identified four categories of screen tourist.[19]

I. The Serendipitous Film Tourist: Those who just happen to be in a destination portrayed by a film.
II. The General Film Tourist: Those who are not specifically drawn to a film location but who participate in film tourism activities while at a film destination.
III. Specific Film Tourist: Those who are actively seeking out places they have seen in a film.
I would argue that there needs to be a fourth category added to this list.
IV. The Accompanying Traveller: Those who are travelling with Tourist(s) I-III but have no interest in the (film) or location.

Quite often those accompanying the screen tourist neither have the inclination nor need to visit the locations. The enthusiasm felt by one traveller cannot really be passed on to others unless they have the shared experience of a film. The needs of these people need to be accounted for, if practicable, within the tourism development template. These people will often travel to one or two locations to satisfy the needs of the film tourist, but if other tourist attractions or historical information can be included in the experience, the needs of these additional travellers are somewhat placated.

The results from this are twofold. Revenue for the tour can be increased and the location/film is exposed to more potential visitors (through social media exposure).

Norway – A Multimedia Case Study[20]

In 2015, I was the unit stills photographer on a film called *Birkebeiner* (released internationally as *The Last King*). The film is an adaptation of the true story of two Norwegians who, in 1217, carry a baby, the heir to the throne, across the mountains of Norway. The feat is celebrated every year in Lillehammer when hundreds of skiers recreate the event. It is part of the Norwegian culture, and it was made into a film (Figure 2.6).

As well as capturing the images for production, the regional tourism groups then commissioned the author to create a self-published guide for visitors. This

FIGURE 2.6 The Last King – telling a 1,000-year-old story and utilizing historic landscapes in Norway. (Left to Right) Skjervald (Jacob Oftebro), Baby Håkon (Jonathan Oskar Dahlgren), and Torstein (Kristofer Hivju).

Source: Image courtesy: Paradox Film/Ian Brodie.

low-budget book sold 1,000 copies in the region and provided an outlet to showcase not only the locations but also the hotels, cafes, and adjacent attractions.

An interesting adaptation of screen tourism was created by one of the local horse stables that provided Icelandic horses for the feature. I was employed by Innovasjon Norway to consult with them regarding opportunities, and the end result was the purchase of some of the original cabins made for the production and placing them on their farm. This location now features a play that is performed in the snow annually in front of more than 5,000 people. It is based on the film, at one of the locations and provides numerous income streams throughout the region.

Opportunities exist to promote regions and film beyond the traditional means. Liaison between all parties provides the chance to expand this which provides multiple benefits – for both the region and the longevity of the film.

As discussed, consultation prior to film production is important for all stakeholders. In 2014, a Norwegian production company prepared to film a mini-series for NRK based on the destruction of the Heavy Water Plant in Rjukan by saboteurs. These events are well known by historians worldwide, and the story is a major part of Norwegian WWII history. In 1965, it was portrayed in the British feature film *The Heroes of Telemark*.

The head of the local tourist office perceived an opportunity to promote the region because of the series, and the author was ultimately commissioned to both be embedded onset (for specific, related images) and create an enhanced e-book. This comprehensive book was subsequently published on the Apple Books and Google Play platforms, and images were used to promote the region as a tourist attraction.

Multimedia production allows film tourism-related material to be used in several different formats allowing multi-pronged promotion of both the film and the locations.

The Future of Film Tourism

In the last few years, I became a co-founder of HIDDEN, which specializes in augmented reality applied to tourism guide applications. Augmented/Virtual Reality and modern mobile technology create many new opportunities not only for the screen tourist but also for the studios. Most productions now feature some form of CGI or, at least, the scenes are rendered initially in pre-visualization. Can these models be easily imported into an augmented reality application that is available to fans and enthusiasts?

Imagine walking through a set from *The Foundation* while visiting Tenerife. See the additions made by the crew and experience the location in augmented reality. The quality of rendered landscapes in computer simulations is now of a quality that can be used by the screen tourist. The future of screen tourism is increasingly portable, personalizable, participation-friendly, and immersive.

Apart from websites, panoramas, and augmented reality, drones and game engines are also adding to our palette of digital tools. I recently utilized Microsoft Flight Simulator 2020 to research and confirm a location in the film *Paris, Texas*. The opening scene portrays Travis Henderson (Harry Dean Stanton) walking in a fugue state through a Texan desert location. Enthusiasts had pinned down this place to "somewhere near here" on private land near to Big Bend National Park.

FIGURE 2.7 Exploring remote film locations in the comfort of your own home. The landscape that appeared in Paris, Texas, as seen from Microsoft Flight Simulator (Xbox).

By flying an aircraft at a low level around the region (Figure 2.7). I was able to accurately pinpoint the location and match it to a frame from the film. Screen media is increasingly varied, flexible, and ubiquitous, with far more powerful technology available at far more accessible prices. The emerging technologies for screen tourism offer up tantalizing prospects not only for screen tourists but also for filmmakers and the tourism industry.

Notes

1 R. Wise, *The Sound of Music* [Film]. 20th Century Fox, 1965, www.britannica.com/topic/The-Sound-of-Music-film-by-Wise.

2 G. Nisbet, Salzburg's Sound of Music: The hills are alive with nostalgia. *West Travel CLub*. Salzburg's Sound of Music: The hills are alive with nostalgia, undated, 7 July. https://westtravelclub.com.au/stories/the-hills-are-alive-with-nostalgia.

3 T. Brueggemann, In 1969, Dennis Hopper and Peter Fonda's 'Easy Rider' Pushed Hollywood Into the '70s, 2019.

4 Internet Movie Database (IMDB) (unknown), *Zabriskie Point*. Accessed 7 July from, www.imdb.com/title/tt0066601/.

5 K. Singh and G. Best, Film-induced tourism: Motivations of visitors to the Hobbiton movie set as featured in The Lord of the Rings. International tourism and media conference proceedings, 2004. https://www.researchgate.net/publication/251790011_Film-Induced_Tourism_Motivations_of_Visitors_to_the_Hobbiton_Movie_Set_as_Featured_in_The_Lord_Of_The_Rings.

6 C. Pinchefsky (2012, 7 July) "The Impact (Economic and Otherwise) of Lord of the Rings/The Hobbit on New Zealand," *Forbes*, www.forbes.com/sites/carolpinchefsky/2012/12/14/the-impact-economic-and-otherwise-of-lord-of-the-ringsthe-hobbit-on-new-zealand/?sh=3f705e6b31b6.

7 New Zealand Tourism Research. (Undated), *INSIGHTS & RESEARCH*. accessed 7 July from, www.tourismnewzealand.com/insights/.

8 I included GPS locations in *The Lord of the Rings Location Guidebook*. This was before the introduction of smartphones with GPS. Many tourists travelled with separate hand-held devices specifically for this purpose.

9 A. Waysdorf and S. Reijnders (2017), "The role of imagination in the film tourist experience: The case of Game of Thrones," *Participations Journal of Audience and Reception Studies*. Accessed July 22, 2022, from www.participations.org/Volume%2014/Issue%201/10.pdf.

10 A NZ-based coach tour company formed after the release of *The Fellowship of the Ring* that specialized in Lord of the Rings tours.

11 Anonymous. (n.d.), *Breaking Bad (Episode 62)*. Internet Movie Database. Accessed 7 July from www.imdb.com/title/tt0903747/.

12 E. Vanderhoof (2018, 7 July), "Ten Years Later, Albuquerque Is Still Breaking Bad's Town," *Vanity Fair*, www.vanityfair.com/hollywood/2018/01/albuquerque-breaking-bad-tourism-10th-anniversary.

13 Film tourism establishes a link between the movie characters, locations and stories, and the film-tourists, who are inspired to immerse themselves and relive again the movie-generated and movie-driven emotions at the location of the movie. M. Gjorgievski and S. Melles Trpkova, "Movie Induced Tourism: A New Tourism Phenomenon," *UTMS Journal of Economics* 3, no. 1 (2012): 97–104.

14 C. William, Thailand's Maya Bay, location for The Beach, to close to tourists. *The Guardian*, July 7, 2018, www.theguardian.com/travel/2018/feb/14/thailand-maya-bay-the-beach-movie-close-to-tourists-leonardo-dicaprio.

15 E. Connolly (2008ibid.), Too long and too many cliches: critics pan Australia, the movie, www.theguardian.com/travel/2008/nov/19/australia-movie-baz-luhrmann-tourism.
16 Great Southern Development Commission (GSDC). (2022, unknown), *H IS FOR HAPPINESS*. Accessed 7 July from www.gsdc.wa.gov.au/our-activities/film-great-southern/filmed-here/h-is-for-happiness/.
17 New Zealand Film Commission. (2021), *The Power of the Dog*. New Zealand Film Commission. Accessed 7 July from www.nzfilm.co.nz/international/showcase/power-dog-0.
18 S. Hudson and J. Ritchie, "Promoting Destinations via Film Tourism: An Empirical Identification of Supporting Marketing Initiatives," *Journal of Travel Research – J TRAVEL RES* 44: 387–396, https://doi.org/10.1177/0047287506286720.
19 N. Macionis, "Understanding the Film-Induced Tourist," International Tourism and Media Conference Proceedings (Melbourne, Australia, 2004).
20 The author has been involved in film tourism in Norway since 2012 when he was invited to be a keynote speaker at a film tourism conference organised by Lillehammer University.

References

Anonymous. (n.d.). "Breaking Bad (Episode 62)." *Internet Movie Database*. www.imdb.com/title/tt0903747/.

Brodie, I. "Rams: A Film Location Guide to the Motion Picture Rams, Filmed in Mount Barker, Western Australia." 2022. https://express.adobe.com/page/VZfcDgJFu5Md5/.

Brueggemann, T. (2019). "In 1969, Dennis Hopper and Peter Fonda's 'Easy Rider' Pushed Hollywood Into the 70s." https://www.indiewire.com/2019/07/easy-rider-1969-dennis-hopper-jack-nicholson-50-year-anniversary-70s-1202157254/.

Connolly, E. "Too Long and Too Many Cliches: Critics Pan Australia, the Movie." *The Guardian*, July 7, 2008. www.theguardian.com/travel/2008/nov/19/australia-movie-baz-luhrmann-tourism.

Gjorgievski, M., and S. Melles Trpkova. "Movie Induced Tourism: A New Tourism Phenomenon." *UTMS Journal of Economics* 3, no. 1 (2012): 97–104.

Great Southern Development Commission (GSDC). "H is for Happiness." 2022. Accessed July 7. www.gsdc.wa.gov.au/our-activities/film-great-southern/filmed-here/h-is-for-happiness/.

Hudson, S., and J. Ritchie. "Promoting Destinations via Film Tourism: An Empirical Identification of Supporting Marketing Initiatives." *Journal of Travel Research* 44 (2006): 387–96. https://doi.org/10.1177/0047287506286720.

Internet Movie Database (IMDB). (unknown). "Zabriskie Point." Accessed July 7. www.imdb.com/title/tt0066601/.

Macionis, N. "Understanding the Film-Induced Tourist." International Tourism and Media Conference Proceedings. Melbourne, Australia, 2004.

New Zealand Film Commission. *The Power of the Dog*. New Zealand Film Commission, 2021. www.nzfilm.co.nz/international/showcase/power-dog-0.

New Zealand Tourism Research. (undated). "Insights & Research." www.tourismnewzealand.com/insights/.

Nisbet, G. "Salzburg's Sound of Music: The Hills Are Alive With Nostalgia." *West Travel Club*, undated, July 7. https://westtravelclub.com.au/stories/the-hills-are-alive-with-nostalgia.

Pinchefsky, C. "The Impact (Economic and Otherwise) of Lord of the Rings/The Hobbit on New Zealand." *Forbes*, July 7, 2012. www.forbes.com/sites/carolpinchefsky/2012/12/14/the-impact-economic-and-otherwise-of-lord-of-the-ringsthe-hobbit-on-new-zealand/?sh=3f705e6b31b6.

Singh, K., and G. Best. "Film-Induced Tourism: Motivations of Visitors to the Hobbiton Movie Set as Featured in The Lord of the Rings." International Tourism and Media Conference Proceedings, 2004. https://www.researchgate.net/publication/251790011_Film-Induced_Tourism_Motivations_of_Visitors_to_the_Hobbiton_Movie_Set_as_Featured_in_The_Lord_Of_The_Rings.

Vanderhoof, E. "Ten Years Later, Albuquerque Is Still Breaking Bad's Town." *Vanity Fair*, July 7, 2018. www.vanityfair.com/hollywood/2018/01/albuquerque-breaking-bad-tourism-10th-anniversary.

Waysdorf, A., and S. Reijnders. "The Role of Imagination in the Film Tourist Experience: The Case of Game of Thrones." *Participations Journal of Audience and Reception Studies* (2017). Accessed July 22, 2022. www.participations.org/Volume%2014/Issue%201/10.pdf.

William, C. "Thailand's Maya Bay, Location for the Beach, to Close to Tourists." *The Guardian*, July 7, 2018. www.theguardian.com/travel/2018/feb/14/thailand-maya-bay-the-beach-movie-close-to-tourists-leonardo-dicaprio.

Wise, R. (1965). *The Sound of Music* [Film]. 20th Century Fox. www.britannica.com/topic/The-Sound-of-Music-film-by-Wise.

3

WINDSHIELD TOURISM GOES VIRAL

On YouTube Scenic Drive Videos of U.S. National Parks

Kornelia Boczkowska

Introduction

Today, scenic drive videos are part of a burgeoning ecomedia[1] genre and are strongly embedded in the discourse of windshield tourism and wilderness, originally proposed by Edward Abbey in 1988.[2] David Louter elaborates on the long-established culture of scenic road and auto-tourism and argues that U.S. national parks have acted as "windshield wilderness, places where the relationship between automobiles and nature seemed to be mutually beneficial."[3]

Videos featuring national parks share a visual strategy with other scenic drive videos as they embrace the road movie's aesthetics, documentary and sensory realism, slowness, duration, anti-narrative and a more artisanal, personal mode of filmmaking, simultaneously negating mainstream culture and commercial media to provide an alternative breathing space in an oversaturated media landscape. Resonating with and generating an embodied spectatorship, defined as "a sense of psychical touching or being touched engendered by an organization of the film image,"[4] such works also invite viewers to perceptually approach the scenic roads at a close proximity so that they can literally attend to the surface detail, texture, shape, or other visual and tactile elements of the medium itself and the on-screen representation of national parks.

Ever since their establishment, national parks have been inextricably bound with what Louter labels windshield tourism. Originally founded as "pleasuring grounds,"[5] America's most spectacular landscapes were preserved for the benefit of the general public by the National Park Service's (NPS's) leaders who embraced automobile tourism to boost park visitation and accommodate a rising flow of motoring travellers. Indeed, the advent of the motor vehicle has significantly contributed not only to the rise of self-drive tourism, characterized by independence,

DOI: 10.4324/9781003327585-3

flexibility, and exploration[6] but also to the growth of national park tourism and many other tourist destinations.

With the emergence of car-based tourism in the early 20th century, which replaced the railroads and increased passenger traffic and ticket sales in U.S. national parks, the automobile has become the major means of transportation for tourists and recreation vehicle users visiting these unique sites of American wilderness. In contrast to destination tourism, windshield or drive-through tourism renders driving and "touring for pleasure" (Pearce 1999)[7] the central part of a visitor's travel experience. This mobilizes a more traditional static tourist gaze, defined as a relational, communal, and socially and technologically mediated practice of gazing, which entails a visual consumption of different places through images and the actual travel to and appreciation of such sites.[8]

This transition exemplifies the changing nature of tourism mobilities and their evolution from a corporeal to a virtual tourism. Building on this assumption, I present a brief historical contextualization of scenic roads in U.S. national parks, followed by a quantitative and qualitative analysis of YouTube's most watched scenic drive videos. The study sheds light on how car-driven leisure and virtual windshield tourism and the wider practices of automobility in national park locations link to the aesthetics of YouTube scenic drive videos.

While echoing the new mobilities paradigm, scenic drive videos heavily rely on some key conventions of the classic road movie and simultaneously fit into an emergent category of YouTube travel videos devoted to conveying and documenting self-drive tourists' national park experience and the larger tradition of touring for pleasure.

Scenic Roads in U.S. National Parks

Ever since the car was invented at the turn of the 20th century, automobile travel has become one of the most popular leisure activities in U.S. national parks, prompted by the construction and development of scenic roads and highways, which have significantly enhanced visitor experience. Encouraged by the expansion of automobile tourism in the interwar period and the promise, grandeur, and mystique of sublime, rugged landscapes and wide-open spaces, many Americans took to the road to encounter parks from the confines of their own cars and demonstrate their patriotism. A more common use of the automobile for nature tourism in the postwar era was possible due to the development of public tourism infrastructure, managed by the National Park Service since 1916, and the Civilian Conservation Corps' provision of cheap labour for road construction during the New Deal, which resulted in the proliferation of scenic roads, parkways and highways, such as the Going to the Sun Road (Glacier, 1933), Rim Drive (Crater Lake, 1941), Trail Ridge Road (Rocky Mountain, 1921), the Wawona Road (Yosemite, 1932), or the Zion-Mount Carmel Highway (Zion, 1930).[9]

By the early 1940s, the National Park Service had completed its flagship road projects, notable for their outstanding engineering and landscape design standards, which served as an embodiment of American progress and "achieved the ultimate goal in scenic preservation" and wildlife protection.[10] However, the construction of two transmountain highways, the Going to the Sun Road and Trail Ridge Road, also marked the end of the era of scenic highways in U.S. national parks, as the project successfully fulfilled Stephen Mather's and Horace Albright's visionary road policy, advanced in the 1920s, especially Mather's concept of national parks organized by motorized mobility, a highway system, and automobile travel.[11]

Propelled by road construction and the emerging highway landscape, windshield tourism and recreational driving created a new category of adventurous travellers lured to specific localities in national parks with stunning landscapes, breathtaking views and the thrill of driving transmountain highways in scenic surroundings. Although today national parks rely on a more extensive infrastructure, including paved roads, parking lots, shuttle bus systems, overnight accommodations, and other commercial facilities, they are primarily considered an auto-friendly setting targeted at middle-class Americans, in which a park visit often becomes a drive-through experience.

With cars still being the primary form of transportation for visiting the NPS areas, recreational driving is an important part of the national park experience and the viewing of car window-framed, postcard-like vistas embody "the freedom and anarchic wandering of the American on-the-road culture."[12] Although national park visits declined between 1988 and 2006 due to the rise in electronic entertainment media and a shift in recreation choices that involved foreign travel,[13] the NPS data indicate that there was a steady increase in visits per year between 2006 (over 272 million) and 2019 (over 237 million) before the pandemic broke out.[14]

I define scenic roads and highways as roadways, which provide an opportunity for recreational travel through a national park and access to vista points, overlook structures, boardwalks, trails, wildlife, and outdoor activities available along the route. Recent findings confirm that the scenic, tranquil, and relaxing value of leisurely driving along with freedom, safety, and convenience of travel are cited as the most frequent reasons why visitors enjoy the national park experience and see the road not necessarily as a means of transportation, but as a means of exploring the area and viewing scenery.[15] As in the 1920s and 1930s, the NPS landscape architects' policy was to integrate the parks' built environment into the natural landscape so that they harmonize with each other, scenic roads were also constructed in a way that their disturbing impact on the surrounding wilderness was kept to a minimum.[16] To fulfil this premise, scenic road construction in national parks was influenced by Andrew Jackson Downing's 19th-century idea of building circular roads in parks and gardens, which was soon adopted by national park designers.

As a result, with the support of the Bureau of Public Roads, numerous loop roads were constructed, which enabled the circular movement of vehicles, created flow through the parks and provided a relatively easy access to the so called "intermediary landscapes"[17] that mediated between the expectations of car tourists and the preservation of pristine, unspoiled nature. Some scenic roads were even designed specifically to enable recreational driving within certain national parks, such as the Blue Ridge Parkway (links the Great Smokey and the Shenandoah), which is currently the most visited NPS site, the Grand Loop Road (Yellowstone), the Teton Park Road (Grand Teton), the Park Loop Road (Acadia), and many others.

Moreover, one of the key factors that influence the quality of driving along national park scenic roads are speed limits, which lower travel speeds and ensure safe sightseeing, aesthetic roadway properties, which provide stopping areas and routes based on scenery and panoramic vistas, and the minimization of features and facilities typical for urban roads, such as traffic signals or signalized intersections, which distract tourists from the scenic aspects of driving for pleasure.[18]

Film Tourism, National Parks, and Scenic Drive Videos

It is well documented that viewing films and TV shows can create and increase visitation to viewed sights.[19] The current research shows that film and media spur and drive tourism through reaching broader audiences more effectively than focused tourism marketing and that 40 percent of visitors are likely to travel to certain destinations as a result of seeing them on screen.[20]

Although no data are presently available that shed light on how Hollywood and independent productions featuring some iconic national park locations affect the development of national park tourism in the United States, it goes without saying that film and media play a crucial role in tourists' choice of travel destinations. The most recent study suggests that NPS official units'[21] contribution to the entertainment, film and television industry is significant (though often overlooked) as they provide some instantly recognisable, long-lasting and memorable images of "grand untouched scenery" for numerous films and TV shows or programs.[22]

Between 1916 and 2013, many NPS units, including national parks and monuments as well as historic sites, were featured in at least one scene in 422 films and TV shows, which is most likely underestimated since the NPS does not keep track of all permits for filming issued since the early 20th century.[23] Some of the biggest hits and blockbusters filmed in the actual park site or its likeness preserved by the NPS unit, which generated several million or billion dollars in profit for studios, include *Butch Cassidy and the Sundance Kid* (Zion, 1969), *ET* (Redwood, 1982), *National Lampoon's Vacation* (Grand Canyon, 1983), *Rocky IV* (Grand Teton, 1985), *Star Trek V* (Yosemite, 1989), *Dances With Wolves* (Badlands, 1990), *Jurassic Park* (Redwood, 1993), *Forrest Gump* (Glacier, 1994), *Brokeback Mountain* (Grand Teton, 2005), and *Into the Wild* (Denali, 2007).[24]

Although the existing statistics indicate certain trends, it is unknown whether the box office success of these productions had any direct impact on the increase in park visitation and ticket sales in the years following their release. For example, according to the University of Montana and Montana Film Office studies, with tourism and hospitality industry being Montana's second largest sector, film-induced tourism has become increasingly significant for the state's economy and as a travel motivator in the last 20 years.[25] Between 1991 and 2019, the number of tourists who visited Montana rose from 7.1 million to 12.6 million, but even though many out-of-state visitors travelled to some major landmarks, including Glacier National Park, it is not indicated if their choice was influenced by cinematic or media representations of this site.[26]

Similarly, the impact of scenic drive videos on national park visitation has never been explored, although the use of YouTube travel videos in tourism and tourism education has become increasingly popular even prior to the Covid-19 outbreak.[27] The most recent study suggests that in 2014, subscriptions for the top YouTube travel channels increased by 106 percent from the previous year.[28] With YouTube becoming the largest video community in the world and the top online streaming platform and video storage, scenic drive videos are among its most-viewed travel videos and have proven to be a vital information, entertainment, and education channel for both individual users and travel and hospitality industry.

Indeed, the last decade has seen a substantial growth in the number of scenic drive videos featuring various locations, which is part of the larger phenomenon of travellers sharing their experiences and seeking inspiration for their next destination. The recent studies conducted by Google with Ipsos MediaCT confirm that YouTube serves as an ideal venue for tourists' exploration of and interaction with video content and that videos posted on the platform influence individual travel decisions as two out of three U.S. consumers view them when considering taking a trip.[29]

Echoing this trend, YouTube scenic drive videos featuring national park locations, produced by actual windshield tourists rather than the NPS or official tourism organizations, can be considered a distinctive object of inquiry and a useful promotional communications tool, which shows a growing significance of the individual consumer in tourism marketing and their power to affect one's social and cultural construction of the world.

While more and more users consume travel-related videos and spend more time viewing them, YouTube data show that 27 percent of travel video views by content type were for personal travel footage and tourist destinations and attractions accounted for 40 percent of the total travel views with viewing up to 179 percent in only over one year.[30] However, there is no research on how users engage with scenic drive videos and what role such videos play in mediating tourism experiences. In contrast to short-form travel videos posted on YouTube, TikTok, Instagram, or Snapchat, scenic drive videos shot in national parks, although also shaped by tourist culture, provide less personalized content and oppose fast-paced modern life or convenient and fast short content consumption.[31]

As the driver typically remains hidden from view, they are less concerned with social identity and self-representation, especially in the context of tourism, and consequently, they are less capable of capturing user attention, as they primarily engage with an environmentally conscious discourse and reflect on lived experiences. In doing so, they fit in with the broader trend of ecomedia and ecocinema to offer an alternative way of engaging with a technologically mediated cinematic representation of the natural world.[32]

Typically digitalized, authorless, ubiquitous, and situated within the visual culture of conservation movements,[33] scenic drive videos function akin to nature videos as they mediate virtual tourists' eco-gaze and affective response to landscape through voyeurism and the conflict between the virtual and the real.[34] As YouTube is currently the world's second largest online search engine after Google with users watching over one billion hours videos per day,[35] it has also become the most popular travel video sharing platform, which effectively converts viewers into virtual tourists.[36]

Scenic Drive Videos Go Viral on YouTube

The first scenic drive videos were posted on YouTube in 2010 and featured timelapse recordings of users' car trips through Arches and Rocky Mountain National Parks, accompanied by the actual soundtrack, with views captured by the camera placed right in front of the windshield.[37] Unsurprisingly, while such videos amounted to only circa 4,110 in 2010, their number has increased by almost 4,000 percent over the last 20 years. For the purpose of this study, I have conducted a search on YouTube for videos tagged with the keywords "scenic drive videos national parks USA" and sorted the search results in descending order by the number of views, which has revealed videos with the highest exposure to the public.

As for January 2022, circa 162,000 videos have been posted on YouTube lasting from a few minutes to over six hours, which feature scenic drives through U.S. national parks. Some of the most iconic videos include (i) over 1,500 videos of Glacier Point Road and over 500 videos of the Wawona Road in Yosemite; (ii) over 1,400 videos of the Trail Ridge Road in Rocky Mountain; (iii) over 1,200 videos of the Zion-Mount Carmel Highway in Zion; iv) over 500 videos of the Going to the Sun Road in Glacier; (v) over 400 videos of the Rim Drive in Crater Lake.

Meanwhile, in terms of views, the three most popular scenic drive videos, shot in real time and accompanied by environmental sounds, were made by user called J Utah, a "frequent traveler with a camera." They include Zion National Park 4K – Scenic Drive – Utah USA (55 minutes; 684,246 views; posted on 2 September 2020), Arches National Park 4K – Scenic Drive – Utah USA (58 minutes; 682,568 views; posted on 23 September 2020), and Drive 4K – America's Most Popular Park – USA, which features Great Smoky Mountains National Park (1 hour 8 minutes; 680,724 views; posted on 26 September 2016). This largely reflects the recent NPS data, which show that the top three most visited parks in 2020 were Great Smoky Mountains National Park (12.1 million recreational visits), Yellowstone National Park (3.8 million recreational visits), and Zion National Park (3.6 million recreational visits).[38]

The majority of scenic drive videos can be characterized by a persistent use of fixed camerawork and single take aesthetics. Despite certain similarities, however, each video depicts a different stretch of the route in a national park and windshield views of the road and adjacent mountains, deserts, canyons, forests, lakes, or meadows. For instance, the most popular scenic drive videos on YouTube, Zion

National Park 4K, Arches National Park 4K and Drive 4K, cover the routes from West Main St, Springdale, UT, to Mt Carmel, UT, from Moab, UT, to Devil's Garden Trailhead, UT, and from Aloe Vera of California Inc., Gatlinburg, TN, to Whittier, NC. Consequently, the trips' highlights include passing through Rockville, Springdale, the Zion-Mount Carmel Tunnel, Petrified Dunes, the Appalachian Trail that runs through the centre of the Great Smoky Mountains National Park, Morton Overlook, and several spectacular vista points. In addition, whereas most videos, including Zion National Park 4K, Arches National Park 4K, and Drive 4K, are shot in real time, feature diegetic and environmental sounds and use the windshield as the key organizing principle in their representation of automobile travel and the driving event, only few are time lapse videos and include the soundtrack and occasionally side window views.

The former category is more prevalent on YouTube and likely to serve the basic functions of online travel videos seen as mediators of tourist experiences in the sense that they "provide mental pleasure to viewers by stimulating fantasies and daydreams," bring back past travel memories and "act as a narrative transportation, providing access to foreign landscapes and socioscapes."[39] At the same time, just like nature cam (movies of nature),[40] relaxation, and YouTube travel videos, they exist within relatively flexible networks of production and circulation and thus lack much aesthetic value.[41] Instead, they serve as an influential source of tourism information and a promotional tool for national park locations that fulfils viewers' desire for a more direct and palpable contact with the wilderness and remote places.

In doing so, most scenic drive videos build intensive time connections between virtual tourists, featured remote sites, and encourage both attentive viewing and place attachment to elicit intimate feelings about and sensuous knowledge of a given locale. With views recorded by dash cams or vehicle dash-mounted cameras, which typically constitute a valuable source of information for road safety planners, the videos present driving along scenic routes mostly through the windshield, seen as a key component of tourist experiences.[42]

From Windshield Wilderness to Virtual Windshield Tourism

In scenic drive videos, the windshield acts as the central framing device, which unfolds passing landscapes as "a narrativized screen space" and converts them into objects of visual pleasure.[43] Echoing William Gilpin's work on picturesque tourism, particularly his landmark *Remarks on Forest Scenery*, highway views tend to transform into a picture, which, once caught in a glass reflection, becomes fixed. Similarly, Dimendberg argues "the highway provides a controlled visual experience analogous to the montage and multiplicity of perspectives afforded by cinema," which he further links to the idea of unrestrained progress and individual freedom.[44] In his study of the Mojave Desert's barren, inhospitable, yet exceptionally stunning landscape, seen from a moving car and known for its scenic beauty, Banham refers to the views as "the alternating sequence of broad slopes" whose affective power lies in "the stately adagio rhythm" that stems from the forward speed of the vehicle.[45] Since scenic drive videos often require an intensely corporeal mode of viewing performed by an embodied rather than disembodied spectator, their apprehension of a changing environment not only is mediated by particular sensations generated by a motorway travel but also emerges in "a confrontation between two worlds and the pictorial packaging of landscape as spectacle."[46] In other words, virtual tourists witness the passing views as "strongly mediated by frames – the car windows and mirrors – and by the sense one has inside the car of being insulated and independent from the natural world."[47]

In particular, the visualities enabled by freeway travelling restrict the windscreen views to the so-called plunging perspective, also known as dromoscopic perception, that is, "a headlong immersion into a free space of movement around which buildings recede."[48] In Zion National Park 4K, Arches National Park 4K, and Drive 4K, the visualities of the roadside scenery are constituted by the unique geology of canyons, mesas, sandstone arches, buttes, mountains, and a variety of tree and plant species. However, unlike driving at speed, travelling in scenic drive videos provides a largely de-dramatized, serene, and tranquil view of national park locales, demands patient viewing, and provides an almost tangible point of contact between the virtual reality and the actual location through imitating the real-time connection.

At the same time, it offers a highly immersive VR and first-person driving experience along the scenic routes seen an affective space, that is the first-person, direct and lived experience of the environment that affects one's emotional states and perception of the world.[49] In scenic drive videos, affective space is usually conveyed through such works' dependence on the chief conventions of road movies, particularly the travelling shot located with the driver or car, and environmental storytelling, which establishes plot points and increases suspense.[50]

In scenic drive videos, driving functions as the primary mode of representation, and the serene first-person travelling shot not only enable virtual tourists to actively retrain their perception and possibly rediscover a mediated connection to nature but also make them part of the larger slow movement. Particularly, due to a rhythmic and non-goal driven nature of the ride, the tracking shot coordinates the surrounding space, simultaneously intensifying and hindering vision, and foregrounds an unemphatic, calm movement, one that lacks special stress or emphasis and is integral to the idea of duration rather than time passing.

As the footage in scenic drive videos is often unedited and running in real time and forward motion, windshield views simply emerge as a sequence of forward-travelling POVs presented from a single and unchangeable setting size, as a result of which the national park scenery forms both an integrated and fragmented vision. Interestingly, in videos shot in real time, motion generates subtle sensory experiences rather than a strong sense of kinaesthetics and inculcates an almost tactile encounter with the mise-en-scène.

In contrast to driving at speed, driving in scenic drive videos is relatively slow due to the speed limits imposed in most national parks, which highlights continuity rather than creative montage, sensitizing the viewer to perceive environmental change as micro-movements and an ongoing source of contemplation. Interestingly, as the videos impose limits on the degree of virtual tourists' interaction with the surrounding environment and their viewing perspective is restricted by the fixed camera position, the video presents a strictly linear experience of moving through the national park and, akin to other drive and relaxation videos, relies on the long take aesthetics and a continuous tracking shot to create the impression of a constant unfolding.

However, despite clearly conveying a sense of exploration, the long-tracking shot prevents virtual tourists from getting the complete view of the surroundings, simultaneously allowing them to literally move with the driver and immerse into the narrative. Since the camera moves smoothly and consistently through space horizontal to the ground in a forward direction, it tends to merge the external space and interiority, the latter of which is represented by the driver's identity and psychological state. In other words, a single dolly shot and longitudinal camera movement act as spatial and temporal metaphors, which project the traveller's psychical state through simplifying the dynamics of driving.

In scenic drive videos, a vehicle-mounted travelling shot functions primarily as an observational tool, as it creates the illusion of an expanding or contracting driving space in national park locations, facilitates the forward movement of virtual tourists, and guides their gaze by bringing them closer to the roadside scenery and putting their mind in a blissful state of heightened relaxation and awareness of the driver's transformed relationship and interaction with nature.

Drawing parallels between windshield tourism and cinema, the camera works to create the feeling of cinematic immersion, which on the one hand gives virtual tourists both an observational and participatory role, but on the other hand restricts their perspective to frontal views only, making them more aware of the direction rather than of the surrounding space. Therefore, although the shot provides some elements of orientation and balance, which reinforces viewers' perceptual engagement with the imagery and their non-judgemental observation, a largely diminished spatial awareness, enhanced by purely diegetic sounds, mostly coming from the car's engine, gives rise to a sense disassociation.

Nevertheless, while the video's long-tracking and single-shot structure destabilizes virtual tourists' spatial relationship to the narrative and prevents their experience of interiority, evident in the fact that they are blinded to the space located outside of the camera's range, the driving event is always paired with an emphatic moment of association with nature. Echoing the recent findings in automobilities research, scenic drive videos articulate an affective experience of highway driving, presenting it as a relaxing, nostalgic, thought-provoking, and all-encompassing experience. In particular, although autospace in national park locations, understood as the scenic road along with the adjacent driving spaces, appears as inherently linear, connected, and predictable, the mobilities of driving and being a passenger are integrated here with what Pearce terms "automotive consciousness"[51] and consequently challenge a dull and desensitizing nature of routine driving.

Whereas the car and, by extension, driving event can serve as a means of escape, meditation, daydreaming, problem-solving, or phenomenological encounters,[52] the practice of driving through national parks may be also conceptualized as a process of becoming or the flow experience, where sensations, fears, desires, memories, and stories constitute the many forms of automobile travel. Therefore, virtual tourists' experience is dependent not only on the mobility of the vehicle or driver car but also on the aesthetics and dynamics of the highway landscape, which may be considered "an extension of the search for the natural sublime,"[53] traditionally associated with the vast, infinite, obscure and fearsome qualities of nature capable of evoking one's intense emotional response to American wilderness.[54] As "the driver absorbs many different items of information from the environment and from the road via the oscillations and vibrations of his vehicle,"[55] driving through national parks becomes a form of corporeal experience, which literally enables one's immersion into the surrounding environment and gives rise to an aesthetically pleasurable spectacle.

The straight-ahead views of the road clearly obstruct virtual tourists' panoramic perception and form an obstacle to the landscape gaze. On the other hand, though, a traveller's alienation from the physical environment creates a distance that directly corresponds to the sublime experience, even though it results from what Desportes labels a fractal mode of perception rather than more holistic ways of seeing.[56] Interestingly, it is both alienated and fragmented perception that augments the corporeality of car travel through building a strong sense of ephemeral landscape, which is subject to the changing weather, traffic conditions, and surroundings.[57]

Although limited, viewpoints provided by scenic drive videos have a significant sensory potential and include framed, panoramic, focused, sequential, and unfolding views. As spectators strongly respond to scenic landscapes on an intuitive and emotional level,[58] driving emerges not as a solely visual experience "reduced to the two-dimensional view through the car windscreen,"[59] but rather as a multi-sensory and kinaesthetic activity that directly links human bodies to national park locations. Since tourism is predominantly defined as a visual practice and way of seeing,[60] many scenic drive videos facilitate a wandering gaze, which shares the penchant for pleasure and spectacle consumption with the tourist gaze. While in such videos, tourism or ecotourism[61] usually organizes places as sights and the tourist gaze, which is now often replaced by virtual gazing, consumes them visually, tourist practices in national park locations involve corporeal movement and proximity.[62]

Although the focus of scenic drive videos is primarily on sightseeing and the virtual tourist's visual encounters with the roadside landscapes and soundscapes, they strongly imply that tourism also embraces the use of other senses and that a pervasive tourist gaze, be it glancing or gazing, is a mediated bodily experience that can be either powerful or powerless, objectifying or neutral, and active or passive.[63]

As the videos create a sense of freedom and cinematic immersion rather than a fleeting and haptic spectacle of cinematic excess, mostly related to the sense of touch, they directly oppose the nature of traditional tourism based on scheduled, fast, and mostly superficial visits to some landmark tourist destinations. Since scenic drive videos operate within the road movie and ecomedia framework, they both foreground and critique a sense of rapid movement crucial for tourism seen as a shallow consumption and pursuit of contrived, inauthentic and prefabricated experiences of being on the move, even if to some extent linked to adventure and venturing into the unknown.

Unlike some YouTube travel videos and vlogs, drive videos clearly work against consumerism and mass culture through their reliance on the imagery that mocks a flashy, instant, and miniaturized tourist world or a scanning, ephemeral tourist travel glance and evokes serenity, pleasure and contemplation rather than anxiety, fatigue, and hastiness. In Zion National Park 4K, Arches National Park 4K, and Drive 4K, the tranquillity effect is brought about by the driver's exposal to some outstanding natural features of national park locations, including rock canyons, towering cliffs, majestic mesas, exceptionally picturesque rock formations, lowland meadows, subalpine forests, and aspen groves, which gradually unwind on screen, dependent on the car's slow, progressive movement.

The fact that the ever-changing surroundings are always captured from the driver's point of view and through the windshield affects one's viewing experience, which not only evokes a sense of pastoral tranquillity but also allows for a contemplative observation and elicits a strong sense of involvement on part of the viewer. While celebrating the pleasures of tourism, the videos draw on cinema's use of modes of embodied experience and activate virtual tourists' imagination through appealing to their senses and encouraging them to develop and practice the art of looking and listening.

However, rather than perpetuating the act of seeing as objectifying, scenic drive videos undermine the power of the gaze, as one's vision is limited to a single viewpoint and consequently imply that travelling on screen also involves the spectator's imaginative activity and creativity. Due to the car's driving style and the driver's perspective restricted mostly or solely to the windshield, the scholarship around such videos is more concerned with exploring the relationship between the visible and invisible and enhancing a multi-sensory, bodily travel experience, which ultimately re-evaluates a superficial and evanescent nature of the tourist glance, replacing sight-seeing with an active, embodied practice of driving through national parks locales.

Conclusion: Scenic Drive Videos and Self-Drive Tourism in National Parks

Scenic drive videos are a relatively recent phenomenon, which constitute a widely accessible and popular form of travel videos and slow ecomedia with the aim of reaching a wider audience. Whereas the videos' fixed frame and extended duration convey a largely unmediated or uninterrupted representation of the national park scenery, the road movie's long-tracking shot and continuity editing facilitate an almost sensory relationship between audience and image, encourage embodied spectatorship, and create the impression of a holistic unity with nature.

The majority of such works use the windshield as the key organizing principle in their representation of automobile travel and the driving event and provides an almost unobstructed first-person perspective of the scenic road or highway with the reference point fixed permanently on horizon, which renders the journey immersive, seamless, and captivating. While the windshield, which acts as the central framing device, and slow ecomedia format unfold passing landscapes, automobility serves to aestheticize the car's external world by perpetuating tourists' panoramic perception and the traditional viewer-as-passenger's schema.

However, in the examined works, driving organizes virtual tourists' perception of the road, which celebrates ways of direct seeing and imposes an ecologically oriented gaze rather than the tourist gaze that typically objectifies national park locations. As they imitate the real-time connection between audience and image and offer an affective experience of the visited locales, scenic drive videos challenge perception as they open up a space, in which virtual tourists dwell on the windshield view and consequently reframe and mediate their national park experience. Through their partly incomplete depiction of a scenic road, which highlights the evasiveness of time and confounds one's spatial awareness, the videos help understand how drive tourism in national parks functions in the broader tradition of ecomedia, as they sustain viewers' attention and commitment to the surrounding environment and encourage them to bond with national park sites, rendering the car-driver an inseparable part of the biotic community.

Despite being significantly underrepresented in critical literature, virtual drive tourism takes different forms, including the 360 VR format, and has recently become the burgeoning and innovative industry sector with the potential to stimulate and facilitate *actual* national park tourism. Likewise, fuelled by technological developments, high levels of Internet and YouTube usage, and the Coronavirus outbreak, scenic drive videos have experienced unprecedented growth in the past few years and are likely to remain popular in the post-pandemic world and can possibly develop into more technologically advanced experiences, such as 360 virtual tours. Popular media have sparked and shaped tourism activities in U.S. national parks since the turn of the 20th century and contributed to invigorating and perpetuating visitors' perception of such places as sights of wonder or exotic, pastoral, picturesque, or romantic landscapes.[64]

Although the impact of scenic drive videos on film tourism has not been determined yet, especially in the context of virtual drive tourism, it seems that the analysed works are likely to facilitate national park visitation, as they reflect a pivotal role of automobile industry in leisure travel in the United States with cars still carrying more tourists than any other mode of transportation.[65] Despite the lack of the official federal scenic byways program,[66] the Federal Highway Administration (FHWA) recognized driving the road as Americans' favourite pastime and a "pleasurable recreational experience."[67]

This coincides with the findings of the 1990 Resources Planning Act that indicated that among all recreational activities in the United States, viewing scenery had the highest participation rate.[68] Today, most tourists still experience national parks within the confines of their own cars,[69] which equates transportation with recreation, and the driving tourist experience is becoming increasingly popular in online streaming platforms, affecting film and popular cultural tourism as it functions in the constantly changing media landscape.

Unsurprisingly, the media are often involved in triggering and promoting various tourist activities, which enthuses the mobility of vision and one's cognitive, imaginative and emotional engagement with the world[70] and consequently encourages virtual travel and post-tourism, the latter of which is understood as an expression of film tourism, one that shifts between the physical and virtual mobility enabled by the omnipresence of technology.[71] As seen, one of the most prominent tourist-generated media products, which convey the national park experience, is YouTube scenic drive videos, which have the potential to affect prospective visitors' understanding of such places through user-generated content and its imaginative force.

As part of the larger phenomenon of YouTube travel videos and amateur travel films, scenic drive videos are also site- and activity-centric and help promote national park tourism as they expand users' imagination through their focus on driving and a detailed coverage of one's trip. Most of them operate within the concept of virtual mobility and provide viewers with a touristic positioning, enabling them to access or literally tour national park scenic roads and highways, consequently generating mental pleasures and stimulating daydreams and reminiscences through imagination. It appears, then, that scenic drive videos can be seen as a powerful tool that intensifies potential travellers' interest in visiting national parks and sheds new light on the role of digital media and technology in assisting, mediating and boosting self-drive tourism experiences on YouTube and elsewhere.[72]

Funding

This work is funded by the National Science Centre, Poland, under the project *Lost highways, forgotten travels: The road movie in the post-war American avant-garde and experimental film through the lens of women and men filmmakers* (grant no. UMO-2018/31/D/HS2/01553).

Notes

1 Ecocinema and ecomedia are a relatively flexible trend in contemporary film and media that "provides an evocation of the experience of being immersed in the natural world," transcends traditional generic classifications and follows no particular conventions but for a widely defined preoccupation with environmental criticism, particularly concerning the experience of cinematic time and environmental change expressed with slow or moralist eco-aesthetics.

2 Edward Abbey, *Desert Solitaire* (Tucson: University of Arizona Press, 1988), 49.

3 David Louter, *Windshield Wilderness: Cars, Roads, and Nature in Washington's National Parks* (Seattle: University of Washington Press, 2006), 165.

4 Annette Kuhn and Guy Westwell, *A Dictionary of Film Studies* (Oxford: Oxford University Press, 2012), 201.
5 Robert Keiter, *To Conserve Unimpaired: The Evolution of the National Park Idea* (Washington, DC: Island Press, 2013), 41.
6 Anne Hardy and Ulrike Gretzel, "Why We Travel This Way: An Exploration Into the Motivations of Recreational Vehicle Users," in *Drive Tourism: Trends and Emerging Markets*, ed. Bruce Prideaux and Dean Carson (London: Routledge, 2011), 194.
7 Philip L. Pearce, "Touring for Pleasure: Studies of the Senior Self-Drive Travel Market," *Tourism Recreation Research* 24, no. 1 (1999).
8 John Urry and Jonas Larsen, *The Tourist Gaze 3.0* (London: Sage Publications, 2011), 201–205.
9 Suzanna Julin, "A Feeling Almost Beyond Description: Scenic Roads in South Dakota's Custer State Park," in *The World Beyond the Windshield: Roads and Landscapes in the United States and Europe*, ed. Christof Mauch and Thomas Zeller (Athens: Ohio University Press, 2008), 81; Louter, *Windshield Wilderness*, x-xi.
10 Louter, *Windshield Wilderness*, 79.
11 Louter, *Windshield Wilderness*, 79–80.
12 Fabio Colonnese and Paolo Rosa, "A Car with a View: Considerations on the Landscape Seen and Represented through the Windshield," in *Digital Draw Connections, Representing Complexity and Contradiction in Landscape*, ed. Fabio Bianconi and Marco Filippucci (Cham: Springer, 2021), 295.
13 Oliver Pergams and Patricia Zaradic, "Is Love of Nature in the US Becoming Love of Electronic Media? 16-year Downtrend in National Park Visits Explained by Watching Movies, Playing Video Games, Internet Use, and Oil Prices," *Journal of Environmental Management* 80 (2006): 387, 393.
14 "Visitation Numbers," *National Park Service*, accessed February 25, 2021, www.nps.gov/aboutus/visitation-numbers.htm.
15 Jeffrey Hallo and Robert E. Manning, "Transportation and Recreation: A Case Study of Visitors Driving for Pleasure at Acadia National Park," *Journal of Transport Geography* 17, no. 6 (2009): 496.
16 Peter Peters, "Roadside Wilderness: U.S. National Park Design in the 1950s and 1960s," in *Observation Points: The Visual Poetics of National Parks*, ed. Thomas Patin (Minneapolis: University of Minnesota Press, 2012), 62.
17 Ethan Carr, *Wilderness by Design: Landscape Architecture and the National Park Service* (Lincoln: University. of Nebraska Press, 1998), 92.
18 Hallo and Manning, "Transportation and Recreation," 497.
19 Sue Beeton, *Film-Induced Tourism* (Clevedon: Channel View Publications, 2002); Timo Thelen, Sangkyun Kim and Elisabeth Scherer, "Film Tourism Impacts: A Multi-stakeholder Longitudinal Approach," *Tourism Recreation Research* 45, no. 3 (2020): 291–306; Noelia Araújo Vila, José Antonio and Fraiz Brea Pablode Carlos. "Film Tourism in Spain: Destination Awareness and Visit Motivation as Determinants to Visit Places Seen in TV Series," *European Research on Management and Business Economics* 27, no. 1 (2021): 21–31.
20 "Kauai Makes List of Top 10 Film Locations," *Pacific Business News*, accessed March 27, 2012, www.bizjournals.com/pacific/news/2012/03/27/kauai-makes-list-of-top-10-film.html.
21 NPS official units include parks, preserves, reserves, sites, monuments, battlefields, parkways, recreation areas, rivers, seashores, lakeshores, scenic trails, etc.
22 Thomas J. Liu, John B. Loomis and Linda J. Bilmes, "Exploring the Contribution of National Parks to the Entertainment Industry's Intellectual Property," in *Valuing U.S. National Parks and Programs: America's Best Investment*, ed. Linda J. Bilmes and John B. Loomis (New York: Routledge, 2020), 94.
23 Liu, Loomis and Bilmes, "Exploring the Contribution of National Parks," 94.
24 Liu, Loomis and Bilmes, "Exploring the Contribution of National Parks," 98–100.

25 "The Economic Impact of Montana Film Production: An Analysis of the Industry and the MEDIA Act, 2019–2020," June 25, 2020 (Helena: Montana Film Office, 2020).
26 "The Economic Impact of Montana Film Production," 30.
27 Denis Tolkacha and Stephen Pratt, "Travel Professors: A YouTube Channel About Tourism Education & Research," *Journal of Hospitality, Leisure, Sport & Tourism Education* 28, no. 5/6 (2021): 1.
28 Hailey Crowel, Haley Gribben, and Jaclyn Loo, "Travel Content Takes Off on YouTube," Think with Google, August 2014, 5, www.thinkwithgoogle.com/consumer-insights/consumer-trends/travel-content-takes-off-on-youtube/.
29 "The 2014 Traveler's Road to Decision," Google/Ipsos MediaCT, June 2014, www.thinkwithgoogle.com/consumer-insights/consumer-trends/2014-travelers-road-to-decision/.
30 Crowel, Gribben and Loo, "Travel Content," 4.
31 Xin Du, Toni Liechty, Carla A. Santos, and Jeongeun Park, "'I Want to Record and Share my Wonderful Journey': Chinese Millennials' Production and Sharing of Short-form Travel Videos on TikTok or Douyin," *Current Issues in Tourism*, 2020, published online, 3.
32 Scott MacDonald, "Toward an Ecocinema." *Interdisciplinary Studies in Literature and Environment* 11, no. 2 (2004): 108–109.
33 Stephanie Lam, "It's About Time: Slow Aesthetics in Experimental Ecocinema and Nature Cam Videos," in *Slow Cinema*, ed. Tiago de Luca (Edinburgh: Edinburgh University Press, 2015), 214.
34 Ike Kamphof, "Webcams to Save Nature: Online Space as Affective and Ethical Space," *Foundations of Science* 16, no. 2–3 (2011): 261.
35 Trine B. Berry, *Videoblogging Before YouTube* (Amsterdam: Institute of Network Cultures, 2018), 9.
36 Sanjoy Kumar Roy, Rinu Fatema, Jamsedul Islam, and Nazmul Huda, "Reviewing YouTube as a Compelling Tool for the Promotion of Tourism," *International Journal of Management and Accounting* 2, no. 5 (2020): 99.
37 See, e.g. Trail Ridge Road, Rocky Mountain National Park Time Lapse Drive (www.youtube.com/watch?v=NbZ3ARHFQL4), Rocky Mountain National Park –Drive to Bear Lake Area (www.youtube.com/watch?v=kZvXyt-qesw), Arches National Park: A Drive in the Park (Time-lapse) (www.youtube.com/watch?v=0GNn7QZw0ek).
38 "Visitation Numbers," *National Park Service*, accessed February 25, 2021, www.nps.gov/aboutus/visitation-numbers.htm.
39 Ilis P. Tussyadiah and Daniel R. Fesenmaier, "Mediating Tourist Experiences: Access to Places via Shared Videos," *Annals of Tourism Research* 36, no. 1 (2009): 24.
40 Nature cam videos typically provide a 24/7 live stream of remotes sites with views recorded by cameras set up in natural landscapes.
41 Lam, "It's About Time," 210.
42 Thomas Zeller, "Staging the Driving Experience: Parkways in Germany and the United States," in *Routes, Roads and Landscapes,* ed. Mari Hvattum, Brita Brenna, Beate Elvebakk, and Janike Kampevold Larsen (Farnham: Ashgate, 2011), 126.
43 Anne Friedberg, "Urban Mobility and Cinematic Visuality: The Screens of Los. Angeles – Endless Cinema or Private Telematics," *Journal of Visual Culture* 1, no. 2 (2002): 184.
44 Edward Dimendberg, "The Will to Motorisation: Cinema, Highways, and Modernity," *October* 73 (1995): 107–108.
45 Reyner Banham, "New Way North," *New Society* 20, no. 4 (1972): 249.
46 Malcolm Andrews, "The View from the Road and the Picturesque," in *The Aesthetics of Human Environments*, ed. Arnold Berleant and Allen Carlson (Peterborough: Broadview Press, 2007), 278.
47 Andrews, "The View from the Road," 278.
48 Mitchell Schwarzer, *Zoomscape: Architecture in Motion and Media* (New York: Princeton Architectural Press, 2004), 99.

49 Federico De Matteis, *Affective Spaces: Architecture and the Living Body* (New York: Routledge, 2021), 6.
50 Susanne Haake and Wolfgang Müller, "Cyberella – Design Issues for Interactive 360 Degree Film," in *Interactivity, Game Creation, Design, Learning, and Innovation*, ed. Anthony L. Brooks, Eva Brooks, and Cristina Sylla (Cham: Springer, 2019), 154.
51 Lynne Pearce, *Drivetime Literary Excursions in Automotive Consciousness* (Edinburgh: Edinburgh University Press, 2016).
52 Pearce, *Drivetime*, 10.
53 David Nye, "Redefining the American Sublime, From Open Road to Interstate," in *Routes, Roads and Landscapes*, ed. Mari Hvattum (Farnham: Ashgate, 2011), 104.
54 See, e.g. Cynthia Freeland, "The Sublime in Cinema," in *Passionate Views: Film, Cognition, and Emotion*, ed. Carl Plantinga and Greg Smith (Baltimore: Johns Hopkins University Press, 1999).
55 Holger Enigk, Uli Foehl, and Verena Wagner, "Haptics Research at Daimler AG," in *Human Haptic Perception: Basics and Applications*, ed. Martin Grunwald (Basel: Birkhauser, 2008), p. 453.
56 Marc Desportes, *Paysages en mouvement. Transports et perception de l'espace, XVIIIe – XXe siècle* (Paris: Gallimard, 2005), 336.
57 Gernot Böhme, "Landscape on the Road," in *Roads, Routes and Landscapes*, ed. Mari Hvattum, Brita Brenna, Beate Elvebakk, and Janike Kampevold Larsen (Farnham: Ashgate, 2011), 235.
58 Janet Swailes, *Field Sketching and the Experience of Landscape* (London: Routledge, 2016), 90–93.
59 John Urry, *Sociology Beyond Societies* (London: Routledge, 2000), 63.
60 John Urry and Jonas Larsen, *The Tourist Gaze 3.0.* (London: Sage Publications, 2011), 14–15.
61 Chris Ryan, Karen Hughes, and Sharon Chirgwin, "The Gaze, Spectacle and Ecotourism," *Annals of Tourism Research* 27, no. 1 (2000): 148–163.
62 John Urry, *The Tourist Gaze*, 2nd ed. (London: Sage, 2002), 154.
63 Jonas Larsen and John Urry, "Gazing and Performing," *Environment and Planning D: Society and Space* 29, no. 6 (2011): 1111–1112.
64 Yolonda Youngs, "Shaping Tourism," in *North American Odyssey: Historical Geographies for the Twenty-First Century*, ed. Craig E. Colten and Geoffrey L. Buckley (Lanham: Rowman & Littlefield, 2014), 309–310.
65 Bruce Prideaux and Dean Carson, "The Structure and Role of Drive Tourism," in *Drive Tourism: Trends and Emerging Markets*, ed. Bruce Prideaux and Dean Carson (London: Routledge, 2011), 3.
66 Rob Draper and Ken Petty, "The National Scenic Byways Program: On the Road to Recreation," *Journal of Physical Education, Recreation & Dance* 72, no. 1 (2001): 27.
67 "National Scenic Byways Program," *Federal Register* 60, no. 96 (1995): 26759.
68 Steven J. Galliano and Gary M. Loeffler, *Scenery Assessment: Scenic Beauty at the Ecoregion Scale* (Portland: US Department of Agriculture, 2000), 9.
69 Jeffrey Hallo and Robert E. Manning, "Analysis of the Social Carrying Capacity of a National Park Scenic Road," *International Journal of Sustainable Transportation* 4, no. 2 (2010): 75.
70 David Crouch, Rhona Jackson, and Felix Thompson, "Introduction: The Media and the Tourist Imagination," in *The Media and the Tourist Imagination: Converging Cultures*, ed. David Crouch, Rhona Jackson, and Felix Thompson (New York: Routledge, 2005), 1.
71 Alfio Leotta, "Navigating Movie (M)apps: Film Locations, Tourism and Digital Mapping Tools," *M/C Journal* 19, no. 3 (2016), https://doi.org/10.5204/mcj.1084.
72 Scott MacDonald, "Toward an Ecocinema." *Interdisciplinary Studies in Literature and Environment* 11, no. 2 (2004): 108.

Bibliography

Abbey, E. *Desert Solitaire*. Tucson: University of Arizona Press, 1988.

Andrews, Malcolm. "The View From the Road and the Picturesque." In *The Aesthetics of Human Environments*, edited by Arnold Berleant and Allen Carlson, 272–90. Peterborough: Broadview Press, 2007.

Banham, Reyner. "New Way North." *New Society* 20, no. 4 (1972): 241–43.

Beeton, Sue. *Film-Induced Tourism*. Clevedon: Channel View Publications, 2002.

Berry, Trine B. *Videoblogging Before YouTube*. Amsterdam: Institute of Network Cultures, 2018.

Böhme, Gernot. "Landscape on the Road." In *Roads, Routes and Landscapes*, edited by Mari Hvattum, Brita Brenna, Beate Elvebakk, and Janike Kampevold Larsen, 229–36. Farnham: Ashgate, 2011.

Carr, Ethan. *Wilderness by Design: Landscape Architecture and the National Park Service*. Lincoln: University of Nebraska Press, 1998.

Colonnese, Fabio, and Paolo Rosa. "A Car with a View": Considerations on the Landscape Seen and Represented Through the Windshield." In *Digital Draw Connections, Representing Complexity and Contradiction in Landscape*, edited by Fabio Bianconi and Marco Filippucci, 295–327. Cham: Springer, 2021.

Crouch, David, Rhona Jackson, and Felix Thompson. "Introduction: The Media and the Tourist Imagination." In *The Media and the Tourist Imagination: Converging Cultures*, edited by David Crouch, Rhona Jackson, and Felix Thompson, 1–13. New York: Routledge, 2005.

Crowel, Hailey, Haley Gribben, and Jaclyn Loo. "Travel Content Takes Off on YouTube." *Think with Google* (August 2014): 1–11. www.thinkwithgoogle.com/consumer-insights/consumer-trends/travel-content-takes-off-on-youtube/.

De Matteis, Federico. *Affective Spaces: Architecture and the Living Body*. New York: Routledge, 2021.

Desportes, Marc. *Paysages en mouvement. Transports et perception de l'espace, XVIIIe – XXe siècle*. Paris: Gallimard, 2005.

Dimendberg, Edward. "The Will to Motorisation: Cinema, Highways, and Modernity." *October* 73 (1995): 90–137.

Draper, Rob, and Ken Petty. "The National Scenic Byways Program: On the Road to Recreation." *Journal of Physical Education, Recreation & Dance* 72, no. 1 (2001): 7–31.

Du, Xin, Toni Liechty, Carla A. Santos, and Jeongeun Park. " 'I Want to Record and Share My Wonderful Journey': Chinese Millennials' Production and Sharing of Short-form Travel Videos on TikTok or Douyin." *Current Issues in Tourism*, 2020, published online.

Enigk, Holger, Uli Foehl, and Verena Wagner. "Haptics Research at Daimler AG." In *Human Haptic Perception: Basics and Applications*, edited by Martin Grunwald, 453–58. Basel: Birkhauser, 2008.

Friedberg, Anne. "Urban Mobility and Cinematic Visuality: The Screens of Los Angeles – Endless Cinema or Private Telematics." *Journal of Visual Culture* 1, no. 2 (2002): 183–204.

Galliano, Steven J., and Gary M. Loeffler. *Scenery Assessment: Scenic Beauty at the Ecoregion Scale*. Portland: US Department of Agriculture, 2000.

Gilpin, Williams. *Remarks on Forest Scenery, and Other Woodland Views (Relative Chiefly to Picturesque Beauty)*. London: R. Blamire, 1791.

Haake, Susanne, and Wolfgang Müller. "Cyberella – Design Issues for Interactive 360 Degree Film." In *Interactivity, Game Creation, Design, Learning, and Innovation*, edited by Anthony L. Brooks, Eva Brooks, and Cristina Sylla, 152–62. Cham: Springer, 2019.

Hallo, Jeffrey, and Robert E. Manning. "Transportation and Recreation: A Case Study of Visitors Driving for Pleasure at Acadia National Park." *Journal of Transport Geography* 17, no. 6 (2009): 491–99.

Hallo, Jeffrey, and Robert E. Manning. "Analysis of the Social Carrying Capacity of a National Park Scenic Road." *International Journal of Sustainable Transportation* 4, no. 2 (2010): 75–94.

Hardy, Anne, and Ulrike Gretzel. "Why We Travel This Way: An Exploration Into the Motivations of Recreational Vehicle Users." In *Drive Tourism: Trends and Emerging Markets*, edited by Bruce Prideaux and Dean Carson, 194–223. London: Routledge, 2011.

Julin, Suzanna. "A Feeling Almost Beyond Description: Scenic Roads in South Dakota's Custer State Park." In *The World Beyond the Windshield: Roads and Landscapes in the United States and Europe*, edited by Christof Mauch and Thomas Zeller, 79–93. Athens: Ohio University Press, 2008.

Kamphof, Ike. "Webcams to Save Nature: Online Space as Affective and Ethical Space." *Foundations of Science* 16, no. 2–3 (2011): 259–74.

"Kauai Makes List of Top 10 Film Locations." *Pacific Business News*, March 27, 2012. www.bizjournals.com/pacific/news/2012/03/27/kauai-makes-list-of-top-10-film.html.

Keiter, Robert. *To Conserve Unimpaired: The Evolution of the National Park Idea*. Washington, DC: Island Press, 2013.

Kern, Stephen. *The Culture of Time and Space 1880–1918*. Cambridge: Harvard University Press, 1983.

Lam, Stephanie. "It's About Time: Slow Aesthetics in Experimental Ecocinema and Nature Cam Videos." In *Slow Cinema*, edited by Tiago de Luca, 207–18. Edinburgh: Edinburgh University Press, 2015.

Larsen, Jonas, and John Urry. "Gazing and Performing." *Environment and Planning D: Society and Space* 29, no. 6 (2011): 1110–25.

Leotta, Alfio. "Navigating Movie (M)apps: Film Locations, Tourism and Digital Mapping Tools." *M/C Journal* 19, no. 3 (2016). https://doi.org/10.5204/mcj.1084.

Liu, Thomas J., John B. Loomis, and Linda J. Bilmes. "Exploring the Contribution of National Parks to the Entertainment Industry's Intellectual Property." In *Valuing U.S. National Parks and Programs: America's Best Investment*, edited by Linda J. Bilmes and John B. Loomis, 93–112. New York: Routledge, 2020.

Louter, David. *Windshield Wilderness: Cars, Roads, and Nature in Washington's National Parks*. Seattle: University of Washington Press, 2006.

MacDonald, Scott. "Toward an Ecocinema." *Interdisciplinary Studies in Literature and Environment* 11, no. 2 (2004): 107–32.

Morse, Margaret. *Virtualities*. Bloomington: Indiana University Press, 1998.

"National Scenic Byways Program." *Federal Register* 60, no. 96 (1995): 26759–62. https://archive.org/stream/federalregister60aunit/federalregister60aunit_djvu.txt.

Nye, David. "Redefining the American Sublime, from Open Road to Interstate." In *Routes, Roads and Landscapes*, edited by Mari Hvattum, 99–112. Farnham: Ashgate, 2011.

Pearce, P. L. "Touring for Pleasure: Studies of the Senior Self-Drive Travel Market." *Tourism Recreation Research* 24, no. 1 (1999): 35–42.

Pearce, Lynne. *Drivetime Literary Excursions in Automotive Consciousness*. Edinburgh: Edinburgh University Press, 2016.

Pergams, Oliver, and Patricia Zaradic. "Is Love of Nature in the US Becoming Love of Electronic Media? 16-Year Downtrend in National Park Visits Explained by Watching Movies, Playing Video Games, Internet Use, and Oil Prices." *Journal of Environmental Management* 80 (2006): 387–89.

Peters, Peter. "Roadside Wilderness: U.S. National Park Design in the 1950s and 1960s." In *Observation Points: The Visual Poetics of National Parks*, edited by Thomas Patin, 55–76. Minneapolis: University of Minnesota Press, 2012.

Prideaux, Bruce, and Dean Carson. "The Structure and Role of Drive Tourism." In *Drive Tourism: Trends and Emerging Markets*, edited by Bruce Prideaux and Dean Carson, 3–13. London: Routledge, 2011.

Roy, Sanjoy Kumar, Rinu Fatema, Jamsedul Islam, and Nazmul Huda. "Reviewing YouTube as a Compelling Tool for the Promotion of Tourism." *International Journal of Management and Accounting* 2, no. 5 (2020): 96–104.

Ryan, Chris, Karen Hughes, and Sharon Chirgwin. "The Gaze, Spectacle and Ecotourism." *Annals of Tourism Research* 27, no. 1 (2000): 148–63.

Schwarzer, Mitchell. *Zoomscape: Architecture in Motion and Media*. New York: Princeton Architectural Press, 2004.

Swailes, Janet. *Field Sketching and the Experience of Landscape*. London: Routledge, 2016.

Montana Film Office. *The Economic Impact of Montana Film Production: An Analysis of the Industry and the MEDIA Act, 2019–2020*. Helena: Montana Film Office, June 25, 2020. https://econsultsolutions.com/wp-content/uploads/2020/07/EcoImpactMTFilmProdReport_2019-2020.pdf.

Thelen, Timo, Sangkyun Kim, and Elisabeth Scherer. "Film Tourism Impacts: A Multi-Stakeholder Longitudinal Approach." *Tourism Recreation Research* 45, no. 3 (2020): 291–306.

Tolkacha, Denis, and Stephen Pratt. "Travel Professors: A YouTube Channel About Tourism Education & Research." *Journal of Hospitality, Leisure, Sport & Tourism Education* 28, no. 5–6 (2021): 1–8.

Tussyadiah, Ilis P., and Daniel R. Fesenmaier. "Mediating Tourist Experiences: Access to Places via Shared Videos." *Annals of Tourism Research* 36, no. 1 (2009): 24–40.

Urry, John. *Sociology Beyond Societies*. London: Routledge, 2000.

———. *The Tourist Gaze*, 2nd ed. London: Sage, 2002.

Urry, John, and Jonas Larsen. *The Tourist Gaze 3.0*. London: Sage Publications, 2011.

Vila, Noelia Araújo, José Antonio, and Fraiz Brea Pablode Carlos. "Film Tourism in Spain: Destination Awareness and Visit Motivation as Determinants to Visit Places Seen in TV Series." *European Research on Management and Business Economics* 27, no. 1 (2021): 21–31.

"Visitation Numbers." *National Park Service*, February 25, 2021. www.nps.gov/aboutus/visitation-numbers.htm.

Youngs, Yolonda. "Shaping Tourism." In *North American Odyssey: Historical Geographies for the Twenty-First Century*, edited by Craig E. Colten and Geoffrey L. Buckley, 309–34. Lanham: Rowman & Littlefield, 2014.

Zeller, Thomas. "Staging the Driving Experience: Parkways in Germany and the United States." In *Routes, Roads and Landscapes*, edited by Mari Hvattum, Brita Brenna, Beate Elvebakk, and Janike Kampevold Larsen, 125–38. Farnham: Ashgate, 2011.

"The 2014 Traveler's Road to Decision." *Google/Ipsos MediaCT*, June 2014. www.thinkwithgoogle.com/consumer-insights/consumer-trends/2014-travelers-road-to-decision/.

4

"FOREVER BALI"

Surf Tourism and *Morning of the Earth* (1972)

Gemma Blackwood

Introduction

Albert Falzon's *Morning of the Earth* (1972) has an enduring legacy as an early Australian documentary of the surf film genre: a free-wheeling Antipodean counterpoint to the breakthrough American surf film *The Endless Summer* (Brown, 1966) that created popular interest in the leisure sport of surfing.[1] Although the style and tone of these films differ, both depict travel quests to far-off lands in search of perfect waves – also known in American surf culture from the 1960s as the "surfing safari" or "surfari."[2] *The Endless Summer* portrayed the world as a series of fun surfer pit stops, and the film hops from nation to nation – from Africa to the Pacific – with amusing commentary from director Bruce Brown whose script developed by touring film footage through surf networks.[3] While there are earlier examples of surfing films that featured travel, *The Endless Summer* can be considered the first surf film that "inexorably linked" surfing with adventurous travel for popular audiences.[4] *Morning of the Earth* added the small Indonesian island of Bali to the international surf touring map. In the film, Bali is privileged as the "purest" surf spot in the movie, and hence represents the film's nirvana-like concept of a "morning of the earth" – a spiritual rebirth in a natural place, away from the problems of modernity.[5]

In this chapter, I am interested in examining the way that Bali is represented in *Morning of the Earth*: in the way that the film has created affective landscapes for imaginative travel, and inspired film tourism to this particular location. Looking at the context of the sports documentary and the surf movie genre, I examine the various ways that travel is imagined within the film. Here, I seek to understand the power of cinematic texts for virtualizing affective travel experiences and creating emotional and social resonances through visual imagery and soundscapes. By reading the film's representation of Bali, I examine how the film helped to create

DOI: 10.4324/9781003327585-4

interest in tourism, and in surfing as an increasingly popular lifestyle sport.[6] Surfing went on to be heavily commodified within the island's emerging tourism industry, and would go on to be a key leisure activity in the emergence of Bali as an exotic playground for the Australian middle classes in the 1980s and beyond.[7] Through a textual analysis of the film, this chapter focuses on the affective spaces of freedom and possibility that were opened up through this sublime representation of surfing travel.

Morning of the Earth is still celebrated in both Australian and international surf cultures, and it is often cited as both an inspirational surf film *and* a travel-inducing film, although its level of influence on tourism decision-making was not quantitatively measured when it was released in the 1970s.[8] More recently, other western films that have depicted Indonesia have demonstrated the potential of film to create popular tourism sites: in Bali, the most well-known example is the increased tourism to the region of Ubud created by the film *Eat, Pray, Love* (2010).[9] Using genre studies and critical tourism studies to examine *Morning of the Earth*, we can understand the film as an historical example of film tourism in the way that it inspired travel to the specific sites and popularizing surf safaris more generally.[10] In a recent article on the best surf movies of all time, *Time Out* suggested that the film "still makes surfers and newbs alike want to quit their jobs, toss their wallets and set up shop on a remote beach."[11] The film created special interest in Bali as a surfing tourism destination and has become famous for having "discovered" the surf break at Uluwatu on the southern peninsula of the island and showing it being surfed for the first time.[12]

In his popular history of Balinese tourism, Phil Jarratt has described the moment in the film when the two surfers, Rusty Miller and Stephen Cooney, first stood on the reef looking out at the surging waves of Uluwatu:

> *Shooting from a lower angle than the first day, Falzon saw an image of profound beauty that would stay with him forever, as it would with so many who later saw the film. The two silhouetted figures were bathed in an eerie yellow light as the high surf crashed around them in this remote and dangerous place. It was an image that captured at once the excitement and trepidation of surf exploration, and in the minds of the pioneer surf explorers, it was an image that would be forever Bali.*[13]

Albert Falzon suggests in an interview that the film dramatically changed patterns of international surf travel after its release:

> *The focus of surfers everywhere at that time was to head to Hawaii. . . . No one looked to Asia for waves – no one – and I think the film changed that. . . . When you went to Hawaii it was already heavily Westernised and not too different from home, but you went to Indonesia and there were different languages, different religions and cultures weaved all through there. You were surfing perfect waves in the midst of this rich culture, and that intoxicated surfers.*[14]

FIGURE 4.1 Surfers Rusty Miller and Stephen Cooney discovering the surf at Uluwatu, Bali.

Source: Image courtesy of Albert Falzon.

My reading of the film aims to discern what kinds of ethical and political stances are contained within the text of *Morning of the Earth* (Figure 4.1). As an Australian sports documentary and surfari, the film utilizes problematic neo-colonial tropes and essentializing modes of otherness in the ways that it depicts the non-surfers: the non-surfers are both the local Balinese residents and visiting hippies of the "Hippie Trail" of South-East Asia who are shown at leisure on the beach. The film's effort to differentiate and "other" the western hippies holidaying in Bali from the surfers also indicates how the surfing subculture is positioned as a very distinctive subcultural practice, indeed very different from other groups that are represented as similar or closely connected in mainstream discourses. Certainly, the way that the surfing subculture is represented in the film seems ideologically linked to national values about masculine heroism, mateship, and individualism connected with Australian sports cultures.[15] As Clifton Evers has noted, "surf culture has long been a privileged site in Australian society, and male surfers have become Australian icons of it."[16]

As Jane Stadler and Peta Mitchell have suggested, an analysis of affect in film can be useful way for understanding the ideological positioning of the film and its potential influence upon audiences. They write,

> [T]he analysis of affect has been a useful critical tool for examining how the mechanisms of cinematic identification that position audience members in relation to the emotional experiences of screen characters also invite the audience to emotionally invest in an ethical and political position with regard to events on screen.[17]

My argument is that the affective positioning of Bali as a utopian site of adventure, freedom, and paradise in the film also means an erasure of local politics, demonstrating a reluctance to understand the socio-political context of Indonesia. Hence, examining the film for its cultural messages also demonstrates how sustained hegemonic ideals can be mapped onto real places, which may go on to shape the way that future tourists imagine or frame local residents and other visitors: in other words, the Australian cultural imaginary of Bali.

Freedom and Possibility: Psychedelic Surf Travel in *Morning of the Earth*

Morning of the Earth's loose travel narrative, groovy folk-rock soundtrack, and lush cinematography suggested that a sublime counter-cultural and trans-oceanic utopia was being created in the simple story of thrill-seeking surfers in search of extreme surfing adventures.[18] This sense of a "utopia" is a generic component of sports documentaries and sports films, as Seán Crosson has suggested:

> *[F]ilm has a powerful ability, inherited and developed by other audiovisual forms such as television and the internet, to preserve, elevate and slow down the utopian moment in sport; not just to capture it for perpetuity, but to transform that moment. . . As a seductive and powerfully influential form, film's fundamentals – including . . . narrative, mise-en-scène, cinematography, editing and sound – provide powerful tools for exploiting utopian sensibilities.*[19]

Sports documentaries are most commonly created for the medium of television, but as an independent and counter-cultural film, *Morning of the Earth* was created for cinematic exhibition, which meant that more focus could be given to aesthetic representation, and great emphasis could be given to the utter extraordinariness and beauty of surfing.[20] The film can also be understood as a "poetic" sports documentary in a mode identified by Ian McDonald, one where "familiar assumptions about sporting practices and settings are disrupted and deconstructed, and the audience is left with the beautiful, bizarre, and barbaric nature of modern sport."[21] Through emphasis on danger and thrill, the presentation of surfing in *Morning of the Earth* connects to extreme sports and is designed to shock and delight its audience as per the characteristic affects of its genre.

Then, as a surfari, *Morning of the Earth* can also be understood as a travelogue or travel documentary, which in Hollywood had enjoyed renewed popularity in the 1950s and 1960s due to the rise of immersive widescreen formats that aimed to take immobile spectators on exciting virtual tours, coinciding with the rise of post-war mass tourism.[22] As with poetic sports documentaries, travelogues were also designed to thrill and awe their audiences with novel and unexpected sights and experiences: they could "broaden audiences' horizons and pique audiences' sense of wonder."[23] *Morning of the Earth* was a sports travelogue with a counter-cultural mien. Using narrative, visual, musical, and aural dimensions, the film creates an affective landscape that draws on visceral emotions such as joy, freedom, peace, a sense of well-being, and the surfer concept of "stoke" – feeling excitement, awe, and thrill after a great time on the waves.

We might consider the film's role as an extension of a broader, subcultural interest in travel in the sport of surfing, which across a range of media platforms were consistently presenting the surfari as an aspirational travel ideal.[24] In *A Political History of Surfing*, Scott Laderman notes that in the 1950s and 1960s, Hollywood developed a "fascination" with Southern California beach culture, "churning out such motion pictures as the Gidget series and the popular comedies of Frankie Avalon and Annette Funicello."[25] One of the earliest surf travel films was American Bud Browne's film *Hawaiian Surfing Movies* (1953), which demonstrated one of the first popular surfing hot spots for U.S. surfers.[26] The sport of surfing ended up being popularized and disseminated globally due to factors such as the international rise of youth culture, the democratization of mass travel, and consumerism, all amplified by mass media.[27] Other than film, media texts, such as American *Surfer* magazine (first established in 1961), helped globally popularize the concept of Californian surf culture, helping to propagate myths about surfing as an "escape" from the difficulties of civilization and modern urbanization.[28] More broadly, Stefan Krause has noted the role of media texts in the production of surfing cultures:

> *Numerous surf magazines, books, films and other industry-related marketing productions make up the "mediascapes" that help to generate images of the surfer lifestyle . . . [t]he history of surfing is thus embodied in the present as surfers negotiate the presented images in the mediascapes with their own lived experiences.*[29]

Both the local Australian and the broader international mediascapes often presented surfing and travel in interconnected ways. By the time of the film's production, surfing and travel were already strongly connected on other media platforms.[30] As an example, Albert Falzon had already created another important surf media text in Australia: the surf magazine *Tracks*.[31] With *Tracks*, Falzon – alongside magazine co-founders John Witzig and David Elfick – had aimed to present surfing as an alternative and eco-conscious lifestyle. *Tracks* brought together counter-cultural stories that included communal living, travel stories, meditation, and yoga, alongside the more traditional sports reportage about surfing that was a conventional aspect of surf magazines. To this end, one might even consider *Morning of the Earth* a transmedia extension of the magazine.[32]

Writing about the affective qualities of *Tracks* magazine, Douglas Booth has argued that it "generates a 'skin-creeping buzz' – it tickles, excites and stimulates the senses and minds of readers; arguably affects are what connect readers to the magazine."[33] Potentially, this description could also be applicable to *Morning of the Earth* (and the surf movie genre more broadly). The film draws upon a set of psychedelic images and sounds that are also in the quest of "buzz," and music-led counter-cultural interludes showing surfers camping, buying home-grown produce from farmers, playing music, and hanging out in exotic places. It is clear from the filmmakers' intentions that the film was made for the same audiences that were interested in the activities described in more quotidian ways in *Tracks*.[34] Hence, the ecstatic and sublime impacts of travel – whether it is the oceanic travel within the surf, or journeys to reach the surf – are similarly built into the very experience of the film.

The film offers its audience the pleasures of virtual travel. In viewing the documentary today, watching the film can transport the viewer back to the 1970s, to a less developed era of surfing culture and travel adventure. In this vein, the promotional copy for the 50th anniversary book of the film suggests that watching *Morning of the Earth* allows one to see "Australia, Bali and Hawaii before rampant development took over." While "last-chance tourism" is a type of real tourism where visitors go to "vanishing destinations" before they are ruined (such as the Great Barrier Reef), we might see watching a film like *The Morning of the Earth* in the current day as a virtual way for reclaiming long-lost spaces.[35] Feeding on the nostalgia for the original film, newly discovered archival footage from the film's original shoot was recently turned into a new 38-minute film called *RE-MOTE: The Lost Reels* (2021), which pairs psychedelic music with sun-tinged travel filmography and surfing shots. This new documentary also draws on the power of seeing early 1970s travel and surf as a key part of its appeal.

The music of the film also plays a role in establishing travel landscapes. The film's folk rock and radio-friendly 1970s soundtrack by G. Wayne Thomas provides another key to its fantastic, affective landscapes, which were particularly aimed for youthful audiences and surfing subculture. In his dissertation on the film's use of music, Tim Gaze describes the movie as a "spontaneously filmed work without use of verbal narration or location sound and only the music soundtrack supporting the imagery and concept."[36] However, I would argue that the lyrics to the songs that feature on the soundtrack serve as a conceptual verbal narration for the film, and they contribute to the affective power of the film to inspire imaginings of travel.

For example, let us consider the introduction of the film. The film starts with an acoustic guitar riff on a black screen, the latter dissolving into the image of sea ripples in deep water. A man is heard beginning to sing: "The forces of the universe and the elements of space/Conjured up your being, your size, your time, your shape." Fluorescent pink light (made with post-production effects) swirls against midnight blue water, creating a split screen effect. Super-imposed shots of rocky coastline move haphazardly in front of shots of water. The images become stark, chiaroscuro: black coastline against bright light. Pink transforms into orange as we see a lustrous sun rise. The light leaks in as fog moves against the sun. "It was the morning of the earth" sings the man as the film's name crystallizes at the top of screen. After this, we see a shot of magnificent surf in a vivid sepia, rolling in slow motion as the camera hypnotically tracks the crest of the falling wave. Then we are told the film is a "fantasy of surfers living in three unspoiled lands and playing in nature's oceans." This specific reference to the fantasy genre instead of the surf movie here demonstrates the film's hybridity as surf documentary. The film leans into many of the fictional and stylistic tropes of the fantasy genre, such as the focus on heroic protagonists with seemingly magical or supernatural powers, the "coming-of-age" narrative and "the quest," and sublime visual landscapes.[37] Here, the music and imagery work together to convey travel as an important spiritual quest.

After this, the film moves more clearly into the generic conventions of the surf movie and the sports documentary, showing surfers and people for the first time. The film primarily consists of shots of surfing, intercut with scenes of life near the waves which acts as the "travelogue" component, as it shows footage of Australia, Bali, and Hawaii. It is these scenes that serve as establishing shots, creating a geographical context for the surf shots and emphasizing the surfari genre. An interesting feature of the film is the way that short shots punctuate longer scenes; often, these shots feature in more detail in later scenes. This play with temporality – seeing glimpses of the future – is the visual equivalent of when a DJ brings in a tiny sample of the next song while the current song is still playing. Overall, the editing techniques of the visuals add a synaesthetic dimension to the sequence as an interconnected music performance. This effect also renders the spectator of the film clairvoyant, as though receiving inexplicable visions from later in the movie. Overall, this lends the film a psychedelic quality, where vision becomes equivalent to stoned hallucination.[38] It also indicates the way that the typical mode of documentary had been fused with "avant-garde techniques" that departed from reality.[39]

The broader concept of a fantasy hero's quest was also resonant with the lysergic, drug-taking counterculture that is tangential to the surfing movement. The mystical sense of taking a "trip" formed yet another concept of a travel journey in the film, and again one virtualized: where mental travel can occur, but actual physical travel does not need to take place. Both surfing and the counterculture shared the concept of a type of travel that meant a change of mental perspective: taking acid or marijuana could lead to an "altered state," while taking your board into swirling blue-green waters could lead to similar altered vistas and a changed type of perception. Laderman suggests this about the film, noting that through its music and its visuals it "dances psychedelically through Australia, Indonesia, and Hawaii; often looking like something out of an acid trip, the smell of burning dope practically wafts from the screen."[40] This psychedelic "trip" is conveyed both through the visual content (e.g. groups of men are shown smoking from exotic pipes) and through the editing and music, in many ways reminiscent of the "Hollywood LSD films" of the late 1960s identified by Harry Benshoff.[41] The psychedelic references work to connect the surfing lifestyle to the film, creating an affective and multi-sensory quality. The music, special effects, and post-production editing of the film – such as the application of vibrant colours and the use of slow-motion – contribute greatly to the film's inspirational powers. In combination with the music and the psychedelic themes, this connects ideas of "travel" with utopian ideals of freedom and possibility.

Analysing Travel to Bali in *Morning of the Earth*

Cathal Kilcline has argued that it is typical in contemporary surfing discourses for surfing protagonists to be presented differently from local residents:

> *Typically, in the discourse of surfing and windsurfing, stereotypes and essentialist characteristics of the local population are recited before the windsurfer retreats into the "cocoon" of his sporting peers and attendant sub-culture. The local populations, if they are recognised at all, are evoked only in static, eternalized form. They are thus disempowered, their lived identity obscured by stereotypes, their places emptied of meaning and thus ripe for appropriation by the dynamic, nomadic surfers.*[42]

Kilcline's research suggests that the local is represented simplistically, by drawing upon stereotypes. This mode of representation within surfing depictions becomes increasingly problematic within the context of cross-cultural travel, particularly with the concern of the neo-colonizing qualities of contemporary tourism. *Morning of the Earth*'s discovery of the surf breaks of Bali occurred as the tropical island was opening up to mass tourism, which would transform its destiny from the 1980s.[43] As Leyla Stevens has summarized, Bali emerged as a popular travel destination in this period due to a range of existing travel associations that tie in with the location's lengthier colonial and political histories:

> *Western imaginations of Bali have long positioned the island as a self-contained paradise, rich with cultural traditions that have withstood global reverberations. Master narratives of Bali as an Eden-like island were initially shaped through colonial policy, when the Dutch positioned Bali as a bastion for an ancient Hindu culture and Balinese as innately artistic and religious people. This move was guided by both the political separation of Bali from Islamic Java and the entrepreneurial development of Bali as a tourist destination.*[44]

Morning of the Earth was released within a decade since Bali and Indonesia had been devastated by the anti-Communist Indonesian mass killings of 1965–1966, which led to the murder of approximately one million people across the country.[45] But you wouldn't know this from watching the film, which presents Bali using paradisiac tropes, including the presentation of its locals as simple, happy villagers and helpful guides. As a sports documentary and surfer film, the film is not concerned with the real-life politics of the area, but rather in capturing wave-based fantasies and showing the "utopian" elements of the activity. As Keith Beattie argues, the film is polemical and like *Tracks* magazine aims to depict "strong environmentalism and an anti-city, back-to-nature stance that endorsed a laid-back lifestyle": it was essential, then, that the locals were shown to be deeply satisfied by their island existence.[46] In the film, the most visible interest in Asian culture is focused on Eastern religions and philosophies: clearly the film's messages seem to draw on a pop cultural version of Buddhism and mystical notions of eastern philosophy particular to the psychedelic culture of the 1960s and 1970s. In its utilization of Kilcline's stereotype of locals as simple and passive within the film's broader narrative, the documentary exhibits a kind of surfer Orientalism over the local Balinese people shown throughout.

The film also depicts a new and potent crossover taking place between Hippie cultures and surfing, which is unsurprising, as Bali was emerging as a key site on the Hippie Trail through southeast Asia in the 1960s and 1970s.[47] Ian Wilson describes how:

> *The sleepy fishing village of Kuta, in particular, became a destination for a new generation of low-budget pleasure seekers . . . Australian surfers, hippies and those on the counter-culture trail slept on the beach or in local homestays, "discovering" the surf spots of Uluwatu and Kuta reef and enjoying magic mushrooms, continuing the tradition of the South Seas island paradise as an escape from conventional morality.*[48]

Mystical Asian religious themes have since become a large part of the visual representation of surfing cultures, going on to influence surfing spots the world over with Yin and Yang symbols and Tibetan prayer flags. Bali became a site of confluence between the Hippie Trail and surf tourism, and both of these travel pursuits went on to influence the island's identity as a mainstream tourist commodity and in particular its position in Australia as an attractive overseas Asian destination for budget travel.[49]

Let us examine more closely the way that the Balinese landscape, the local residents, and the surfing visitors are represented in *Morning of the Earth.* The Bali sequence of the documentary starts after a first act that focuses on the east coast of Australia. Bali emerges after an abrupt sonic shift, from folk ballad to gentle flute instrumental. An establishing shot shows a mountainous island, dramatic cloudscapes, the glistening ocean, and the silhouettes of trees: this is a return to the sublime, fantastic landscapes shown in the opening sequence of the film. By doing this, Bali is clearly indexed as the ultimate "morning of the earth" of the film. Into this landscape, Balinese villagers are shown walking along paths, some with baskets on their head, others carrying children. Two young, white surfers are then shown, walking like the villagers through palm tree-clad streets, carrying their surf boards the way that the Balinese carry their children and goods. The camera focuses right on the sun to emphasize the intensity of the tropical heat. This is a clear example of a surfari (in the colonizing sense of the word "safari"), as soon the surfers are in quest mode, walking through a barren, tree-less wasteland. Then, we see a shot of the giant surf of Uluwatu. The surfers look down at it from the clifftops. What follows is a spectacular slow-motion scene as the men walk on the reef towards the crashing waves at Uluwatu: the iconic image of "forever Bali" that is one of the film's most famous moments. Finally, we see the surfers trying out these spectacular waves. Alex Leonard has suggested that the film's narrative is quite typical of the surfer film genre *and* real-life surfing stories, writing:

> *This origin story of Uluwatu's discovery can be said to be exemplary of Western surfers' stories of discovery throughout the world, as it contains such generic elements as some initial struggle, confusion and frustration . . . a turn-off the beaten track, a chance sighting of an unusual landmark . . . a display of the protagonists' courage, curiosity and desire, and the discovery itself, of a previously undiscovered treasure, whose strange worth the protagonists subsequently reveal to its previously unenlightened custodians and thus somehow transform them (Uluwatu locals, some of whom later take up surfing).*[50]

Here, we see a translation taking place: the "discovery" mode of surfing stories as a real-life physical activity is being replicated in cinematic representation. Potentially, this replication may be one of the reasons that the film has such an effect on its viewers: it invites a virtual experience of surfing discovery that accords with the real-life affective dimensions of thrill, joy and stoke. At the same time, this scene draws upon neo-colonialist motifs as it presents the surfers entering into this landscape, which is typical of the travelogue documentary which often contained "anthropological" moments of cultural discovery.[51] Certainly, the focus ultimately remains on surfers and the surf, and it draws upon their affective experiences of joy and the edgework of surfing as a dangerous and thrilling activity.

Other Balinese scenes show both locals and tourists represented on the island. All scenes take place outdoors, which was a deliberate choice made by the filmmakers to emphasize the beauty of outdoors and nature.[52] The difficulty of travelling with surf equipment is stressed, with the surfers needing to use local transport to get around, even hitching a ride with their boards on horse and cart. Local men sit with the surfers in a cave at Uluwatu, laughing and cheering as they take turns smoking from a pipe (in the credits they are called the "cave crew"). Then, a local man takes a traditional boat out to the surf, towing the surfers behind him. At sunrise, Balinese men return to shore with their fishing boats, dragging a small shark onto the beach. Later, a filmed cockfighting scene run by Balinese men seems to be used as an analogy for the difficulty of ordinary life, as the surfers respond by walking away towards the surf. Again, the lyrics of the song that plays directly after the cockfighting seems to narrate the film's message in a concrete way: "Wake my friend/ There's nothing in this place/ You're living in dream/ Free yourself from what you want to be/ Coz there is no reality." Here, the lyrics suggest the importance of leaving the ordinary and/or mundane and also emphasize the power of personal reinvention, which is exemplified in the film through travel and experiencing the "stoke" of the waves. The Balinese people in the film are not privileged to have this ability to travel: either virtually (through the psychedelic "trip") or physically through surfing and touring.

Therefore, the power to travel in this film is only connected to the surfers, which creates neo-colonial associations of possession and discovery as they discover the surf of Uluwatu and southern Bali. In the film, the Balinese are largely represented as guides and/or helpers or in the case of the cockfighting scene are used to represent unthinking and unimaginative humanity, caught in cycles of endless violence. Of course, the real-life context of the recent human massacres in Indonesia does not get a mention here: it does not fit in with the humorous and positive tropes of the surf film, or with the documentary's connection to fantasy and psychedelia and modernist notions of the individual's ability to "free yourself" that remains its core message.

Morning of the Earth and the Hippie Trail

As I have mentioned, this surfari film *Morning of the Earth* shows an intersection between the south-east Asian Hippie Trail and surfing cultures, so the representation of the travelling Hippie is another vital aspect of Bali's representation in the film. As a documentary, the film observes the rising popularity of Kuta's beach as nexus for alternative Western travel in Asia, showing its popularity as an international stop on the south-east Asian Hippie Trail (or "the Overland"), which first emerged in the late 1960s.[53] Agnieszka Sobocinska has outlined the geographical boundaries of this novel form of cross-continental land travel, suggesting that European visitors would travel to Iraq, Iran, Afghanistan, and Pakistan to India and Nepal and home again. Australians and New Zealanders would typically travel in a line to Indonesia and Singapore, and then either fly to India or travel through Malaysia, Thailand, and Burma to re-join the main Hippie Trail on their way to London.[54] Sobocinska argues that despite the iconoclastic ideals of the Hippies, the travel route and the type of travel still conformed with conventional Western fantasies about travel to Asia, noting that on the trail "Orientalist images and imperial behaviours were augmented to suit a new generation that liked to think of itself as radically breaking from the past."[55] Therefore, it was a group that appeared revolutionary but remained fundamentally enmeshed within the same cultural frameworks as they apprehended and engaged with foreign places and cultures.

Morning of the Earth demonstrates this renewed interest from Western tourists in the early 1970s, and they are named the "Hippies" in the credits. But it is important to note that they are presented differently from the two travelling surfers – the heroes – who are there to discover the waves. The film uses the anthropological gaze of the travelogue as it examines the Hippies: they are presented as local oddities, similar to how *The Endless Summer* narrates hanging out with the local "natives" of Africa and the Pacific. Here, the viewer is shown a young couple and their infant, frolicking in the waves; as well as groups of naked Hippies sunbaking in a large, communal group. The Hippies are only shown on the beach at Kuta. There is a humorous tone throughout: one particularly funny moment is a full-frontal, nude shot that slowly pans up a man's body as he plays with a yo-yo on the beach (he is named the "Yo-Yo Man" in the credits). Indeed, the film presents the

FIGURE 4.2 Representing a Balinese local in Morning of the Earth.

Source: Image courtesy of Albert Falzon.

western Hippies to be equally the objects of the tourist gaze, in fact as exotic and strange to the surfers as the local Balinese people (Figure 4.2).

While this scene presages the dominance that western bodies would go on to have on Balinese beaches – and soon after the natural landscape of Kuta Beach would start to be altered beyond recognition as international resort and party-town – in the film, the impact of this representation of Hippie culture consolidates the importance of the surfing subculture as distinctive.[56] In direct contrast to the Hippies lounging on the beach, the surfers are presented as the heroic figures, looking for epic adventures on the waves. Hence, the representation of the Hippies in the film serves to present the surfing visitors as the "real" adventurers and heroes of the film. They are shown as active and valiant, while the Hippies are languid and more passive, as they hang out on the beautiful beaches of Bali. Here there seems to be a distinction made between the "traveller" and the "tourist," with the surfer representing all the wild freedoms of the "traveller" and the Hippie taking on the less active concept of "tourist."

The Hippie Trail of the late 1960s and early 1970s was ultimately transformative for the Balinese tourism industry.[57] It was at this time that the exotic, religious, and historical associations of the island as a reason for travel started to switch more towards the "sun, sea and sand" concept of a lazy beach holiday, focused on Kuta Beach and nearby beaches on the southern peninsula. The idea of Balinese beach culture was transformative for the destiny of the island as tourist destination. Real estate on the cliffs and beaches of the island – until then treated by local Balinese people as wasteland because of the lack of agricultural potential – started becoming extremely valuable for tourism development.[58] Michel Picard notes that between 1970 and 1980, "the number of foreign visitors to Bali multiplied from fewer than 30,000 to over 300,000 a year, reaching around 1 million in 1990 and up to around 2 million in 1997."[59] In the early 1980s, mass tourism intensified and changed the urban and cultural landscape of the region. Now, the coastline of Bali – especially Kuta – looks markedly different.[60] There have been waves of capitalist development that have made the island's coastline particularly susceptible to tourist construction. Aiming to characterize the contemporary position of Bali, Amanda Achmadi has suggested that it is now "an island increasingly dominated by growing sprawl and unplanned urban environment."[61] Ultimately, there have been large social consequences from the development of Bali as a major tourist hub. Therefore, considering the long-term social and political consequences of international travel on this location, there may be some damaging effects that have been at least in part caused by its paradisical and singular position and its prominent place in the western cultural imaginary, as depicted in films such as *Morning of the Earth*.

Conclusion

This chapter has examined the potent ways by which travel, tourism, and surfing journeys have been imagined in the ground-breaking Australian surf documentary *Morning of the Earth*, and it has reflected on the power of this surf film as an affective mode for staging both virtual and actual forms of travel for its viewers. As I have investigated in this chapter, through a close reading of the film, it is clearly the surfers as main protagonists who are most linked to forms of imaginative travel, whether it is local travel, international travel, taking a psychedelic "trip," or else even just the movement and sense of adventure that comes from riding the surging waves. The film is able to draw upon powerful emotional resonance through both the imagery and the musical soundtrack, and special effects and other post-production techniques such as slow motion and enhanced colours work to draw out a sense of thrill and "stoke," especially connected to the small figures that are shown surfing in such overpowering and epic landscapes.

As I have shown in this chapter, the concept of Bali as static, eternal, and unchanging – or, "forever Bali" as it appears in *Morning of the Earth* – invites critique. Achmadi has suggested that "twentieth-century colonial, orientalist and travel discourses on Bali have constructed, perpetuated and objectified the notion of 'real Bali' through the twentieth century."[62] One Australian-Balinese artist – Leyla Stevens – has recently sought to find ways of disrupting the film's static modes of cultural representation, especially critiquing the way that gender, race, and culture was presented in the film. By doing so, the artist raises urgent questions about how Balinese people and places were depicted in the original documentary. Her multi-channel film called *A Line in the Sea* (2019) is a deliberate response to *Morning of the Earth*. Retelling the iconic scene of "forever Bali" that saw Rusty Miller and Steven Cooney striding towards the surf at Uluwatu, in *A Line in the Sea* we see two young Balinese women as they sit on the beach in their surfing gear, looking out at the undulating sea. Mixed with static shots of the Balinese jungle, *A Line in the Sea* features no dialogue and in doing so evokes the silent, untold stories of the victims of the Indonesian massacres, who were often buried in mass unmarked graves. Stevens' film questions the cultural ideals of the original film and rewrites it from a feminist and postcolonial position.

Now that *Morning of the Earth* has celebrated its 50th anniversary and continues to receive acclaim as a ground-breaking surf movie and travelogue, critical interest in the text can reveal Australian cultural imaginaries. By examining this surf documentary's presentation of travel, this chapter has demonstrated the affective ways that films can present narratives and elicit audience responses. Some bigger questions are raised: if a documentary is to inspire tourism, what ethical responsibility does it have to engage with the socio-political realities of the geographical location it aims to represent? If it is predominantly a sports documentary focused on the art of surfing, is there room for imaginative takes on travel, and how should such a film draw upon potent pre-existing concepts about foreign cultures? Ultimately, this chapter has raised bigger questions about the influence of film genre and form to be used ideologically, and the unintentional consequences that this may have on real places when there are real media effects and a "tourist gaze" on a given destination, which in *Morning of the Earth* turned to the beaches of southern Bali.[63]

Notes

1 In 2022, the documentary celebrated its 50th year with the release of a 4K restoration and digital remaster of the original film. The film's soundtrack was also remastered and *Tracks* magazine presented a screening tour of the film across major cities in Australia.

2 Keith Beattie has argued that the promotional advertisements for *Morning of the Earth* clearly drew upon visual similarities to *The Endless Summer* in order to maximise their connection as surf movies. See Keith Beattie, "Morning of the Earth," *Metro Magazine: Media & Education Magazine*, no. 169 (2011): 118–125.

3 Joan Ormrod, "*Endless Summer* (1964): Consuming Waves and Surfing the Frontier," *Film & History* 35, no. 1 (2005): 39.

4 Cathal Kilcline, "California Dreaming: Surfing Culture in Mediterranean France," in *Sport, Representation and Evolving Identities in Europe*, ed. Philip Dine and Seán Crosson (Oxford: Peter Lang, 2010), 3.

5 *Morning of the Earth* was made independently: Falzon filmed the documentary in his spare time, taking his friends and driving up to the northern Australian coast on weekend trips and holidays from his home in Sydney. The film's Australian locations include Kirra, Lennox Head, Broken Head, Angourie and Whale Beach in Queensland and New South Wales. In Hawaii – at the time considered to be the world's essential surfing destination – the production filmed at Rocky Point and North Shore. Bali in Indonesia became the third location, especially the beach at Kuta and Uluwatu on the southern peninsula of the island. "It didn't really have any plans, it just unfolded . . . we didn't have a storyboard, we didn't have any money, we just had an idea" said Falzon in the making-of film *MOTE: 50 Years Around the Sun*. This was a go-with-the-flow philosophy that matched the surfie perspective and it gave a loose freedom to the film's structure.

6 Belinda Wheaton, "Introducing the Consumption and Representation of Lifestyle Sports," in *The Consumption and Representation of Lifestyle Sports*, ed. Belinda Wheaton (Oxon: Routledge, 2013), 2.

7 Ian Wilson, "'Outlaw's Paradise': Australian Outlaw Bikers, Pre-crime Regimes and the Appeal of Bali," in *Linking People: Connections and Encounters between Australians and Indonesians*, ed. Antje Missbach and Jemma Purdey (Berlin: Regiospectra, 2015): 255.
8 Sue Beeton, *Film-induced Tourism* (Clevedon, UK: Channel View Publications, 2005).
9 Eerang Park, "The Impacts of Film Tourism on Place Change and Tourist Experience: A Lesson from *Eat Pray Love* in Bali, Indonesia," in *Film Tourism in Asia: Evolution, Transformation and Trajectory*, ed. Sangkyun Kim and Stijn Rejinders (Singapore: Springer, 2018).
10 Beeton, *Film-induced Tourism.*
11 Kate Wertheimer, "The 15 Most Epic Surf Movies," *Time Out*, December 8, 2021, www.timeout.com/film/the-15-most-epic-surf-movies.
12 Mark Bannerman, "It's consistently ranked as one of the top five surf movies of all time, but *Morning of the Earth* was once seen as a gamble," *ABC Online*, February 20, 2022, www.abc.net.au/news/2022-02-20/50-year-anniversary-morning-of-the-earth-still-relevant/100837084.
13 Phil Jarratt, *Bali: Heaven and Hell* (Melbourne: Hardie Grant Publishing, 2014).
14 Sean Doherty, "Surf Pioneer Albe Falzon: 'I'm so grateful to be living this life,'" *The Weekend Australian Magazine*, July 25, 2020, www.theaustralian.com.au/weekend-australian-magazine/surf-pioneer-albe-falzon-im-so-grateful-to-be-living-this-life/news-story/edcfd0de4c235f54cbf6d5f59e78c09d.
15 Kay Saunders, "'Specimens of Superb Manhood': The Lifesaver as National Icon," *Journal of Australian Studies* 22, no. 56 (1998): 96.
16 Clifton Evers, "'The Point': Surfing, Geography and a Sensual Life of Men and Masculinity on the Gold Coast, Australia," *Social & Cultural Geography* 10, no.8 (2009): 893.
17 Jane Stadler and Peta Mitchell, "Never-Never Land: Affective Landscapes, the Touristic Gaze and Heterotopic Space in *Australia*," *Studies in Australasian Cinema* 4, no. 2 (2010): 175.
18 The soundtrack by G. Wayne Thomas became the country's first gold album for an Australian-produced soundtrack, and the album has remained a significant text in Australian popular culture. See Tim Gaze, "Sun, Surf and Sound: The Role of Surf Music in the Development of Australian Popular Culture. An Investigation of the Iconic Surf Film, *Morning of the Earth*, as a Medium" (Professional Doctorate diss., Queensland University of Technology, 2020).
19 Seán Crosson, *Sport and Film* (Oxon: Routledge, 2013), 8.
20 Ian McDonald. "Situating the Sport Documentary," *Journal of Sport & Social Issues* 31, no. 3 (2007): 209.
21 McDonald, "Situating," 211.
22 Ellen Strain, *Public Places, Private Journeys: Ethnography, Entertainment and the Tourist Gaze* (New Brunswick, NJ: Rutgers University Press, 2003), 133.
23 Jennifer Lynn Peterson, *Education in the School of Dreams: Travelogues and Early Nonfiction Film* (Durham, NC: Duke University Press, 2013), 21.
24 Ormrod, "*Endless Summer.*"
25 Scott Laderman, *Empire in Waves: A Political History of Surfing* (Berkeley: University of California Press, 2014), 3.
26 Davis Jones, "History of Surfing: Bud Browne Goes to the Movies," *Surfer*, April 25, 2017, www.surfer.com/features/history-of-surfing-bud-browne-goes-to-the-movies/.
27 Joan Ormrod, "Expressions of Nation and Place in British Surfing Identities" (PhD diss., Manchester Metropolitan University, 2007), 2.
28 Kilcline, "California Dreaming," 2.
29 Stefan Krause, "Pilgrimage to the Playas: Surf Tourism in Costa Rica," *Anthropology in Action* 19, no. 3 (2012): 38.
30 Beattie, "Morning of the Earth," 119–120.
31 *Tracks* magazine was founded in 1970. The magazine still exists today, albeit in a very different format from its original incarnation.

32 Douglas Booth, "(Re)reading the Surfers' Bible: The Affects of *Tracks*," *Continuum* 21, no. 1 (2012): 20.
33 Booth, "(Re)reading," 20.
34 Beattie, "Morning of the Earth," 119–120.
35 Harvey Lemelin et al., "Last-chance Tourism: The Boom, Doom, and Gloom of Visiting Vanishing Destinations," *Current Issues in Tourism* 13, no. 5 (2010).
36 Gaze, *Sun, Surf and Sound*, 7–8.
37 In many ways, this is indicative of the broader sports film genre, which develops "dramatic narratives to appeal beyond specialist audiences, treating themes often latent in the topoi of modern sports." Bruce Babington, *The Sports Film: Games People Play* (New York: Wallflower Press, 2014), 6.
38 Harry Benshoff, "The Short-lived Life of the Hollywood LSD film," *Velvet Light Trap* 47 (2001).
39 Beattie, "Morning of the Earth," 122.
40 Scott Laderman, "Wanderlust: Surfing, Modernization, and Cultural Diplomacy in the Long 1970s," *Diplomatic History* 40, no. 5 (2016).
41 Benshoff, "The Short-lived Life."
42 Kilcline, "California Dreaming," 8.
43 Wilson, "Outlaw's paradise," 255.
44 Leyla Stevens, "Afterimages: Constructing a Counter-archive of South Bali Histories Through Moving Image" (D. Creative Arts diss., University of Technology Sydney, 2021), 109.
45 Vannessa Hearman, *Unmarked Graves: Death and Survival in the Anti-communist Violence in East Java, Indonesia* (Singapore: National University of Singapore Press, 2018).
46 Beattie, "Morning of the Earth," 119.
47 Agnieszka Sobocinska, "Following the 'Hippie Sahibs': Colonial Cultures of Travel and the Hippie Trail," *Journal of Colonialism and Colonial History* 15, no. 2 (2014).
48 Wilson, "Outlaw's Paradise," 255.
49 Wilson, "Outlaw's Paradise," 255.
50 Alex Leonard. "Learning to surf in Kuta, Bali," *Review of Indonesian and Malaysian Affairs* 41, no. 1 (2007): 27.
51 Strain, *Public Places*, 133.
52 Bannerman, "*Morning of the Earth* was Once Seen as a Gamble."
53 Sobocinska, "Following the 'Hippie Sahibs.'"
54 Sobocinska, "Following," par. 8.
55 Sobocinska, "Following," par. 8.
56 The film was first exhibited just prior to Bali's new prominence as Australia's pre-eminent beach resort getaway and international hotspot. In the following decades, Bali emerged as an attractive and exotic destination for international tourists. See Michel Picard, "From 'Kebalian' to 'Ajeg Bali': Tourism and Balinese Identity in the Aftermath of the Kuta Bombing" in *Tourism in Southeast Asia: Challenges and New Directions*, ed. Michael Hitchcock, Victor T. King, and Michael Parnwell (Copenhagen: NIAS Press, 2008), 101.
57 Agnieszka Sobocinska. "Innocence Lost and Paradise Regained: Tourism to Bali and Australian Perceptions of Asia." *History Australia* 8, no. 2 (2011): 199–222.
58 Amanda Achmadi. "Reading Urban Bali: Untold History, Unwanted Urbanism," *Review of Indonesian and Malaysian Affairs* 44, no. 2 (2010): 149–178.
59 Picard, "From 'Kebalian' to 'Ajeg Bali,'" 101.
60 Kadek Wiweka and Komang Arcana. "The Impact of Tourist Accommodation Development Toward the Socio-cultural Aspects in the Seminyak Village, District of Kuta, Regency of Badung, Bali (in Perspective of the Local Community)," in *Asia Tourism Forum 2016-the 12th Biennial Conference of Hospitality and Tourism Industry in Asia* (Atlantis Press, 2016), 216.

61 Achmadi, "Reading Urban Bali," 150.
62 Achmadi, "Reading Urban Bali," 152.
63 John Urry. *The Tourist Gaze: Leisure and Travel in Contemporary Societies* (London: Sage, 1990).

References

Achmadi, Amanda. "Reading Urban Bali: Untold History, Unwanted Urbanism." *RIMA: Review of Indonesian and Malaysian Affairs* 44, no. 2 (2010): 149–78.

Babington, Bruce. *The Sports Film: Games People Play*. New York: Wallflower Press, 2014.

Bannerman, Mark. "It's Consistently Ranked as One of the Top Five Surf Movies of All Time, but *Morning of the Earth* Was Once Seen as a Gamble." *ABC Online*, February 20, 2022. www.abc.net.au/news/2022-02-20/50-year-anniversary-morning-of-the-earth-still-relevant/100837084.

Beattie, Keith. "Morning of the Earth." *Metro Magazine: Media & Education Magazine*, no. 169 (2011): 118–25.

Beeton, Sue. *Film-Induced Tourism*. Clevedon: Channel View Publications, 2005.

Benshoff, Harry. "The Short-Lived Life of the Hollywood LSD Film." *Velvet Light Trap*, 47 (2001).

Booth, Douglas. "(Re)reading *the Surfers' Bible*: The Affects of *Tracks*." *Continuum* 21, no. 1 (2012): 17–35. https://doi.org/10.1080/10304310701676285.

Crosson, Seán. *Sport and Film*. Oxon: Routledge, 2013.

Doherty, Sean. "Surf pioneer Albe Falzon: 'I'm so Grateful to Be Living This Life.'" *The Weekend Australian Magazine*, July 25, 2020. www.theaustralian.com.au/weekend-australian-magazine/surf-pioneer-albe-falzon-im-so-grateful-to-be-living-this-life/news-story/edcfd0de4c235f54cbf6d5f59e78c09d.

Evers, Clifton. "The Point: Surfing, Geography and a Sensual Life of Men and Masculinity on the Gold Coast, Australia." *Social & Cultural Geography* 10, no. 8 (2009): 893–908. https://doi.org/10.1080/14649360903305783.

Gaze, Tim. "Sun, Surf and Sound: The Role of Surf Music in the Development of Australian Popular Culture. An Investigation of the Iconic Surf Film *Morning of the Earth*, as a Medium." Professional Doctorate diss., Queensland University of Technology, 2020.

Hearman, Vannessa. *Unmarked Graves: Death and Survival in the Anti-Communist Violence in East Java, Indonesia*. Singapore: National University of Singapore Press, 2018.

Jarratt, Phil. *Bali: Heaven and Hell*. Melbourne: Hardie Grant Publishing, 2014.

Jones, Davis. "History of Surfing: Bud Browne Goes to the Movies." *Surfer*, April 25, 2017. www.surfer.com/features/history-of-surfing-bud-browne-goes-to-the-movies/.

Kilcline, Cathal. "California Dreaming: Surfing Culture in Mediterranean France." In *Sport, Representation and Evolving Identities in Europe*, edited by Philip Dine and Seán Crosson, 1–18. Oxford: Peter Lang, 2010.

Krause, Stefan. "Pilgrimage to the Playas: Surf Tourism in Costa Rica." *Anthropology in Action* 19, no. 3 (2012): 37–48. https://doi.org/10.3167/aia.2012.190304.

Laderman, Scott. *Empire in Waves: A Political History of Surfing*. Berkeley: University of California Press, 2014.

———. "Wanderlust: Surfing, Modernization, and Cultural Diplomacy in the Long 1970s." *Diplomatic History* 40, no. 5 (2016): 865–884. https://doi.org/10.1093/dh/dhw030.

Lemelin, Harvey, Jackie Dawson, Emma J. Stewart, Pat Maher, and Michael Lueck. "Last-Chance Tourism: The Boom, Doom, and Gloom of Visiting Vanishing Destinations." *Current Issues in Tourism* 13, no. 5 (2010): 477–493. https://doi.org/10.1080/13683500903406367.

Leonard, Alex. "Learning to Surf in Kuta, Bali." *RIMA: Review of Indonesian and Malaysian Affairs* 41, no. 1 (2007): 3–32.

McDonald, Ian. "Situating the Sport Documentary." *Journal of Sport & Social Issues* 31, no. 3 (2007): 208–25.

Ormrod, Joan. "*Endless Summer* (1964): Consuming Waves and Surfing the Frontier." *Film & History* 35, no. 1 (2005): 39–51. https://doi.org/10.1353/flm.2005.0022.

———. "Expressions of Nation and Place in British Surfing Identities." PhD diss., Manchester Metropolitan University, 2007.

Park, Eerang. "The Impacts of Film Tourism on Place Change and Tourist Experience: A Lesson from *Eat Pray Love* in Bali, Indonesia." In *Film Tourism in Asia: Evolution, Transformation and Trajectory*, edited by Sangkyun Kim and Stijn Rejinders. Singapore: Springer, 2018.

Peterson, Jennifer Lynn. *Education in the School of Dreams: Travelogues and Early Nonfiction Film*. Durham, NC: Duke University Press, 2013.

Picard, Michel. "From 'Kebalian' to 'Ajeg Bali': Tourism and Balinese Identity in the Aftermath of the Kuta Bombing." In *Tourism in Southeast Asia: Challenges and New Directions*, edited by Michael Hitchcock, Victor T. King, and Michael Parnwell, 99–131. Copenhagen: NIAS Press, 2008.

Saunders, Kay. "'Specimens of Superb Manhood': The Lifesaver as National Icon." *Journal of Australian Studies* 22, no. 56 (1998): 96–105. https://doi.org/10.1080/14443059809387364.

Sobocinska, Agnieszka. "Innocence Lost and Paradise Regained: Tourism to Bali and Australian Perceptions of Asia." *History Australia* 8, no. 2 (2011): 199–222. https://doi.org/10.1080/14490854.2011.11668380.

———. "Following the 'Hippie Sahibs': Colonial Cultures of Travel and the Hippie Trail." *Journal of Colonialism and Colonial History* 15, no. 2 (2014): 1–10. https://doi.org/10.1353/cch.2014.0024.

Stadler, Jane, and Peta Mitchell. "Never-Never Land: Affective Landscapes, the Touristic Gaze and Heterotopic Space in *Australia*." *Studies in Australasian Cinema* 4, no. 2 (2010): 173–187. https://doi.org/10.1386/sac.4.2.173_1.

Stevens, Leyla. "Afterimages: Constructing a Counter-archive of South Bali Histories Through Moving Image." Doctor of Creative Arts diss., University of Technology Sydney, 2021.

Strain, Ellen. *Public Places, Private Journeys: Ethnography, Entertainment and the Tourist Gaze*. New Brunswick, NJ: Rutgers University Press, 2003.

Urry, John. *The Tourist Gaze: Leisure and Travel in Contemporary Societies*. London: Sage, 1990.

Wertheimer, Kate. "The 15 Most Epic Surf Movies." *Time Out*, December 8, 2021. www.timeout.com/film/the-15-most-epic-surf-movies.

Wheaton, Belinda. "Introducing the Consumption and Representation of Lifestyle Sports." In *The Consumption and Representation of Lifestyle Sports*, edited by Belinda Wheaton, 1–25. Oxon: Routledge, 2013.

Wilson, Ian. "'Outlaw's Paradise': Australian Outlaw Bikers, Pre-Crime Regimes and the Appeal of Bali." In *Linking People: Connections and Encounters Between Australians and Indonesians*, edited by Antje Missbach and Jemma Purdey, 253–70. Berlin: Regiospectra, 2015.

Wiweka, Kadek, and Komang Arcana. "The Impact of Tourist Accommodation Development Toward the Socio-Cultural Aspects in the Seminyak Village, District of Kuta, Regency of Badung, Bali (in Perspective of the Local Community)." In *Asia Tourism Forum 2016-the 12th Biennial Conference of Hospitality and Tourism Industry in Asia*, 216–25. Amsterdam: Atlantis Press, 2016.

5

LOCATING FELLINI

Affect, Cinecittà, and the Cinematic Pilgrimage

Alberto Zambenedetti

Introduction

In 2020, as the Covid-19 pandemic claimed the lives of many, exposed entrenched systemic inequalities, and restructured our entire way of socializing, cinephiles all around the world pressed on with the celebrations of Federico Fellini's centenary. While in-person retrospectives, exhibitions, and scholarly gatherings were postponed or held online, documentaries, books, and re-releases of his classic films poured into the marketplace, confirming that Fellini remains one of the most celebrated directors and profitable brands in all of cinema's history, and that the public's appetite for Fellini often spills out of his filmic universe.[1] In this chapter, I explore the manifestations of Fellinian affects in objects, places, and art in which the maestro's presence is felt, but remains intangible, ultimately arguing that Fellini's endurance transcends the confines of his films and is sustained by a complex articulation of cinematic memories, public discourse, and media.[2] Ranging from marketing to tourism, I examine three different invocations of Fellini, connecting them to their corresponding films and emotions. Specifically, in the first part, I examine how nostalgia anchors the launch of an Italian electric car, focusing on the mobilization, commodification, and wilful misinterpretation of *La dolce vita* (1960) and *Amarcord* (1973). In the second section, I discuss melancholia in relation to Cinecittà as a site of tourism for cinematic pilgrims, underscoring how the studio's present emptiness remains haunted by the production *Il Casanova di Federico Fellini* (1976), one of the director's most melancholic characters. Finally, in the third example I focus on *saudade* and the practice of sceneframing, which I connect to Fellini's own repurposing of Fellinian imagery in *Intervista* (1987).[3] Ultimately, what I seek to address is how, even without direct invocation of his films, his own words, or even his images, Fellini continues to elicit affective responses long after his passing.

DOI: 10.4324/9781003327585-5

Nostalgia for the Future: On Fellini and Electric Cars

In August 2020, FIAT brand CEO Olivier François introduced the all-new 500 La Prima, the marque's "first" fully electric subcompact car.[4] In a launch video posted on the company's official YouTube channel, the manager asks: "Have you ever seen the Fellini movies from the '60s? The soundtracks by Nino Rota, to me, are the sound of the dolce vita." As the new model pulls up next to an original 1957 FIAT 500 convertible and a 2007 reissue of the iconic car, François confidently turns to the camera and exclaims: "Look. Nothing has changed. But in fact, everything has changed."[5] The invocation of the automaker's history and tradition alongside Federico Fellini's cinema is not gratuitous – albeit erroneous in chronology and lax in interpretation. In fact, while the expression "la dolce vita" does translate to "the sweet life," Fellini's eponymous film is anything but a celebration of the kind of stylish lounging the words seem to evoke. *La dolce vita* is both a document of the Italian economic boom, of which FIAT was a protagonist and a driving force, and a reflection on the customs of a rudderless generation dangerously lost in unbridled hedonism. Moreover, La Prima does not play the theme from *La dolce vita*, which is decidedly more sombre, but the serene music from a film that was released in 1973, when the hobnobbing and jet-setting trends of the 1960s were decidedly over.

Due to their quiet engines, in the majority of markets, electric automobiles are mandated to be equipped with acoustic vehicle alerting systems (AVAS) to signal their presence to nearby pedestrians when travelling at low speeds. While most companies opt for futuristic hums or chimes, the FIAT 500 La Prima emits the theme from *Amarcord* by legendary composer and frequent Fellini collaborator Nino Rota. Anchoring the 500 to Italian cultural icons of global renown is a recurrent promotional strategy for Fiat Chrysler Automobiles (FCA). In 2012, when the brand returned to North American markets after a 30-year hiatus, the group partnered with fashion house Gucci to create a convertible 500 emblazoned with the designer's classic red and green stripes. Similarly, for its first issue of the electric La Prima, FCA collaborated with Giorgio Armani, Bvlgari, and Kartell to produce three limited-edition vehicles with premium features such as upgraded paint, bejewelled steering wheels, upholstered dashboards, and custom exterior mirror caps and grilles, all of it using primarily recycled or sustainable materials. As I argued elsewhere, FIAT's campaigns often situate their products in the 500 line-up (500, 500L, and 500X) within imagined lifestyles that, while glamorous, remain attainable to the masses.[6] Rather than extolling the cars' virtues, the advertisements position them as the playful, vibrant, cheerful alternatives to the dull utilitarianism of their American, German, or East Asian counterparts. Moreover, the styling of new 500 models strikes a balance between past and present, often ignoring current design trends in favour of retro-chic solutions such as returning to FIAT's mid-century curved lines and supple volumes. In other words, the underpinnings of FCA's messaging are entirely predicated on affects such as joy, youthfulness, cheekiness, sexiness, coolness, and, most importantly, nostalgia.

In semiotic terms, while the specific interventions of Gucci, Armani, Bvlgari, and Kartell on the 500 are indexical, in that their customizations establish a physical relationship between the car and the brands' respective products, the evocation of Fellini's cinema in the launch campaign of La Prima is entirely symbolic: in order to grasp its secondary meaning (that is, beyond that of being alerted to the presence of a nearby vehicle), the pedestrian hearing the melody emanating from the car needs to be familiar with the auteur's work. The automaker's assumption, then, is that the descending quarter notes of the first eight beats of *Amarcord*'s theme are not only a recognizable melody but also a widely shared cultural reference that transcends cinematic boundaries. Of course, marketing applications of the Fellini name (and by extension, images and sounds) have been afoot at least since the 1980s, as Rebecca Bauman has convincingly argued, and they continue to hold "a rich trove of possible associations for consumers, including craftsmanship, creativity, italianità, hedonism, and fun."[7]

Consumerism is predicated on the sustained acquisition of goods that are meant to bring pleasure to their owner, in whatever shape or form that might take; it is only natural for brands to harness the power of a name that, at least since the release of *La dolce vita* in 1960, has become synonymous (perhaps inaccurately) with a certain kind of joie de vivre.[8] However, Bauman continues,

> The naming of products and establishments after Fellini following the director's death has only multiplied and no doubt will continue. The farther we get from the Fellini era, the deeper the *nostalgia* for the historical and cultural trappings of the time periods he represented.[9]

In this light, *Amarcord* emerges as a singularly appropriate film for the launch of a "new" FIAT 500, a product whose very existence in the 21st century is predicated on the invocation of a bygone era; the film's title, as is well known, is the univerbation of the phrase *a m'arcord*, which means "I remember" in Romagnol, the regional language spoken in Fellini's hometown of Rimini. John Agnew bristles at the idea that Fellini mobilized nostalgia in his films about Rimini (*I vitelloni*, *Amarcord*) and Rome (*Le notti di Cabiria*, *La dolce vita*, *Roma*, *Intervista*), arguing

> There is nothing nostalgic or celebratory about Fellini's sense of place, no yearning for better times or a world we have lost. Rather the sense of the past is colored by a scepticism that planned interventions, such as those associated with Fascism, ever turn out as intended.[10]

Nevertheless, as Bauman's argument makes salient, anchoring a product to Fellini means capitalizing on the unquestionable presence of this affect (and variations thereof) in his cinematic universe. In my view, both Bauman and Agnew are correct, and turning to Svetlana Boym's pioneering work can help explain how nostalgia is not a singularly romanticizing or idealizing feeling. Returning to the etymology of the word, the late scholar distinguished between the two kinds of affect commonly associated with nostalgia:

> Restorative nostalgia stresses nostos and attempts a transhistorical reconstruction of the lost home. Reflective nostalgia thrives in algia, the longing itself, and delays the homecoming – wistfully, ironically, desperately. Restorative nostalgia does not think of itself as nostalgia, but rather as truth and tradition. Reflective nostalgia dwells on the ambivalences of human longing and belonging and does not shy away from the contradictions of modernity. Restorative nostalgia protects the absolute truth, which reflective nostalgia calls it into doubt.[11]

In Fellini's case, the distinction outlined by Boym is extremely helpful; in fact, Fellini did not romanticize the past, but he sought to understand its relationship with the present, and often projected its teachings into the future with uncanny foresight. If there is nostalgia in his work, it is of the reflective variety. Thusly understood, the association of Fellini's *Amarcord* with La Prima appears to be less flippant than previously imagined; hearing the new FIAT drive by does not evoke nostalgia for a generically construed *la dolce vita,* as François reductively seems to suggest, but rather this aural perception activates a sophisticated network of affects anchoring a reflection that ranges from tradition to legacy in the Anthropocene, from the (desirable) demise of petroculture to the (long-awaited) rise of sustainable industrial and commercial, but also touristic and cinematic practices in the face of environmental catastrophe.

In fact, Fellini gestured to the ecological disasters caused by human activity as early as 1972 in the extended sequence set on the Grande Raccordo Anulare (GRA) in *Roma.* The ring-shaped highway encircling the extensive metropolitan area of the Italian capital city is depicted in his experimental feature as a genuine hellscape. Congested, loud, flooded, muddy, surrounded by dirty construction sites and factories spewing thick columns of black smoke into the air, this major piece of modern infrastructure is the theatre of dreadful accidents.

Bloodied cattle strewn on the asphalt, fire engulfing overthrown lorries, protestors confronting policemen in riot gear. Everything is chaos, and all of it is documented by Fellini's crew, which also contributes to the gridlock and the pollution with its diegetic and extradiegetic trucks and cranes. Beginning at a tollbooth and ending (nonsensically) at the Colosseum, the sequence suggests that perhaps our civilization is destined to extinction, and that our monuments will be such hideous highways, a far cry from the majestic ruins left behind by the ancient Romans, now reduced to traffic islands strangled by coils of dirty vehicles. Perhaps, among the many artefacts we will leave behind, is also a small, anachronistically styled electric car emitting *Amarcord*'s soundtrack; a calculated, yet desperate last attempt to use restorative nostalgia to paper over Fellini's pessimism about the future of humanity.

A Journey Into Melancholia: Cinecittà and the Cinematic Pilgrimage

Comprising vignettes culled from childhood memories (either personal or borrowed), on the surface, *Amarcord* appears to be (but as I sought to demonstrate, it is not) steeped in a restorative nostalgia that attracted criticism from many, including pioneering critic and Fellini's long-time friend Renzo Renzi, who underscored the film's problematic enmeshment of history, memory, and imagination. In Renzi's crosshairs are precisely *Amarcord*'s dominant affects, which in his opinion do not adequately expose the horrors of fascism and its troubling latency in Italian culture and beyond. In this open letter to the director, Renzi writes: "*Amarcord*, instead of disquieting spectators, in my opinion, in the end rewards their habit to slip into *nostalgic* 'rêverie' which is corrected by some passing refusal, against your stated intention."[12] As Giuliana Minghelli notes, Renzi's essay is accompanied by 25 images (23 photographs and 2 stills from a LUCE newsreel) that counterbalance Fellini's re-imagining of the fascist period with the hard truths of photojournalism, propagandist as it might have been.[13]

Contributing to the film's atmosphere of political and historical "disembeddedness" is the lack of a physical relationship between its setting and the actual (few) locations existing in the real world. In fact, as is well known, since *La dolce vita*, the director began to progressively retreat into the studio, where he asked his collaborators to build increasingly abstract versions of the places where his films were set; many of the streets, squares, houses, and monuments of *Amarcord*'s "Rimini" are no exception.[14] Marco Bertozzi argues that

> Fellini's films knit together the different skeins of Rimini: the realistic, like a description of a road, a café or square; the impressionistic, such as a bright sky, details of sound and touch, perceptions linked to the "flavour" of the town; and, lastly, the imponderable and personal: flashes of the past, hallucinations, fragments of inner life, phantasms of the memory in which one wanders bewildered and enchanted.[15]

As the scholar notes, Fellini's reconstruction of Rimini becomes intelligible not simply as a city, but as a visual and aural recollection of the time and space of his childhood. From a narrative standpoint, many of Fellini's films are loose collections of episodes rather than causally linked chains of events, and *Amarcord* is no exception. However, while impressionistic in nature, the film also takes a holistic approach to the senses, abounding in references to corporeality – bodily excretions being almost celebrated as a site of resistance to fascism's efforts to "sanitize" the nation's biopolitics.

Yet, if *Amarcord* decidedly invites the body into its critique of fascism, it also withholds any direct condemnation of its physical manifestations in architecture and urban planning, and Fellini's complex relationship with Cinecittà speaks to this point. Since the 1960s the sprawling studio complex, and soundstage (Teatro) number 5 in particular, has become synonymous with Fellini. However, Cinecittà's long history is also inextricably linked to fascism. When it was inaugurated in 1937, a triumphant Benito Mussolini famously declared that cinema was the regime's strongest weapon, suggesting that government intervention and investment in the film industry, at the very least, would result in productions that expressed views favourable to his administration.[16] Consequently, cinema would help him garner the people's acquiescence, if not their consent. In his invective against *Amarcord*'s perceived edulcoration of fascism, Renzi directs his attention to the sequence depicting the 21 April celebrations[17]:

> It expresses the "rejection of politics" typical of the petit-bourgeois spirit. . . . In this way, a past that is rebuilt using a good part of the culture of those days, which seems far away, is in fact really close, because it appeals to (and obtains) our *emotional* participation, showing that the two eras are connected; there is no fracture, we left nothing behind the fence, even if we told ourselves that we did. Your film, in other words, is about such "continuity."[18]

It is unclear whether Renzi included Cinecittà itself among the unchallenged manifestations of 1930s culture he detected in the film (and against which he railed). However, the studio complex is thinly disguised in the sequence he attacks, which begins near Cinecittà's immediately recognizable main entrance on Via Tuscolana 1055. In the film, the Rationalist building serves as a makeshift railway station where the arrival of a gaggle of fascist echelons is announced by a cloud of thick smoke – presumably exhaled by a fatigued locomotive travelling from Rome to the small coastal town. As the fascists begin their parade through *Amarcord*'s "Rimini," they leave the physical reality of the storied entrance behind, and enter the papier mâché world of Fellini's nostalgia, his (problematic) digression into childhood.

Franco Monteleone writes "for many spectators, and for many authors, Cinecittà . . . is the custodian of memories, *emotions*, images that the neighbourhood theatres, in the intimacy of the dark room, have delivered for decades."[19] As the scholar notes, among the variety of loci tied to the history of Italian cinema, Cinecittà is the one locus that established a clear affective link connecting production with exhibition; in other words, the fondness that Italian cinema goers experience for this juggernaut transcends its ideologically fraught beginnings, because "Cinecittà continues to be the *myth*, but overall the *place*, as we said, capable of transforming itself into anywhere in the world, the antonomasia of collective dream and imagination."[20] It is precisely this ability to become any place (and any time) that enticed Fellini, who took advantage of these endless possibilities to materialize his own subconscious. The development of the Cinecittà-inspired theme park Cinecittà World in Rome's Castel Romano suburbs is a tribute to this enduring legacy. As further evidence of the continuing commodification of Fellini's name and imaginary explored by Bauman, the exhibition *Pezzi da Oscar* features a bust of the director, and the small circus tent at the centre of the vast theme park is named after Fellini.[21]

Inside the gates of Cinecittà is the spacious courtyard that leads to the complex's open-air sets and pool, production and storage facilities, and 19 soundstages. Today, the rectangular lawn is adorned with the head of Venusia, a large artefact crafted by scenographer Giantito Burchiellaro for the opening and closing scenes of *Il Casanova di Federico Fellini*, which was made entirely on the premises. The sculpture is not the only sign of Fellini's profound and prolonged relationship with the place, far from it; a low-slung, peach-coloured building on the grounds has been renamed Palazzina Fellini in honour of the director, and now contains a permanent exhibition of paraphernalia from his films. As Andrea David notes in her account of her visit to the hallowed Italian film studio, it is the first stop on the self-guided tour of Cinecittà's historic buildings.[22] In one of the longest-running websites devoted to this practice, FilmTourismus.de, the German travel writer asks: "What is film tourism?" According to David, it is a phenomenon strictly related to, and caused by, the habit of moviegoing, except that "one skilfully plays with illusions, the other promises real experiences."[23] Identified by the neologism "set jetting," the custom of travelling the world to visit film locations is a growing industry to which a fair amount of scholarly literature has been devoted.[24] However, the affective dimensions of this particular manifestation of cinephilic tourism remain underexplored.[25]

David's post on her visit to Cinecittà focuses not on the films, but on physical structures such as the Palazzina Fellini and the studio's open-air sets, and closes with a shot of an empty Teatro 5. It is in this cavernous space that Fellini and his collaborators were able to breathe filmic life into his most dream-like titles, often barely signifying real places by means of cardboard cut-outs, plastic sheets, and light effects.[26] Nevertheless, a pilgrimage to this vacant hangar-like building inspires awe and reverence for the art that was created therein. As Rob Stone writes, this is a quality that the studio does not share with other tourist sites:

> Unlike the crowded and exploited tourist areas of the nearby metropolis, Cinecittà remains a sacred site whose inaccessibility and relative obscurity only adds to its appeal. It is not nostalgia that powers the quest but a kind of faith in film. . . . That is to say, the locations where something was filmed and the extant evidence of that happening many years later are treated as sacred by the pilgrims to Cinecittà.[27]

These positive affects (return and belonging) are the antidote to the void (or lack) upon which nostalgia is predicated, and are made possible by the unique relationship between tourist/audience and place/film. In other words, visiting Cinecittà is akin to what Roger C. Aden termed "symbolic pilgrimages," which "feature individuals ritualistically revisiting powerful places that are symbolically envisioned through the interaction of story and individual imagination."[28] Aden intends the pilgrimage as located in "the metaphorical terrain of culture rather than in its material manifestation."[29] Yet, if the location is vacant, as is the case for the empty Teatro 5, then the pilgrimage is necessarily both material *and* symbolic, precisely because the "place" built by means of cinematic technique has never truly existed as such; it is illusion, as David noted. In other words, when considering the notion of place in Fellini's late films, one must realize that they are iconic, not indexical, and must be discussed not in terms of locations, which they are not, but as networks of affects.

But if Olivier François's uncritical evocation of a generically understood (restorative) "nostalgia" sends us in the wrong direction, what are the affects that capture Fellini's invocations of places (real or imagined) from his favourite soundstage? Thinking about the "looking into the past" photographs may be useful here. In his writing on this cultural phenomenon – whereby images from the past are juxtaposed with the location where they were shot, generating a new picture containing the old one – Ryan Lizardi argues that these photographs:

> encourage a mourning that learns from and releases the object of past desire once we have learned from it, as opposed to *melancholia* where the past is lingered and dwelled on indefinitely. The importance of this distinction is that in one form of past engagement, mourning, a difference is felt for the missing object of the past where one can understand that the object is gone or not as it once was, whereas in the other form, *melancholia*, the object takes on an assumed permanence that can never actually be achieved.[30]

Building on the distinctions drawn by Boym in the realm of nostalgia, Lizardi identifies mourning as the healthier of the two affects he examines, since the emotional closure it provides allows for recovery and growth to take root. He argues that "through these images we are encouraged to mnemonically mourn by looking directly at the loss, learning from it, and moving on."[31] Conversely, he sees melancholia as a sustained inability to process the passing of time, which is made all the more unfathomable by the illusory permanence of the object (the film) while its physical manifestation (the set) is long gone.

When thinking about Fellini's relationship with Cinecittà, mourning and melancholia emerge as the affects capable of bringing into sharp relief the crucial differences between *Amarcord* – the film that best encapsulates Fellini's relationship with history, memory, and place – and *Il Casanova di Federico Fellini*. If the former can be understood as mourning the loss of youth, innocence, and home, the latter presents the unresolved lament of a character afflicted by uprootedness and melancholia. At the film's end, the elderly protagonist (Donald Sutherland) realizes he might never return to his native Venice. In the solitude of his chambers in faraway Bohemia, the man slips into a reverie and the film cuts to a dream sequence in which his younger self walks on his hometown's frozen waters and detects the submerged head of Venusia – the same artifact that now welcomes visitors to Cinecittà. Characters from his past appear to him like evanescent memories, until a stable form manifests in the night. It is Rosalba (Angela Lojodice), the dancing automaton

Casanova had romanced at the court of the Duke of Wurtemberg (Dudley Sutton) in one of the film's previous chapters. At the time, the gentlemanly Venetian mistook her for a real woman and sprang into action to defend her honour against the salacious behaviour of the Duke's many drunken guests. In his last vision Casanova approaches her, and Rosalba begins her familiar mechanical dance. Soon the two figures glide smoothly over the plastic sheets that signify Venice's canals, accompanied by Nino Rota's ethereal *La Poupée Automate (II versione)*. Eventually, the discreet long shot is interrupted by a close-up of the aged Casanova's wrinkled eyes, his expression betraying a surge of emotion as he fantasizes about this impossible reunion. The film's concluding shot returns to the dream, which pictures Rosalba and the young Casanova clasped in a tender but frozen embrace, revolving like the statuettes atop a music box.

As Fellini biographer and scholar Tullio Kezich wrote, Casanova is "construed to be a completely unhappy character, more baroque than 18th century, deplorable, a symbol of emptiness fated to remain empty."[32] Yet this finale returns the melancholic lothario to the "place" of his youth, where the film also begins: a nocturnal Venice barely signified by a hollowed-out replica of the Rialto Bridge propped up by visible scaffolding. Everything is still in this quiet landscape, permanently embalmed in the grips of a fictional winter. Rosalba and Giacomo will twirl on

FIGURE 5.1 Photograph of the open-air Cinecittà pool.

Source: Courtesy of Alberto Zambenedetti.

forever, the historical figure obfuscated by the myth of himself he expertly crafted in his magnum opus, his multivolume memoirs.

No physical evidence of this scene is left in the empty soundstage photographed by David during her visit to Cinecittà, only the actual facilities, such as the open-air pool where the film's festive opening scene was shot (Figure 5.1). But the head of Venusia, a monument to the entire production, and to Fellini's presence at large, remains to guard the gates of the compound, welcoming visitors to this cathedral of Italian filmmaking. The cinematic pilgrimage, in this case, is akin to that experienced by believers who travel to the place where the spirit is said to have manifested itself, yet they have no hope to see it with their own eyes. While Venusia points to the materiality of their journey to Cinecittà, their pilgrimage remains largely symbolic, full of longing for places that never were (the Rimini of *Amarcord*, Casanova's Venice, or even the Rome of *La dolce vita*), existing only in their cinematic evocations, and continuing to reverberate through time and generations of cinephiles.

Wistfully Travelling Through Cinema: Sceneframing Fellini's *Saudade*

At the midpoint of *Fellini's Intervista* (1987), the director and Marcello Mastroianni (dressed as one of Fellini's favourite comic-book characters, Mandrake the Magician) travel to Anita Ekberg's secluded villa in the Roman countryside, where they are welcomed by the actress with genuine affection. Once they settle in with their entourage, which includes the fictional Japanese crew interviewing Fellini, Mastroianni waves Mandrake's cane into the air, a cigarette between his lips, commanding it to bring back the good old days. With a puff of smoke, a white sheet unfurls in the middle of the living room, hanging magically from the ceiling. Wistfully, Mastroianni/Mandrake whispers the word "music," and the *Intervista: Marcello Mandrake Suite* by Nicola Piovani begins to play.

Inspired by the score to *La dolce vita*, Piovani's suite is both a celebration of Nino Rota's accompaniment to the 1960 film and a variation on its primary themes. Slowly, the Magician begins to swing his hips, and gently shuffles behind the sheet where, in silhouette, he is joined by Ekberg. They begin to slow dance, their shadows embracing. The image fades to black and the two acting legends quickly reappear as their younger selves, projected onto the white sheet, dancing cheek-to-cheek at the Baths of Caracalla in *La dolce vita*. In a two-shot, 1987 Ekberg and Mastroianni, now sitting down, gaze at the iconic scene with misty eyes, the camera slowly closing on them. Mastroianni jokingly asks for some grappa, Ekberg fights back a tear, and eventually they toast to their monumental achievement: their performance in *La dolce vita* has transcended the history of cinema; it has become myth, legend, pure emotion. Another wave of the magic wand and the sheet/screen disappears in a puff of smoke.

In his theorization of Fellini as a 20th-century incarnation of the Greek god Hermes, a figure of mediation, interpretation, and circulation of knowledge, Ivan

Pintor Iranzo comments on Fellini's own revision of *La dolce vita*'s famous scene, which was itself based on a 1958 spread in the magazine *Il Tempo*. The scholar writes:

> *Intervista* turns into what is practically an essay on the recycling of images, introducing a third stage of interpretation of these sequences and opening them up to their conversion into icons and their reconstruction as Instagram pictures or parades like the Fendi brand's runway show in 2016 in the same location.[33]

Cast in this light, Fellini's ability to remediate and recirculate images might evoke certain postmodern practices based on appropriation and re-signification. However, while it is undoubtedly true that the director was engaged in the business of repurposing objects of popular culture since before his debut behind the camera, his practice involved a profound imbuing of the resulting images with affects attributable to his own artistic sensibility. In fact, Fellini's films abound with scenes that not only convey emotion but also are structured to create it: for instance, the many brusque transitions from crowded celebrations in full swing to empty squares littered with stragglers (*La strada, Le notti di Cabiria, La dolce vita, Roma*); or the enigmatic endings following characters that are experiencing (or have just experienced) enormous distress (*La strada, Il bidone, Le notti di Cabiria*); or the many goodbyes and farewells bid by major or minor players, often straight to the camera (*Roma, Amarcord, E la nave va, Ginger e Fred, Intervista, La voce della luna*); or the use of music cues and transitions moving fluidly in and out the diegesis (*La strada, 8 ½*). Precisely because it revisits a handful of iconic shots from *La dolce vita*, but re-inscribes them within the director's on-and-off screen iconography and associations (Cinecittà, Mandrake, sudden bursts of crass humour, etc.), the Ekberg-Mastroianni reunion in *Intervista* is not simply an instance of postmodern metalinguistic referentiality, but an expression of quintessentially Fellinian *saudade*, an affective quality that the Instagram and Fendi appropriations mentioned by Pintor Iranzo attempt to capitalize on, much like the FIAT 500 campaign built on *Amarcord*'s (putative) nostalgia.

An emotion neighbouring and overlapping with nostalgia and melancholia, *saudade* is similarly characterized by longing, desire, and perhaps sadness. However, as Zuzanna Bułat Silva writes:

> There is something very bad and something very good about *saudade*. One can be killed or tortured by it, but it is also a precious jewel everybody desires. *Saudade* . . . is also an overwhelming feeling: you can't not think about it. What is very important in the meaning of *saudade* is the element of "thinking," recalling the past and also the positive evaluation of that state of mind – it is good to feel *saudade*.[34]

Unlike melancholia and (restorative) nostalgia, but similar to mourning, *saudade* can be a positive way to engage with the past and the sense of lack emerging from the absence (of things, people, places) caused by the passing of time. Moreover, while the first two affects are generally associated with space and/or time, *saudade* is itself borne out of experiencing an emotion, and therefore is implicitly self-referential. If Fellini's return to his own imaginary is in fact an instance of *saudade*, then so are the many sceneframes images produced by cinematic pilgrims. Formally identical to the aforementioned "looking into the past" photographs, these ubiquitous sceneframes function both as memento of a visit to the loci of cinema (much like Andrea David's blog), and as recall of something as fleeting as an emotion sparked by the viewing of a film – once again, an "illusion," as David put it.

Existing at the nexus between what W.J.T. Mitchell has described as western society's "pictorial turn" – a return to pictures as sites for the interrogation of the relationship between people, institutions, power, and the very acts of looking – and the emergence of John Urry's "mobilities paradigm" – a renewed attention to the connections between different forms of travel, social organizations, and infrastructures – sceneframes engender a unique relationship with the moving image not only in spatial and temporal terms but also in the way they subtend participatory affects such as joy, awe, or *saudade*.[35] In other words, while the "looking into the past" photographs enabled a mourning of what is no longer there, sceneframes celebrate the emotions experienced through cinema, and they project them onto the location, effectively making it into a distinctly "cinematic place" that exists only *because* said film was shot there. Whether the location was already historically or architecturally important is entirely irrelevant to the affective value of the sceneframe.

In his ongoing series FILMography, Canadian photographer Christopher Moloney produced seven sceneframes of Rome's Trevi Fountain, using stills from *To Rome With Love* (Woody Allen, 2012), *Three Coins in the Fountain* (Jean Negulesco, 1954), *Roman Holiday* (William Wyler, 1953), the episode *Play It Again, Brian* from the animated TV show *Family Guy* (Season 6, Episode 10, 2008), and, of course, *La dolce vita*, the ur-text of all Trevi Fountain cinematic events (Figure 5.2).[36] Cropped, often out of focus, and largely obscured by black-and-white printouts of relevant frames from the aforementioned titles, the neoclassical masterpiece by architect Nicola Salvi is barely intelligible in Moloney's images, whose primary purpose is not to document the monument, but to signal the photographer's presence in the place where the movie cameras once rolled – that is, where something related to the history of cinema once happened. In other words, Moloney's visit to the Trevi Fountain is a cinematic pilgrimage animated by the symbolic relationship with those titles, and as such it may override the location's original artistic significance.

As Rob Stone explains, "instead of referring to the history of the actual building, monument or landscape that is visited, the cinematic pilgrimage to the filming location relocates its meaning in an intra-history of viewing experiences."[37] Consequently, I argue that the cinematic pilgrim's production of a sceneframe is not simply the reframing of an existing image, which might seem as an intrinsically postmodern practice. It is, above all, a positive engagement with one's emotions regarding the experience of a cultural event, and therefore in-and-of-itself an act

FIGURE 5.2 Sceneframe of Trevi Fountain.

Source: Courtesy of Christopher Moloney.

of intertextual, spatializing *saudade* – a lesson taught by Fellini in *Intervista*'s proudly self-referential celebration of *La dolce vita* not only as monumental achievement in the history of cinema but also as a personal memory of a time long gone.

Ultimately, if *Amarcord* is tinged with problematic nostalgia, and *Il Casanova di Federico Fellini* engenders a melancholic relationship with absence – not only of a person from a place but also of someone's affection for someone else – then, as Marco Bertozzi argues, *Intervista* functions as a synecdoche for Fellini's entire artistic journey, and for cinema at large, exemplifying that

> Well beyond the nostalgia for the past, of the dreams of an artist adrift in his *rêveries* (a favourite aspect of fellini-ism), Fellini's films lead to emotional and cognitive experiences capable of challenging our functional discernment, allowing the foundational concept that cinema is an experiential journey, without projections, of the world.[38]

In conclusion, returning to those parcels of physical world that have been immortalized by cinema, even if they are no longer there, there can be a powerful emotional affair. Hearing a car drive by can make us daydream of a bygone era that a skilled director has immortalized (and partly created), and that has come to represent Italy itself for so many people around the world. Walking a self-guided tour of an empty studio lot can fill us with awe, and can help us process loss. And finally, taking a photograph in a place that we have seen "at the movies" can make us feel one with them, even if both are, the film and the emotion we felt, just an illusion.

Notes

1 As Corsi, Nicoli, and Venturini note, "Fellini as a brand was built first by leveraging the idea of his authorship and then by elevating him to the rank of producer. His legend as a producer was started by the film critic Tullio Kezich and publisher Angelo Rizzoli. Fellini himself then cultivated it through interviews and books as well as by constantly investing in his celebrity. He was not exploited merely because he had become a valuable brand but also because he was personally engaged in creating his own stardom." Barbara Corsi, Marina Nicoli, and Alfonso Venturini, "Fellini the Founder? The Fellini Brand in Film Production," *Journal of Italian Cinema & Media Studies* 9, no. 1 (2021): 134–5.

2 Rather bitterly, Centorrino, Dalla Gassa, and Minuz write that "There are restaurants called 'La Dolce Vita' all over the world. Words like paparazzi have made it into dictionaries. Airports, streets, piazzas and schools are named after Federico Fellini. And yet, as cinema scholars understand, Fellini's films are not nearly as well-known as one might expect." Centorrino, Clizia, Marco Dalla Gassa, and Andrea Minuz, "With a Hat and a Red Scarf: The Construction of Federico Fellini's Public Image," *Journal of Italian Cinema & Media Studies* 9, no. 1 (2021): 3.

3 *Saudade* is a feeling of longing, melancholy, or nostalgia that is supposedly characteristic of the Portuguese or Brazilian temperament. Sceneframes are photographs of film stills produced on the location where the films were originally shot, often aligning them with the surrounding landscape. I borrow the definition from the Finnish cinematic tourism website Fangirl Quest, www.fangirlquest.com/sceneframing/. See also Stepping Through Film. www.steppingthroughfilm.co.uk.

4 Despite its name and marketing campaign, La Prima is not, in fact, FIAT's first fully electric 500. A prototype named 500 Elettra BEV was presented at the 2010 Detroit North American Auto Show and went into production as 500e in 2012. It was sold in select locations in North America from 2013 to 2019.

5 "New Fiat 500," YouTube, www.youtube.com/watch?v=EBl7oWnw_ZU&list=RDCMUC_nBvBIV0K6P5C3tA27Hwrg&start_radio=1&t=295. In a separate video, the CEO compares the sounds from different automakers, arguing that the Italian car "is not the voice of a computer. It is the voice of the Italian creative spirit." See "Fiat 500 la Prima," YouTube, accessed April 10, 2022, www.youtube.com/watch?v=ruFvmcfa7Vc. In the Italian-language version of the launch video, François's words are slightly different, but the substance remains essentially unchanged. " 'Il giorno de La Prima' con Olivier Francois," YouTube, accessed April 10, 2022, www.youtube.com/watch?v=E_JaXUvjoLI&list=RDCMUC_nBvBIV0K6P5C3tA27Hwrg&index=14.

6 See Alberto Zambenedetti, "Making the Old New Again: FIAT's (North American) Second Coming," in *Print to Digital: Aspects of a Cultural Discomfort*, ed. Francesco Guardiani (Firenze: Franco Cesati Editore, 2017), 237–48.

7 Rebecca Bauman, "The Fellini Brand: Marketing Appropriation of the Fellini Name," in *A Companion to Federico Fellini*, ed. Frank Burke, Marguerite Waller, and Marita Gubareva (Malden: Wiley-Blackwell, 2020), 393.

8 On the linguistic history of the expression *la dolce vita* and its semantic shifts over time, see Massimo Arcangeli, ed., *Itabolario. L'Italia in 150 parole* (Roma: Carocci Editore, 2011), 210–11.

9 Bauman, "The Fellini Brand," 401; my emphasis.

10 John Agnew, "Fellini's Sense of Place," in *A Companion to Federico Fellini*, ed. Frank Burke, Marguerite Waller, and Marita Gubareva (Malden: Wiley-Blackwell, 2020), 122. And yet, the discourse surrounding his films is one of enduring nostalgia. For instance, see Juliet Rix's account of a visit to Fellini's hometown tinged with this

affect. Juliet Rix. "Inspired by Rimini: A Tour of Fellini's Magical Home Town," *The Guardian*, January 16, 2020. www.theguardian.com/travel/2020/jan/16/rimini-tour-fellini-centenary-exhibition-home-town.

11 Svetlana Boym, *The Future of Nostalgia* (New York: Basic Books, 2001), XVIII.

12 Renzo Renzi, "Il fascismo involontario (a proposito di 'Amarcord')," in *Il fascismo involontario e altri scritti* (Bologna: Universale Cappelli, 1975), 131–81, 167; my emphasis. Translations from the original Italian are mine.

13 While Renzi's essay does not discuss the photographs directly, his pages implicitly invite a visual counter-reading of these images. See Giuliana Minghelli, "Amarcord/Am'l'arcord. Fascism, Memory, and the Visual in Federico Fellini and Renzo Renzi," *Italica* 97, no. 4 (2020): 848–78.

14 During long shoots, Fellini even set up living quarters in a small building attached to the enormous soundstage no. 5. Strangely, the paucity of real-world locations one can visit may augment the significance of affects such as nostalgia when embarking in cinematic pilgrimages related to Fellini, since tourists must turn to structures such as Cinecittà, the theme park Cinecittà World, and the new Fellini museum in Rimini.

15 Marco Bertozzi, "Media Architecture: Fellini's Rimini, a Town of the Imagination," *Journal of Italian Cinema & Media Studies* 9, no. 1 (January 2021): 19. On Fellini's "invention" of Rimini in Cinecittà, Bertozzi also writes: "The Minghini collection at the Rimini Gambalunga Library now possesses a precious trove of some thousand photos of real Rimini buildings and of Cinecittà, where stage designer Danilo Donati made the sets for the Romagna townscape. One can see Corso d'Augusto, Fontana della Pigna, Piazza Cavour, the bar Commercio, the Fulgor Cinema, etc., allowing a comparison of images that transport the cinema-town relationship onto a level beyond mere iconographic reference." Bertozzi, "Media architecture," 18. For an extensive discussion of the intricate relationship between Fellini and his hometown, both in terms of memory and *urbs*, see Marco Bertozzi, *L'Italia di Fellini. Immagini, paesaggi, forme di vita* (Venice: Marsilio, 2021), 21–46.

16 "La cinematografia è l'arma più forte." YouTube, https://youtu.be/Q_6yglVyaaU.

17 April 21 is the putative date of Rome's founding in 753 BC. It was proclaimed a fascist official holiday by Mussolini on April 3, 1921, thereby establishing an ideological link between ancient Rome and his regime.

18 Renzi, "Il fascismo involontario," 178; my emphasis.

19 Franco Monteleone, "L'albergo più amato di Federico Fellini: Cinecittà," *Studi Novecenteschi* 35, no. 75 (2008): 286; my emphasis.

20 Franco Monteleone, "L'albergo più amato," 288.

21 "Pezzi da Oscar," Cinecittà World, www.cinecittaworld.it/it/parco/attrazioni/pezzi-da-oscar.

22 "Cinecittà Studios Rom," Filmturismus, Translations from the original German are mine. Accessed April 11, 2022, www.filmtourismus.de/cinecitta-studios-rom/.

23 "Was is filmturismus?" Filmturismum, accessed April 11, 2022, www.filmtourismus.de/was-ist-filmtourismus/.

24 See Sue Beeton's *Travel, Tourism, and the Moving Image* (Bristol: Channel View Publications, 2015) for an influential overview of set jetting research.

25 One of the few contributions to this phenomenon is by Jane Stadler and Peta Mitchell, "Never-Never Land: Affective Landscapes, the Touristic Gaze and Heterotopic Space in *Australia*," *Studies in Australasian Cinema* 4, no. 2 (2010): 173–87. Also of note is the work of Rodanthi Tzanelli, especially *Cinematic Tourist Mobilities and the Plight of Development: on Atmospheres, Affects, and Environments* (Abingdon: Routledge, 2019).

26 Fellini confessed that "In the case of a film director like myself, the studio is not merely a physical space where I work, a place that cannot be exchanged for another place or abolished altogether, but a kind of mental and psychological dwelling-place, the very means of expression for me." Federico Fellini, *Fellini on Fellini* (New York: Delta, 1976), 111.

27 Rob Stone, "Cinematic Pilgrimages: Postmodern Heritage Cinema," in *Screening European Heritage: Creating and Consuming Film*, ed. by Paul Cooke and Rob Stone (London: Palgrave Macmillan, 2016), 266.

28 Roger C. Aden, *Popular Stories and Promised Lands. Fan Cultures and Symbolic Pilgrimages* (Tuscaloosa and London: The University of Alabama Press, 1999), 10.
29 Aden, *Popular Stories*, 11.
30 Ryan Lizardi, "The 'Looking Into the Past' Photographs: Comparing and Sharing the Past," *Visual Communication Quarterly* 23 (2016): 139; my emphasis.
31 Ryan Lizardi, "The 'Looking Into the Past' Photographs," 139.
32 Tullio Kezich, *Federico Fellini. His Life and Work* (New York: Farrar, Strauss & Giroux, 2006), 327.
33 Ivan Pintor Iranzo, "The Gesture of Hermes: Federico Fellini as an Interpreter and Circulating Agent of Images," *Journal of Italian Cinema & Media Studies* 9, no. 1 (January 2021): 89.
34 Zuzanna Bułat Silva, "Saudade – a Key Portuguese Emotion," *Emotion Review* 4, no. 2 (April 2012): 207.
35 For the foundational theorisations of the respective "turns," see W.J.T. Mitchell, *Picture Theory: Essays on Verbal and Visual Representation* (Chicago: University of Chicago Press, 1994), and John Urry, *Mobilities* (Cambridge, UK: Polity Press, 2007).
36 FILMography, accessed April 12, 2022, https://philmfotos.tumblr.com/search/trevi.
37 Rob Stone, "Cinematic Pilgrimages: Postmodern Heritage Cinema," in *Screening European Heritage: Creating and Consuming Film*, ed. by Paul Cooke and Rob Stone (London: Palgrave Macmillan, 2016), 259.
38 Marco Bertozzi, *L'Italia di Fellini. Immagini, paesaggi, forme di vita* (Venice: Marsilio, 2021), 161.

Bibliography

Aden, Roger C. *Popular Stories and Promised Lands: Fan Cultures and Symbolic Pilgrimages.* Tuscaloosa and London: The University of Alabama Press, 1999.

Agnew, John. "Fellini's Sense of Place." In *A Companion to Federico Fellini*, edited by Frank Burke, Marguerite Waller, and Marita Gubareva, 117–27. Malden: Wiley-Blackwell, 2020.

Arcangeli, Massimo, ed. *Itabolario. L'Italia in 150 parole.* Roma: Carocci Editore, 2011.

Bauman, Rebecca. "The Fellini Brand: Marketing Appropriation of the Fellini Name." In *A Companion to Federico Fellini*, edited by Frank Burke, Marguerite Waller, and Marita Gubareva, 391–402. Malden: Wiley-Blackwell, 2020.

Beeton, Sue. *Travel, Tourism, and the Moving Image.* Bristol: Channel View Publications, 2015.

Bertozzi, Marco. "Media Architecture: Fellini's Rimini, a Town of the Imagination." *Journal of Italian Cinema & Media Studies* 9, no. 1 (January 2021): 11–26. https://doi.org/10.1386/jicms_00048_1.

———. *L'Italia di Fellini: Immagini, paesaggi, forme di vita.* Venice: Marsilio, 2021.

Boym, Svetlana. *The Future of Nostalgia.* New York: Basic Books, 2001.

Centorrino, Clizia, Marco Dalla Gassa, and Andrea Minuz. "With a Hat and a Red Scarf: The Construction of Federico Fellini's Public Image." *Journal of Italian Cinema & Media Studies* 9, no. 1 (January 2021): 3–9. https://doi.org/10.1386/jicms_00047_2.

Corsi, Barbara, Marina Nicoli, and Alfonso Venturini. "Fellini the Founder? The Fellini Brand in Film Production." *Journal of Italian Cinema & Media Studies* 9, no. 1 (2021): 133–48. https://doi.org/10.1386/jicms_00055_1.

Farrell, Patrick M. "Portuguese Saudade and Other Emotions of Absence and Longing." In *Semantic Primes and Universal Grammar: Empirical Evidence from the Romance Languages*, edited by Bert Peeters, 235–258. Amsterdam, The Netherlands: John Benjamins, 2006.

Fellini, Federico. *Fellini on Fellini.* New York: Delta, 1976.

Iranzo, Ivan Pintor. "The Gesture of Hermes: Federico Fellini as an Interpreter and Circulating Agent of Images." *Journal of Italian Cinema & Media Studies* 9, no. 1 (January 2021): 83–100. https://doi.org/10.1386/jicms_00052_1.

Kezich, Tullio. *Federico Fellini: His Life and Work*. New York: Farrar, Strauss & Giroux, 2006.

Lizardi, Ryan. "The 'Looking into the Past' Photographs: Comparing and Sharing the Past." *Visual Communication Quarterly* 23 (2016): 135–46.

Minghelli, Giuliana. "Amarcord/Am'l'arcord. Fascism, Memory, and the Visual in Federico Fellini and Renzo Renzi." *Italica* 97, no. 4 (2020): 848–78.

Mitchell, W. J. T. *Picture Theory: Essays on Verbal and Visual Representation*. Chicago: University of Chicago Press, 1994.

Monteleone, Franco. "L'albergo più amato di Federico Fellini: Cinecittà." *Studi Novecenteschi* 35, no. 75 (2008): 281–91.

Renzi, Renzo. "Il fascismo involontario (a proposito di 'Amarcord')." In *Il fascismo involontario e altri scritti*, 131–81. Bologna: Universale Cappelli, 1975.

Rix, Juliet. "Inspired by Rimini: A Tour of Fellini's Magical Home Town." *The Guardian*, January 16, 2020. www.theguardian.com/travel/2020/jan/16/rimini-tour-fellini-centenary-exhibition-home-town.

Roesch, Stefan. *The Experiences of Film Location Tourists*. Bristol: Channel View Publications, 2009.

Silva, Zuzanna Bułat. "Saudade – a Key Portuguese Emotion." *Emotion Review* 4, no. 2 (April 2012): 203–11. https://doi.org/10.1177/1754073911430727.

Stone, Rob. "Cinematic Pilgrimages: Postmodern Heritage Cinema." In *Screening European Heritage: Creating and Consuming Film*, edited by Paul Cooke and Rob Stone, 257–278. London: Palgrave Macmillan, 2016.

Tzanelli, Rodanthi. *Cinematic Tourist Mobilities and the Plight of Development: On Atmospheres, Affects, and Environments*. Abingdon: Routledge, 2019.

Urry, John. *Mobilities*. Cambridge, UK: Polity Press, 2007.

Zambenedetti, Alberto. "Making the Old New Again: FIAT's (North American) Second Coming." In *Print to Digital: Aspects of a Cultural Discomfort*, edited by Francesco Guardiani, 237–48. Firenze: Franco Cesati Editore, 2017.

———. "Emplacing Time: Photography, Location, and the Cinematic Pilgrimage." *Space and Culture* 23, no. 4 (2020): 548–60.

6

WALKING IN CARY GRANT'S FOOTSTEPS

The *Looking for Archie* Walking Tour

Charlotte Crofts

Introduction

Looking for Archie: Cary Grant's Bristol is a guided walking tour of Hollywood actor Cary Grant's hometown Bristol, UK, where he was born as Archibald Leach in 1904. "Archie" emigrated to New York in 1920 with a troupe of acrobats and was signed by Paramount in 1931 when he changed his name to Cary Grant.[1] The locus for the tour originates from my wider curatorial practice as the director of the biennial *Cary Comes Home Festival*, established by myself and Anna Farthing in 2014.[2] The festival aims to celebrate Grant's Bristol roots, develop new audiences for his work, and recreate the golden age of cinema-going by programming immersive cinematic experiences in significant locations which featured in Grant's life and work. These include a gala screening at Bristol Hippodrome, where Archie worked backstage as a lad; a cream tea at the Avon Gorge hotel where he often stayed; and a screening of *Bringing Up Baby* (1938), in which Grant plays palaeontologist David Huxley, next to the dinosaur exhibit in Bristol Museum. These aims "hinge on the embodied, haptic, located experience of attending in person," with the "festival's stimulus to cinematic tourism and pilgrimage and the importance of place and 'emplacement' are pivotal."[3]

Originally developed in 2017 with funding from the *Being Human Festival*, the theme of which was "Lost and Found," we framed the tour around uncovering Bristol's "hidden cinema history" through the lens of one of the city's most iconic figures.[4] The tour has since had several iterations, but for the purposes of this chapter I base most of my discussion on the original 2017 tour and point towards how it has developed and changed in response to audience feedback.[5] A souvenir guide of the live walk was produced (see Figure 6.1), one side mapping the points of interest and the other providing a brief description and image of each map point.

DOI: 10.4324/9781003327585-6

The process of creating the map involved identifying key points that could be navigated within walking-distance of Bristol city centre. There are many other significant points in the city that could not be accommodated, such as Grant's birthplace in Horfield, the Glenside Hospital in Fishponds, or the Avon Gorge hotel in Clifton, because they are too dispersed across the city and not navigable within the timescale of a walking tour. As well as the process of selecting points on the map, there was also the job of scripting a narrative for use in the tour, which would enable these disparate locations to be joined up in an emotionally satisfying way. According to Rebecca Solnit, the act of mapping itself creates meaning, imagining "the particulars of place" makes it "much more immediate, visceral and affecting."[6] Mapping both Archie's and Grant's locations across the city centre illuminates not only the intersecting geography of his biography and stardom but also the shifting identities of the city itself.

FIGURE 6.1 The Looking for Archie map, designed by Eleanor Elliot-Rathbone.

Source: Photograph: Eleanor Elliott-Rathbone.

This fascination with site-specific content has been a sustained feature of my previous practice research around local screen heritage, in particular two place-making heritage apps, the *Curzon Memories App* and *The Lost Cinemas of Castle Park App*, both of which explore the relationship between contemporary physical geographical locations and the past.[7] Both smartphone applications were developed for the Apple iOS platform using Calvium's AppFurnace platform that harnesses the affordances of (then emerging) pervasive media including QR codes and GPS to automatically trigger context-sensitive content in significant locations.

One of the key aims of these locative heritage apps was to bring cinema history to life in the places where it actually happened, using "digital culture to enhance the public's relationship with the physical, material world of the cinema and its history by encouraging them to visit the site and experience it anew through creative media technologies."[8] These projects emphasized "thinking about heritage in terms of material experience, not just digital archive," and the importance of real-life geographical locations to the experience.[9] Both apps have an "arm-chair" mode where the content can be experienced remotely. However, based on the iterative design process and user evaluation feedback, it became apparent that the apps were most compelling when used in situ: cinema history truly comes alive when experienced on location.

Another of the key research findings was the need to factor the visitors' imagination into the user experiences: namely the necessary oscillation between hearing about the past while being immersed in the present location and having to make that imaginative leap to fill in the gaps between what was, and what is. For example, writing about their GPS-triggered "Mediascape," *Riot 1831*, Calvium's Jo Reid et al. discuss those "*magic moments* that are deemed to be both moving and memorable."[10] This relates to the location-based frisson of being there where history unfolded, triggered when the experience hits that sweet spot in which imagination, media, and real-life intersect. This is something that is at play in *Looking for Archie*, bringing Grant's Bristol to life and encouraging people to experience it on site.

For Reid et al. "one of the magical aspects of mediascapes" is "their relation to the physical world, the process of walking through that digital layer and interacting with digital media which can take you into a parallel world."[11] This is equally valid when using low-tech ways to bring the past to bear on the present, such as simply holding up a photograph in situ, or looking at a physical map (see Figure 6.2).

The tour covers two-and-a-half miles of the city centre, both featuring places that young Archie used to frequent growing up, as well as the locations at which Grant was photographed by the *Bristol Evening Post* on his many visits home after he had become famous. Although many people are aware of his Britishness, Grant is still most often figured as an American actor. While it is common knowledge that he was born in Bristol, fewer are aware that he continued an enduring relationship with the city, returning often, well into his seventies. The tour aims to raise awareness of these long-lasting connections to the city of his birth. In an audience feedback survey conducted in 2017 most participants already knew that Grant was

FIGURE 6.2 Participants looking at maps next to Bristol Cathedral on College Green.
Source: Photograph by Eleanor Elliott-Rathbone.

from Bristol (91.4 percent), but fewer were aware of his ongoing relationship with the city (74.3 percent). 22.9 percent had not realized that he returned regularly to Bristol until well into his old age, and 2.9 percent were only partially aware of this.[12] Another participant claimed, "the tour has made me appreciate how special a place Bristol is and how little I know about it."[13]

Stardom, Celebrity, and Star-Induced Tourism

The emphasis of most academic literature about film-induced and cinematic tourism is on how film and TV *texts* stimulate the tourist impulse,[14] but stardom, celebrity and fandom can also inform the impetus for travel. Rather than set-jetting, and the desire to relive the fantasy of film locations, this is motivated by the yearning to experience an "auratic"[15] co-presence, with the object of adulation, what I term "*star*-induced" tourism. Indeed, Grant's star identity and his associations with the city are used as a call to visit the city on both the VisitBritain and Visit Bristol websites and he is cited as one of the things that makes Bristol a UNESCO city of film.[16] In their analysis of tourists' motivations to visit film sites, Niki Macionis and Beverley Anne Sparks found that some participants' personal motivations were highly influenced by wanting to "feel close to my favorite (sic) actor."[17]

Returning to the festival from the United States for the third time in 2018, superfan Colleen Zwack described her desire to "get a closer connection to Cary. . . . Seeing and being in the actual places that were important to a young Archie Leach and later, Cary Grant, is just something extremely special."[18] Another participant in the 2018 tour, visiting from the United States for the first time, enjoyed the way in which the tour played with the "juxtaposition of who he was in life and on film."[19] The ethos behind *Looking for Archie* is the importance of recognizing this productive tension between Cary Grant and Archibald Leach in our understanding of Grant's star persona. Grant himself often drew attention to his former identity: "I have spent the greater part of my life fluctuating between Archie Leach and Cary Grant; unsure of either, suspecting each. Only recently have I begun to unify them into one person."[20]

Richard Dyer emphasizes the "structured polysemy" of the star image, "the finite multiplicity of meanings and affects they embody."[21] Grant's star identity is both complex and fascinating, oscillating as it does between the suave, debonair, global film icon, and the working-class Bristolian boy. This duality between Archie and Grant and the different Bristols that they inhabit is drawn out in the *Looking for Archie* tour, which uncovers layers of interlinked and little-known local history through a place-based journey of discovery.

Each layer offers new perspectives on and engagement with the complex interrelationship between the present and historic physical environment, Grant's persona, his life, and the tour participant's lived experience. For example, this overlapping of Grant's biography and his star image is brought into focus by a publicity shot of Grant on the roof of Bristol Cathedral in 1965, helping to raise funds for the repairs. This photograph was shown outside the cathedral, which was the starting point for the first edition of the tour in 2017; the venue deliberately chosen to tie in with the Christmas screening of *The Bishop's Wife* (1947) the following weekend. In that film, Grant plays Dudley, a guardian angel who answers the prayer of Reverend Henry Biggs (David Niven) to help him build a cathedral. The realization that life has imitated art, with the plot of the film so closely paralleling Grant's later support for the real cathedral, has the potential to provide an affective thrill where the intersection between Grant's biography and star persona is foregrounded.

Another "magic moment" occurred sharing a photo of Grant on College Green in front of City Hall, taken shortly after it was built in the 1950s when he was at the height of his stardom. These magic moments can be intensified if there is an "unexpected connection between physical and the virtual worlds" or when "the shared context evokes a gesture or feeling of social bonding giving a shared private movement within a public space."[22]

For the participants, standing on College Green in exactly the same spot as Grant creates a frisson between these different palimpsestic layers: imagining the Bristol Archie grew up in, before City Hall was built; comparing the freshly planted saplings in the photograph to the fully grown trees now obscuring the municipal architecture. Later in the tour, participants are asked to observe how Grant's confident stride in the photograph has been used as the template for the pose of the Cary Grant statue.

In both examples, the experience of encountering site-specific knowledge about Grant can also have a defamiliarizing effect, resulting in participants "looking at the world with new eyes."[23] In the 2017 audience survey, over half of the respondents agreed that the walk had changed their view of Bristol (60 percent), with an additional 25.7 percent saying their view of Bristol had partly been changed. A large majority of people felt that the tour had changed their understanding of Grant (74.3 percent) or partly changed (11.4 percent) it.

The most significant way in which participants' view of Grant changed was in gaining a better understanding of his ongoing connections to Bristol, and the warmth and affection he felt for the city of his birth (38.5 percent). A total of 19.2 percent of respondents learnt that he was more "down to earth" than they had realized, saying the tour "made him more human" and 11.5 percent learnt that his life was more "complex" than they had realized. The tour also gave participants a new perspective on Grant's early life, his family and mother in particular (34.6 percent), his marriages (7.7 percent), and the possibility that he may have been bisexual (3.9 percent). People learnt about "places of significance" and his connections to Bristol theatres (7.7 percent).

In her discussion of the Manhattan TV Tour, Leshu Torchin outlines how the tour director "overlays a series of landscapes over the physical terrain . . . causing each site to resonate with a compounded spatial complexity."[24] *Looking for Archie* similarly explores the ways in which we experience the city as a palimpsestic layered space: Archie's childhood growing up in Bristol, contrasting with his experience returning as an adult having found fame; the contemporary landscape, overlaid with the traces of how the built environment has developed and changed through time.

Another tension is between absence and presence, between our physical experience of place as it is now and with our imagined understanding of how it used to be or might have been. Torchin identifies this experience as "a constant negotiation of space for the tourist," acknowledging the complicated relationship between a location and its representation, as well as the affective labour of the cinematic tourist to make sense of that relationship.[25]

Emplaced Interaction and the Bristol Blitz

One of the key ways we can see this affective labour is in how the legacy of World War 2 plays out across the city and weaves its way through Grant's life story. Throughout the walk, we encounter evidence of the Bristol Blitz and its devastating impact on the Leach family. Bristol was the fifth most bombed city in the UK during the Second World War due to the strategic importance of its harbour as a major port for American supply ships and its aircraft factories. In the winter of 1940–1941, there were six major bombing raids in which the city suffered heavy loss of life and destruction of the built environment, which is still visible today in the city's eclectic architecture.[26] According to the *Western Daily Press*, Grant lost relatives in a direct hit in one of these raids: "Mr and Mrs John Henry Leach, uncle and aunt of Grant, the Bristol-born film star, their daughter, son-in-law and grandson were buried by a direct hit on their house in a recent air raid."[27] An entire family was wiped out in a single incident.

Participants first experience this theme early in the tour when we encounter the Bristol Basin plaque on the centre, opposite the Bristol Hippodrome. This is a replica of the original plaque erected in New York City in 1942 by the English-Speaking Union (ESU) of the United States to "commemorate the sacrifices made by the people of Bristol, England, during World War II."[28] Rubble from the Bristol Blitz was used as ballast for returning U.S. supply ships and dumped in the East River between 23rd and 34th Streets as landfill forming the foundations for what is now Franklin D Roosevelt Drive. Stephen Vincent Benetto was commissioned to write a poem: "these/Fragments that were once homes/Shall testify while men love/Freedom to the resolution and/Fortitude of the people of Britain."[29]

The plaque ends: "And broad-based under all/Is planted England's oaken-hearted mood,/As rich in fortitude/As e'er went worldward from the island-wall." This last section is a quote from another poem called *National Ode* by Bayard Taylor, which was written for the Centennial International Exhibition of 1876, celebrating the Declaration of Independence, underscoring the long and complex relationship between Britain and the United States.

In December 1974, Grant re-dedicated the plaque when it was relocated due to the construction of the Waterside Plaza. In his speech, Grant was clearly moved, stating "I have a deep-seated emotion about this ceremony."[30] He revealed that "a great part of my family were wiped out in the war."[31] In May 2022, when the tour ran as part of Bristol WalkFest, one of the participants visiting from New York revealed that he had recently visited the original plaque at Waterside Plaza and had brought a fragment of rubble with him to return to Bristol. I photographed him holding up a selfie on his phone pictured next to the original plaque in New York City, standing in front of the Bristol replica (see Figure 6.3).

Later we encounter the bombed-out ruin of St Peter's Church in Castle Park, left standing as a testimony to the Blitz. A metal plaque is dedicated to the "Civilians and Auxiliary Personnel who lost their lives in the Greater Bristol Area during the air raids carried out between June 1940 and May 1944. They live in the memory of the living." Listed among the dead are several Leaches, including Grant's uncle John Henry and aunt Rose, his cousin Dorothy Ellen Marsh and her husband Francis Richard Bruiss Marsh, the family mentioned in the newspaper, and another uncle, Frederick Leach. According to Mark Glancy, "the impact of the bombs was so intense that their bodies were never recovered."[32]

The experience of encountering this memorial and seeing their names is affective. Earlier on in the walk, we learnt about the Bristol Basin plaque and intellectually comprehend that he lost relatives, but there is something about recognizing their names engraved on the plaque on the ruin of a shell of the church that

FIGURE 6.3 Participant holding up an image of himself in front of the Bristol Basin plaque in New York, next to the replica plaque in Bristol, UK. Looking for Archie, May 2022.

Source: Photograph by Charlotte Crofts.

brings home this abstract understanding into a more embodied experience (see Figures 6.4 and 6.5).

It is useful to turn to Aston's concept of "emplaced interaction" here, which is a useful paradigm with which to move beyond the mind/body relationship of "embodiment," reinserting us back into the physical world around us, and "bringing our bodies and minds into direct interplay with the wider environment."[33] Aston's argument for a new way of working with interactive documentary could equally apply to how we might think about screen tourism and affective landscapes

FIGURE 6.4 Participants looking at the plaque commemorating civilian loss of life. Looking for Archie, May 2022.

Source: Photographs by Charlotte Crofts.

FIGURE 6.5 Close-up of Leaches lost in the Blitz. Looking for Archie, May 2022.

Source: Photographs by Charlotte Crofts.

in the digital age. As Aston suggests, it is becoming increasingly important to return to the physical, "to make us feel 'alive' through maintaining some direct physical contact with our fellow humans and with the world at large."[34] In *Looking for Archie*, this "emplaced interaction" with the physical memorial and the specific example of Grant's personal loss enables us to grasp the enormity of the lasting legacy of the Blitz on Bristol as a whole, resonating with contemporary historical events in deepening our understanding of the horrors of war and their impact on civilian populations.

One of the most emotional points on the tour is the moment when we hear about Archie getting lost in a Marks and Spencer department store as a young child, as recounted in Maureen Donaldson's memoir.[35] When his mother Elsie finds him, she punishes him by saying "You see, Archie, nobody wants you. Nobody came for you, did they? I'm the only one – *the only one* – who cares about you. Nobody else. And don't you ever forget it, because the next time you let go of my hand and wander off, I won't come back."[36] Elsie and Elias Leach had lost their first-born son, John, to septicaemia when he was a baby and Elsie never recovered from the guilt. Grant remembers her as a controlling mother, exacting in her expectations of his manners and aspirational in her hopes for his future.[37]

This is a point of anagnorisis for Donaldson, as Grant's lover, a critical moment of recognition about the effect that his relationship with his mother had on all his subsequent relationships with women. This resonates because earlier in the tour we have heard about how Archie came home from primary school in February 1915 to find Elsie gone. He was told she had gone to recuperate in a nearby seaside town, but she never came back, and he was led to believe that she was dead. She had been committed to a mental hospital, which Grant did not discover until over 20 years later after he had become famous.

The landscape of Castle Park creates an affective backdrop for this story. Participants are required to draw on their imagination to picture what is effectively now a concrete path through a grassy park as a bustling high street, and to imagine the anguish of Elsie thinking she had lost Archie and punishing him, and the adult Grant's long-held pain because of that punishment. Indeed, that moment of anagnorisis is extended further because what used to be the busiest shopping street in Bristol was razed to the ground after suffering severe bomb damage during the War, so the pathos we feel for Archie and Elsie is once again mingled with a greater sense of loss for the city itself. According to the 2017 survey, people's view of Bristol was changed by a better understanding of local history and buildings (6.7 percent), the effect of War on Bristol (13.33 percent) and Castle Park in particular (23.33 percent), and new knowledge about Bristol's cinema history (23.33 percent). Several people felt that they had seen another side of the city either offering "a different point of view on familiar places" (10 percent) or introducing them to places that they had not seen before (23.33 percent).[38]

In 2017, this was even more emotionally resonant because two separate branches of Elsie Kingdon's family (Grant's mother) met each other for the first time on one of the tours. One participant stated that the tour "renewed my interest in possible family connections," and another asserted that as "Kingdons, put us in touch with another family member." In a personal follow up email these family connections were further elaborated:

> My sister . . . and I, thoroughly enjoyed the "Looking for Archie" walking tour and can't wait to attend the next event. . . . Your passion and enthusiasm, definitely left us both wanting to learn more about Archie Leach, our family connection and the history of Cinema in Bristol.[39]

Bristol as Port City: Transatlantic Travel and the Tourist Gaze

It could be argued that Grant's trajectory, from Horfield to Hollywood, would not have been possible without, but also performs, what John Urry has coined the "tourist gaze."[40] Indeed, Bristol has been instrumental in the development of transatlantic travel and globalization, on which tourism itself is predicated. Italian navigator and explorer John Cabot (Giovani Caboto) sailed from Bristol in the Matthew, funded by Bristol merchants, in 1497, and discovered North America.[41] Young Archie would have been well aware of this history, as Cabot Tower – a prominent feature of the Bristol skyline – was erected to commemorate the 400th anniversary of Cabot's expedition in 1897, shortly before his birth.[42] Archie's life, then, was shaped by his relationship to the sea, and Bristol's role as a port town in the early 20th century. He wrote about frequenting the docks, where he found solace from an unhappy childhood, fantasizing about sailing away on one of the tall ships, which still sailed right into the city centre:

> *I regularly haunted the Bristol wharves where in those days, schooners and steamships came right up the Avon River into the center (sic) of town; and on weekends, when most of my school friends were playing cricket, I sat alone for hours watching the ships come and go, sailing with them to far places on the tide of my imagination, trying to release myself from the emotional tensions which disarranged my thoughts.*[43]

This extract is heard near Pero's bridge, named after Pero Jones, an enslaved African who lived in Bristol, within sight of a statue of John Cabot which overlooks the floating harbour, opposite the quay where the statue of slave trader Edward Colston was unceremoniously dumped into the docks in the summer of 2020. The River Frome, which used to run right into Bristol's centre, is now culverted and the expansion of Avonmouth and Portbury Docks at the mouth of the River Avon, together with post-war, redevelopment has transformed the working harbour into Bristol's leisure quarter; cargo ships have been replaced by pleasure boats. Multiple layers of Bristol's naval history and its relationship to the transatlantic slave trade coexist. The penultimate stop was the SS Great Britain, where we have photographs of Grant visiting the wreck shortly after it returned to Bristol in the early 1970s. Built in 1843 by another iconic Bristolian, Isambard Kingdom Brunel, she was designed as the first great ocean liner between Bristol and New York, demonstrating the longevity of the relationship between the city of his birth and the country of his future citizenship.

Archie was not to cross the Atlantic until July 1920 when he emigrated with Bob Pender's Troupe of Knockabout Boys to New York City, where he worked his way up from Vaudeville to Broadway, and then made his way to Hollywood and reinvented himself as Cary Grant. Archie's first voyage on the RMS Olympic – the Titanic's sister ship and an exact replica – epitomizes the height of modernity and the "golden age" of transatlantic travel – with its stratified accommodation. As Urry argues, the late 19th and early 20th centuries are the period in which "mass tourism has become widespread within much of Europe and North America."[44] Archie and the Penders travelled Second Class, whereas film stars Douglas Fairbanks and Mary Pickford who he met on the voyage, travelled in First.[45] Seeing America in the movies propelled him on his transatlantic journey. Grant vividly remembers his first glimpse of the New York skyline, arriving at dawn on 28 July 1920:

> *Manhattan Island. That skyline in the early-morning July sunshine. New York City. There it was; but was I there? Was I actually there at the ship's rail, neatly scrubbed and polished, standing with a small, solitary band of Pender-troupe boys – none of whom had slept all night for fear of missing the first glimpse of America? The excitement. Those skyscrapers I had seen so many times before. Oh my, yes. In England. In Bristol. In the films.*[46]

Not only did Bristol's status as a port inform Archie's desire to travel and escape from his difficult childhood, but so did the representations of "elsewhere" in the films he watched on the Bristol cinema screens. Tourism is strongly motivated by the desire to experience locations that one has already seen in images. In his discussion of the "tourist gaze," Urry suggests the "anticipation . . . of intense pleasures" is "constructed and sustained through a variety of non-tourist practices, such as film, TV, literature, magazines, records and videos."[47]

It is useful to extend this concept to include what I call "film-induced *migration*." Once in New York, Archie's cinema-going continued to inform his trajectory, as Orry-Kelly attests, "whenever Archie had any spare money or time, he'd go to the movies to study technique," learning from the likes of Stanley Lupino, Stan Laurel and Harry Langdon.[48] As Urry argues, "the gaze is constructed through signs, and tourism involves the collection of signs."[49] Of course, Grant went on to play numerous naval characters, and boats feature in many of his films, from his first appearance as First Sailor in *Singapore Sue* (1932) to *Father Goose* (1964).

So, the tourist gaze comes full circle. Archie Leach watched movies in Bristol and studied them in New York; these in turn motivated him to travel to Hollywood and informed the cultivation of Grant's paradoxical star persona, as a British-born actor who represented the apex of American masculinity. With his new-found wealth he was able to travel widely, returning to his hometown regularly to see his mother and even after her death. His many visits home were assiduously documented by *Bristol Post* – they were given exclusive photo shoots in return for leaving him alone for the rest of his stay. Images of Grant perpetually circulate on social media, pictured on glamorous cruise ships (both in life – he met his third wife Betsy Drake on a cruise – and in art – as in his film, *An Affair to Remember* [1957]); his star image symbolizes the golden age of transatlantic travel, in turn motivating cinematic tourism today.

From Virtual Communitas to Making the Pilgrimage

Given the contemporary ubiquity of Grant's image circulating on the Internet and the accessibility of his films on Video On Demand platforms such as Netflix and Amazon Prime, why do some fans still feel the need to visit his birthplace? Is there something even more compelling about physical pilgrimage in the digital age?

The concept of pilgrimage and the related concept of communitas have been explored not only in tourism studies but also in fandom and celebrity studies, as well as in festival studies.[50] As Urry argues,

> Like the pilgrim the tourist moves from a familiar place to a far place and then returns to the familiar place. At the far place, both the pilgrim and the tourist engage in "worship" of shrines which are sacred, albeit in different ways, and as a result gain some kind of uplifting experience.[51]

Drawing on Victor Turner's theorization of pilgrimage, Will Brooker outlines the concept of fan pilgrimage as a three-stage journey from "habitus" to the "promised land" and back again.[52]

Brooker complicates this in his discussion of "inbetweeness" – that liminal space between the fantasy world of the text (or star) and the reality of the actual location, arguing that "pilgrims never quite make the sacred connection but instead hover on the borders between actual and fictional, holding both in a double-vision of alternate realities."[53] This "double-vision" relates to the duality between Archie and Cary identified earlier; the frisson between then and now which has to be negotiated by the fan/tourist. Such sites, Brooker argues, "demand a significant amount of emotional commitment and imaginative work on the pilgrim's part to even approach a sense of communion with the fictional text."[54] This fan labour to transform the everyday into the sacred can also be applied to celebrity and stardom: "mundane houses that become special only when a fan drives past and projects upon them the knowledge that their idol once lived there."[55]

In *The Tourist Gaze 3.0*, Urry and Larsen identify an "upsurge in 'media pilgrimage'" in the digital age.[56] Cardoso et al. also note the impact of "information and communication technologies" in informing tourist expectations, their practices, and the sharing of tourist experiences.[57] In the early days of the Internet, an email list for Grant fans, The Email Warbrides (named after *I Was A Male Warbride* [1949]) was established in 1996 and a website curated by Debbie Dunlap was set up in 1998. Members of the list held "Caryventions," including a visit to Bristol in 1999, helped raise money for the Cary Grant statue and visited the city when it was unveiled in 2001.[58] Lincoln Geraghty et al. highlight how fan loyalty can also contribute to the cyclical development of tourism in the digital age of globalization and mediatization, through word of mouth and sharing on social media which "make it possible for them to exchange experiences and stories before, during, and after visiting tourist spaces."[59] Indeed, two 2018 visitors from the United States were mobilized to visit via their membership of the Cary Comes Home Facebook group.[60] Today Grant's image spreads via social media platforms, engaging a new generation, with stan accounts and memes promulgating on Facebook, Twitter, Instagram, and more recently TikTok, contributing to the longevity of his star image.[61]

Dressing Up, Selfies, and Playful Fan Practices

The longevity of Grant as a star is not simply maintained by the passive circulation of his star image, but via the ongoing inter/active engagement of fans. Geraghty et al. argue that popular cultural tourism is motivated not only by the desire to escape everyday life but also by fans' desire to interact with "icons and symbols."[62] As Paul McDonald suggests, it is important to recognize the "many practical actions which stardom motivates in society," calling for a "pragmatics of star practices" to supplement the "semiotics of star meanings."[63] Recognizing the "social activity of stardom,"[64] enables us to think about the agency of the spectator or fan, actively negotiating the institution of stardom, rather than passively consuming it.

Similarly, Christine Geraghty calls for a re-examination of "how we understand the audience's activity in relation to stars" moving beyond Richard Dyer's emphasis on stars "as a means of exploring social identity" to an evaluation of the "extratextual work of fans."[65] Discussing "cult geographies," Hills describes this as an "affective-interpretative process."[66] This affective, extratextual work is emotionally loaded. Rebecca Williams suggests that "media fans often have strong emotional interests in finding and visiting sites related to their favorite (sic) films, TV shows or celebrities."[67] As Lincoln Geraghty et al. point out, fan practices are the expression of people's "deep involvement" with the object of adoration which often "permeates several aspects of their lives," including not only travel but also other features, such as fashion.[68] This is borne out by the fact that some participants on the *Looking for Archie* tour emulate Grant in their choice of attire, wearing suits and hats in homage to Grant as style god (see Figure 6.6).

Dressing up in vintage style, or as characters from Grant's films, is positively encouraged during the *Cary Comes Home Festival*, with prizes for the best dressed at our gala screenings.[69] Dedicated fans take this further and dress up for the whole festival. For the walking tour, participants are encouraged to dress appropriately, whatever the weather. The fact that some participants choose to don outfits that would not look out of place in a Cary Grant film represents a playful approach to engaging with Bristol's screen heritage.

FIGURE 6.6 Participant dressing the part with tour guide Pam Beddard on the right. Looking for Archie, December 2021.

Source: Photograph by Charlotte Crofts.

This playful, performative mode particularly comes to the fore in the way that participants engage with the Cary Grant statue in Millennium Square. All iterations of the tour traditionally end here. In November 2017, and in subsequent years, the statue was incorporated into the snug of the Aprés Bar as part of the Christmas Fair. We usually have an arrangement with the bar to secure the area immediately surrounding the statue and participants were invited to toast "our Archie" with a complimentary glass of prosecco, while filling in our audience surveys. Ending the tour at the statue with a toast contributes towards a sense of communitas and provides an opportunity for participants to reflect on their experience of the tour. But what is most fascinating is the way in which participants tend to spontaneously interact with the statue, touching it, even hugging and kissing it, posing for photographs which they then share on social media (see Figure 6.7).

Urry and Larsen note a "'performance turn' within tourist studies" which acknowledges the experience of place as embodied "in more multi-sensuous ways,

FIGURE 6.7 Tweet of participant posing with a map in front of the Cary Grant statue in Millennium Square. Looking for Archie, November 2018.

Source: Photograph by Clare Sheridan.

touching, tasting, smelling, hearing."[70] This performative sense of play, this tactile, haptic engagement with Grant's memorial demonstrates a shift from the tourist as audience to performer.[71] It also speaks to the carnivalesque potential of fan practices, spilling out of the everyday and becoming transformative, creating what Urry and Larsen identify as a " 'liminoid' situatio[n] where everyday obligations are suspended or inverted."[72] Pilgrimage becomes a creative act in which, as Will Brooker points out "performance, disguise and carnival . . . symbolically transforms the location in question, temporarily inverting social structures and making the city into a liberating, playful space."[73] By personifying the statue at the end of the tour, we give permission for "playful 'non-serious' behaviour and the encouragement of a relatively unconstrained 'communitas' or social togetherness."[74] That "double-vision" between Archie/Cary is maintained right until the last.

Future Directions

The *Looking for Archie* walking tour has contributed to the wider audience development for the *Cary Comes Home Festival*. The challenge is how to continue to develop both the festival and the tour to reach a wider and more diverse audience. We are currently exploring how to enhance the tour, collaborating with Show of Strength Theatre company – who operate several Bristol-based "theatre walks" – to explore incorporating elements of itinerant physical theatre, such as circus performance, to pay homage to Grant's acrobatic past.[75]

There is also potential to develop a *Looking for Archie* app. A basic prototype has already been produced, but completion of the full app is currently on hold, partly because it was interrupted by the pandemic but also because, having experienced the short shelf-life of my previous iOS apps, I am reluctant to continue with a native mobile app (one that uses the device's mobile operating system). Both *Curzon Memories* and *Lost Cinemas* are now no longer available on the App Store, having become defunct due to Apple's ever-evolving Application Programming Interface (API) and the high cost of updating and maintaining them.

Instead, I intend to develop a web-based virtual tour using the ArcGIS StoryMap platform which will be accessible from any Internet browser.[76] This will enable the tour not only to expand to include further Bristol locations not accessible within the spatial-temporal confines of one walking tour but also to incorporate other international locations pertinent to Grant's life and work, such as New York. The advantage of a web-based online tour is wider-audience reach and longevity. However, at present the StoryMap platform does not have context-aware functionality which means that it would not be truly "locative."

Future directions for this research could explore how screen cultures impact on tourists' understandings of and engagements with place and the ways in which emerging technologies affect our relation to landscapes. How might the next evolution of the Internet drive cinematic tourism using XR (eXtended Reality) to enhance our view of the physical environment? However, while we have become so reliant on the Internet as a way to relate to the world, even more so post-pandemic, it is nevertheless important not to lose touch with the physical and tactile world around us.[77] Experiencing the tour remotely using the "armchair mode" or immersed through the lens of XR may detract from the affective, embodied, and emplaced interaction with the physical landscape and the act of pilgrimage which makes the original walking tour so compelling.

Notes

1 Throughout this chapter I will refer to him as "Archie" (as he refers to himself in his autobiography (Cary Grant, "Archie Leach." *Ladies Home Journal* Part 1 (January–February 1963); Part 2 (March 1963); Part 3 (April 1963))), when referencing the period before he had become Cary Grant, and Grant when referencing the period after he had changed his name.

2 See Cary Comes Home Website for documentation of past events, www.carycomeshome.co.uk.

3 Charlotte Crofts, "Bristol Fashion: Reclaiming Cary Grant for Bristol – Film Heritage, Screen Tourism and Curating the Cary Comes Home Festival," *Open Screens* 4, no. 2 (December 31, 2021): 17, https://doi.org/10.16995/OS.8018.

4 *Being Human* is a national festival promoting public engagement with humanities research led by the School of Advanced Study, University of London, in partnership with the Arts and Humanities Research Council and the British Academy. Being Human, *Lost and Found Festival Programme* (2017), 56, https://issuu.com/schoolofadvancedstudy/docs/being_human_2017_low_res_proof_for_/58.

5 The tour was reprised at the 2018 festival where it sold out and was featured on BBC Online, "Bristol Cary Grant film festival celebrates third year" (November 23, 2018) www.bbc.co.uk/news/uk-england-bristol-46178044 and ITV West Country News (November 23, 2018) https://vimeo.com/505041536. We then partnered with Bristol WalkFest in May 2019. After a hiatus due to the Pandemic, in December 2021 we ran the tour as part of our 20th anniversary celebration of the unveiling of the Cary Grant statue in Millennium Square and in May 2022 we again partnered with Bristol WalkFest.

6 Chris Gondek, "Infinite City – Interview with author Rebecca Solnit and UC Press Art Director Lia Tjandra," UC Press Podcast, December 8, 2010, https://soundcloud.com/uc-press/infinite-city-interview-with-author-rebecca-solnit-and-uc-press-art-director-lia-tjandra.

7 Charlotte Crofts, *Curzon Memories App,* 2012 (which won Best Multimedia Award, Learning on Screen, 2013) celebrates the Curzon Community Cinema, Clevedon, UK; and Charlotte Crofts, *The Lost Cinemas of Castle Park App*, 2013, uncovers 17 hidden Bristol city centre cinemas. See also Charlotte Crofts, "City Strata," Poster at the *Media, Communication and Cultural Studies Association Annual Conference*, 2013.

8 Charlotte Crofts, "Technologies of Seeing the Past: The Curzon Memories App," in Electronic Visualisation and the Arts (EVA 2011) (EVA), July 2011, 164, https://doi.org/10.14236/ewic/EVA2011.29.

9 Crofts, 2011, 164.

10 Josephine Reid, Richard Hull, Kirsten Cater, and Constance Fleuriot, "Magic Moments in Situated Mediascapes," in *Proceedings of the 2005 ACM SIGCHI International Conference on Advances in Computer Entertainment Technology* (emphasis in the original) (New York: Association for Computing Machinery, 2005), 290, https://doi.org/10.1145/1178477.1178529.

11 Reid et al., 2005, 290.

12 Charlotte Crofts, "Looking for Archie Being Human Evaluation Report" (January 2021), there were 36 responses (out of roughly 40 who attended the two walks).

13 Anonymous participant, Personal email to author, November 20, 2017.

14 See Sue Beeton, "The Advance of Film Tourism," *Tourism and Hospitality Planning & Development* 7, no. 1 (February 1, 2010): 1–6; Lucília Cardoso, et al., "Film-Induced Tourism; a Systematic Literature Review," *Tourism and Management Studies* 13, no. 3 (September 2017): 23–30.
15 Walter Benjamin, "The Work of Art in the Age of Mechanical Reproduction," Trans. JA Underwood (Penguin Books, 2008).
16 Visit Bristol, "The Best of Bristol – Discover the City's Oscar-Winning Connections," https://visitbristol.co.uk/blog/read/2016/02/the-best-of-bristol-discoverthe-citys-oscar-winning-connections-b241 and VisitBritain, "Visit Bristol Before the Year Ends," www.visitbritain.com/au/en/8-reasons-visit-bristol-year-ends-0; Bristol City of Film, "What Makes Bristol A City of Film?," http://bristolcityoffilm.co.uk/about/what-makes-bristol-a-city-of-film.
17 Niki Macionis and Beverley Anne Sparks, "Film-induced Tourism: An Incidental Experience Article," in *Tourism Review International* 13 (October 2009): 97, https://doi.org/10.3727/154427209789604598.
18 BBC Online, "Bristol Cary Grant," 2018.
19 Cary Comes Home, "Cary Comes Home 2018 US Visitors' Vox Pops," https://vimeo.com/manage/videos/505034998.
20 Cary Grant, "Archie Leach," *Ladies Home Journal*, Part 1, January/February; Part 2, March; Part 3, April 1963. Accessed June 8, 2022, www.carygrant.net/autobiography/.
21 Richard Dyer and Paul McDonald. *Stars*. Bloomsbury Academic, 1998, 3.
22 Reid et al., 2005, 291–2.
23 Reid et al., 2005, 291.
24 Leshu Torchin, "Location, Location, Location the Destination of the Manhattan TV Tour," *Tourist Studies* 2, no. 3 (2002): 248.
25 Torchin, 2002, 247–8.
26 Eugene Byrne, "The Night That Changed Bristol Forever – by the People Who Were There," *Bristol Post*, November 24, 2020, www.bristolpost.co.uk/news/bristol-news/night-changed-bristol-forever-people-4731571.
27 "Cary Grant Lost Relatives," *The Western Daily Press and Bristol Mirror*, January 27, 1941, 5.
28 Waterside Plaza, "Cary Grant Links Waterside to Bristol," *Waterside Views* (April 2006), 1, www.watersidewire.com/residentservices/documents/NewsletterPDFApril06_000.pdf.
29 Waterside Plaza, 2006, 1.
30 Waterside Plaza, 2006, 1.
31 BBC Online, "How Bricks From Bristol Were used to Build Part of New York," October 21, 2017, www.bbc.co.uk/news/av/uk-england-bristol-41700547.
32 Mark Glancy, *Cary Grant,* A Press: 2021, 222.
33 Judith Aston, "Interactive Documentary and Live Performance From Embodied to Emplaced Interaction," in *i-Docs: The Evolving Practices of Interactive Documentary,* ed. Judith Aston et al. (New York: Wallflower Press, 2018), 232–4.
34 Aston, 2017, 233.
35 Maureen Donaldson and William Royce, *An Affair to Remember: My Life With Cary Grant* (New York: Putnam, 1989), 152–4.
36 Donaldson and Royce, 1989, 153. Emphasis in the original.
37 Grant, "Archie Leach," 1963.
38 Crofts, "Evaluation Report," 2021.
39 Personal email, 2017.
40 John Urry, *The Tourist Gaze: Leisure and Travel in Contemporary Societies* (London: Sage, 1990) and John Urry and Jonas Larsen, *The Tourist Gaze 3.0* (London: Sage, 2011).
41 Evan T. Jones and Margaret M. Condon, *Cabot and Bristol's Age of Discovery: The Bristol Discovery Voyages 1480–1508* (University of Bristol, November 23, 2016). The foundation stone was laid in 1897 and work was completed in 1898, just six years before Archie's birth.

42 Historic England, "Cabot Tower," https://historicengland.org.uk/listing/the-list/list-entry/1202009.
43 Cary Grant, "Archie Leach," 1963.
44 Urry, 1990, 4. Scrutinising the passenger manifests initially for Archie Leach, and later for Grant's many transatlantic sea-voyages, it is instructive to note that the passenger records cease from the late 1950s when air travel takes over and transatlantic travel becomes more accessible.
45 Archie clearly somehow inveigled his way onto the upper deck: "Among the fellow passengers were newlyweds Douglas Fairbanks Sr. and Mary Pickford, the world's most popular honeymooners and the first film stars I ever met. They were gracious and patient in face of constant harassment, by people with cameras and autograph books, whenever they appeared on deck; and once even I found myself being photographed with Mr. Fairbanks during a game of shuffleboard."
46 Cary Grant, "Archie Leach," 1963.
47 Urry, 1990, 3.
48 Orry-Kelly, *Women I've Undressed* (London: Allen & Unwin, 2016), 93.
49 Urry, 1990, 3.
50 On fandom see Will Brooker, "The Blade Runner Experience: Pilgrimage and Liminal Space," in *The Blade Runner Experience: the Legacy of a Science Fiction Classic*, ed. Will Brooker (London: Wallflower Press, 2004), 11–30 and Matt Hills, *Fan Cultures* (London and New York: Routledge, 2003); on festivals, see Marijke de Valck, *Film Festivals: From European Geopolitics to Global Cinephilia,* 1st ed. (Amsterdam: Amsterdam University Press, 2007), http://doi.org/10.1017/9789048506729.
51 Urry, 1990, 359.
52 Brooker, 2004, 12.
53 Brooker, 2004, 13.
54 Brooker, 2004, 13.
55 Brooker, 2004, 14.
56 Urry and Larson, 2011, 116.
57 Cardoso et al., 2017, 23–4.
58 Email Warbrides, accessed June 9, 2022, www.carygrant.net/warbrides.html. A life-sized bronze statue of Grant by Graham Ibbeson was erected and unveiled by Grant's fifth wife, Barbara Harris-Grant on December 7, 2001, following a world-wide campaign led by David Long which raised over £60,000. Cary Comes Home commemorated the twentieth anniversary of the statue with a tour in 2021 and an online event: "The Cary Grant Statue: 20 Years – a conversation with David Long," www.carycomeshome.co.uk/events/the-cary-grant-statue-20-years.
59 Lincoln Geraghty et al., "Guest Editorial: Exploring the Popular Culture and Tourism Place Making Nexus." *The Journal of Popular Culture* 52, no. 6 (2019): 1241–4.
60 Cary Comes Home. "Vox Pops."
61 Lucy Bolton and Julie Lobalzo Wright, eds., *Lasting Screen Stars: Images That Fade and Personas That Endure* (London: Palgrave Macmillan, 2016).
62 Lincoln Geraghty et al., "Guest Editorial: Exploring the Popular Culture and Tourism Place Making Nexus," *The Journal of Popular Culture* 52, no. 6 (2019): 1241–4.
63 Dyer and McDonald, 1998, 200.
64 Dyer and McDonald, 1998, 200.
65 Geraghty, Christine, "Re-Examining Stardom: Questions of Texts, Bodies and Performance," in *Reinventing Film Studies*, ed. Christine Gledhill and Linda Williams (London: Arnold, 2000), 105, http://eprints.gla.ac.uk/49612/.
66 Hills, 2003, 210.
67 Rebecca Williams, "Fan Tourism and Pilgrimage," in *The Routledge Companion to Media Fandom* (Milton Park, Abingdon and New York: Routledge, 2017), 98.
68 Lincoln Geraghty et al., "Guest Editorial: Exploring the Popular Culture and Tourism Placemaking Nexus," *Journal of Popular Culture* 52, no. 6 (2019): 1243.

69 As documented on the Cary Comes Home Facebook photos albums.
70 Urry and Larsen, 2011, 14.
71 Urry, 1990, 10.
72 Urry and Larsen, 2011, 12.
73 Will Brooker, "Everywhere and Nowhere: Vancouver, Fan Pilgrimage and the Urban Imaginary," *International Journal of Cultural Studies* 10, no. 4 (2007): 423.
74 Urry and Larsen, 2011, 12.
75 Show Of Strength perform in 'found' spaces in and around Bristol, including itinerant "theatre walks" themed around local culture, such as *Treasure Island* and *Frankenstein*, https://showofstrength.org.uk/.
76 For a successful example of the StoryMap platform, see Lisa Stead, *Reframing Vivien Leigh: South West Star Histories*, https://reframingvivienleigh.exeter.ac.uk/storymap/.
77 Aston, 2017, 233.

Bibliography

Aston, Judith. "Interactive Documentary and Live Performance from Embodied to Emplaced Interaction." In *i-Docs: The Evolving Practices of Interactive Documentary*, edited by Judith Aston, Sandra Gaudenzi, and Mandy Rose, 222–36. New York: Wallflower Press, 2018.

BBC Online. "How Bricks from Bristol Were Used to Build Part of New York." *BBC*, October 21, 2017. www.bbc.co.uk/news/av/uk-england-bristol-41700547.

———. "Bristol Cary Grant Film Festival Celebrates Third Year." *BBC*, November 23, 2018. www.bbc.co.uk/news/uk-england-bristol-46178044.

Beeton, Sue. "The Advance of Film Tourism." *Tourism and Hospitality Planning & Development* 7, no. 1 (February 1, 2010): 1–6. https://doi.org/10.1080/14790530903522572.

Being Human. "Lost and Found Festival Programme." 2017. https://issuu.com/schoolofadvancedstudy/docs/being_human_2017_low_res_proof_for_/58.

Benjamin, Walter. *The Work of Art in the Age of Mechanical Reproduction*. Translated by J. A. Underwood. London: Penguin Books, 2008.

Bolton, Lucy, and Julie Lobalzo Wright, eds. *Lasting Screen Stars: Images That Fade and Personas That Endure*. London: Palgrave Macmillan, 2016. https://doi.org/10.1057/978-1-137-40733-7.

Bristol City of Film. "What Makes Bristol A City of Film?" Accessed June 8, 2022. http://bristolcityoffilm.co.uk/about/what-makes-bristol-a-city-of-film.

Brooker, Will. *The Blade Runner Experience: The Legacy of a Science Fiction Classic*. New York: Columbia University Press, 2005.

———. "Everywhere and Nowhere: Vancouver, Fan Pilgrimage and the Urban Imaginary." *International Journal of Cultural Studies* 10, no. 4 (December 1, 2007): 423–44. https://doi.org/10.1177/1367877907083078.

Byrne, Eugene. "The Night that Changed Bristol Forever – by the People Who Were There." *Bristol Post*. November 24, 2020. www.bristolpost.co.uk/news/bristol-news/night-changed-bristol-forever-people-4731571.

Cardoso, Lucília, Cristina Estevão, Cristina Fernandes, and Helena Alves. "Turismo Cinematográfico: Revisão Sistemática Da Literatura." *Tourism and Management Studies* 13, no. 3 (September 2017): 23–30.

Cary Comes Home. "Cary Comes Home 2018 US Visitors' Vox Pops." November 2018. https://vimeo.com/manage/videos/505034998.

———. "The Cary Grant Statue: 20 Years – A Conversation with David Long." December 7, 2020. www.carycomeshome.co.uk/events/the-cary-grant-statue-20-years.

———. "Facebook." www.facebook.com/carycomeshome.

———. "Instagram." www.instagram.com/carycomeshome.

———. "Twitter." www.twitter.com/carycomeshome.

———. "Website." www.carycomeshome.co.uk.

Crofts, Charlotte. "Technologies of Seeing the Past: The Curzon Memories App." *Conference Proceedings of Electronic Visualisation and the Arts (EVA 2011)* (July 2011): 163–70. https://doi.org/10.14236/ewic/EVA2011.29.

———. "City Strata." Poster at the Media, Communication and Cultural Studies Association Annual Conference, January 2013. https://curzonproject.files.wordpress.com/2012/10/city-strata-poster-final-draft.pdf.

———. "Looking for Archie." *Walking Tour*, November 2017. https://uwe-repository.worktribe.com/output/6993516/looking-for-archie-a-walking-tour-of-cary-grants-bristol.

———. "Looking for Archie Being Human Evaluation Report." January 2021. https://uwe-repository.worktribe.com/OutputFile/7094711.

———. "Bristol Fashion: Reclaiming Cary Grant for Bristol – Film Heritage, Screen Tourism and Curating the Cary Comes Home Festival." *Open Screens* 4, no. 2 (December 31, 2021). https://doi.org/10.16995/OS.8018.

de Valck, Marijke. *Film Festivals: From European Geopolitics to Global Cinephilia*, 1st ed. Amsterdam: Amsterdam University Press, 2007. http://doi.org/10.1017/9789048506729.

Donaldson, Maureen, and William Royce. *An Affair to Remember: My Life with Cary Grant*. New York: Putnam, 1989.

Dyer, Richard, and Paul McDonald. *Stars*. London: Bloomsbury Academic, 1998.

Email Warbrides. Accessed June 9, 2022. www.carygrant.net/warbrides.html.

Geraghty, Christine. "Re-Examining Stardom: Questions of Texts, Bodies and Performance." In *Reinventing Film Studies*, edited by C. Gledhill and L. Williams, 183–201. London: Arnold, 2000. http://eprints.gla.ac.uk/49612/.

Geraghty, Lincoln, Vassilios Ziakas, and Christine Lundberg. "Guest Editorial: Exploring the Popular Culture and Tourism Place Making Nexus." *The Journal of Popular Culture* 52, no. 6 (2019): 1241–49. https://doi.org/10.1111/jpcu.12867.

Glancy, Mark. *Cary Grant, the Making of a Hollywood Legend*. Oxford and New York: Oxford University Press, 2021.

Gondek, Chris. "Infinite City – Interview with Author Rebecca Solnit and UC Press Art Director Lia Tjandra." *University of California Press Podcast*, December 8, 2010. https://soundcloud.com/uc-press/infinite-city-interview-with-author-rebecca-solnit-and-uc-press-art-director-lia-tjandra.

Grant, Cary. "Archie Leach." *Ladies Home Journal* Part 1 (January–February 1963); Part 2 (March 1963); Part 3 (April 1963). Accessed June 8, 2022. www.carygrant.net/autobiography.

Hills, Matt. *Fan Cultures*. London and New York: Routledge, 2003.

Historic England. "Cabot Tower." https://historicengland.org.uk/listing/the-list/list-entry/1202009.

Jones, Evan T., and Margaret M. Condon. *Cabot and Bristol's Age of Discovery: The Bristol Discovery Voyages 1480–1508*. Bristol: University of Bristol, 2016.

Macionis, Niki, and Beverley Anne Sparks. "Film-induced Tourism: An Incidental Experience Article." *Tourism Review International* 13 (October 2009): 93–101. https://doi.org/10.3727/154427209789604598.

Orry-Kelly. *Women I've Undressed*. London: Allen & Unwin, 2016.

Reid, Josephine, Richard Hull, Kirsten Cater, and Constance Fleuriot. "Magic Moments in Situated Mediascapes." In *Proceedings of the 2005 ACM SIGCHI International Conference on Advances in Computer Entertainment Technology*, 290–93. New York: Association for Computing Machinery, 2005. https://doi.org/10.1145/1178477.1178529.

Stead, Lisa. "Reframing Vivien Leigh: South West Star Histories." https://reframingvivienleigh.exeter.ac.uk/storymap/.

Torchin, Leshu. "Location, Location, Location the Destination of the Manhattan TV Tour." *Tourist Studies* 2, no. 3 (2002): 247–66.

Urry, John. *The Tourist Gaze: Leisure and Travel in Contemporary Societies*. London: Sage, 1990.

Urry, John, and Jonas Larsen. *The Tourist Gaze 3.0*. London: Sage, 2011.

Visit Bristol. "The Best of Bristol – Discover the City's Oscar-Winning Connections." Accessed June 8, 2022. https://visitbristol.co.uk/blog/read/2016/02/the-best-of-bristol-discoverthe-citys-oscar-winning-connections-b241.

Visit Britain. "Visit Bristol Before the Year Ends." Accessed June 8, 2022. www.visitbritain.com/au/en/8-reasons-visit-bristol-year-ends-0.

Waterside Plaza. "Cary Grant Links Waterside to Bristol." *Waterside Views*, April 2006. www.watersidewire.com/residentservices/documents/NewsletterPDFApril06_000.pdf.

The Western Daily Press and Bristol Mirror. "Cary Grant Lost Relatives." January 27, 1941, 5.

Williams, Rebecca. "Fan Tourism and Pilgrimage." In *The Routledge Companion to Media Fandom*, 98–106. Milton Park, Abingdon and New York: Routledge, 2017. https://doi.org/10.4324/9781315637518-13.

7

VANCOUVER UNMOORED

Hollywood North as a Site of Spectres

Erin Ashenhurst

Introduction

The city of Vancouver sits at the edge of the Pacific Ocean, cradled by mountains (Figure 7.1). The ancestral, unceded territory of the Squamish, Musqueam, and Tsleil-Waututh peoples, the landscape was once carpeted with temperate rainforest. Now, weekend hikers clad in costly rain gear drive out to roam among the massive cedars and Douglas firs of the North Shore. These forests still feel ancient, easily cast as mystical or menacing, depending on the light. In contrast, the downtown core of the city is a raccoon's collection of treasures, slick and shiny surfaces, towers growing like crystals at odd angles, grit, and rot mixed with sea glass. New architecture is unanchored to a sense of place, with edges easily reframed as Seattle, as Portland, as San Francisco, New York, or even a crowded street in Mumbai in *Mission Impossible: Ghost Protocol*. On the Eastside, older buildings can be imparted with any imagined history, stately pasts, or more often the noir shadows of crime dramas. As a popular stage for film productions, Vancouver – known by the industry as "Hollywood North" – is a seasoned actor. Through the stories of television and cinema, the city becomes a membrane of alternate dimensions, the true nature of which is porous and elusive. Here, ghosts of fictions haunt the streets.

While there is some thrill in seeing *Deadpool* filming on one's commute to work, or a grinning Joshua Jackson crossing the street fresh off the set of *Fringe*, recognizing the city on screen as it shape-shifts through narratives can be disorienting. Vancouverites are shown their city in disparate pieces, a changeling with iconic markers digitally removed and surfaces stickered over with American flags. For the viewer, landmarks trigger recollections fraught with hybrids as fragments of fabricated stories muddle with memories of lived experience. The gaze hesitates in these spaces that are known, and yet rendered unknowable. "Hey, I've been

DOI: 10.4324/9781003327585-7

there," one insists, pointing at the television, but that place does not exist. That place is a composite, an exercise in deception. With watching, there is a sensation like dreaming.

In this chapter, I employ an autoethnographic approach to convey the lure of narrative, especially in the creation of emotive stories linked to landscape, and to illustrate the sense of ambivalence in navigating instances where fictions played out on screen tangle with lived experience. To follow are a series of scenes, instances when the familiar is infused with strangeness. In the early days of the Covid-19 pandemic, people were encouraged to stay close to home. We searched screens for information, or escapism. With shops closed and downtown storefronts boarded up, some urbanites binged on television for stimulus, savouring glimpses of identifiable places from the "time before." But so often they are the sites of crime (*Criminal Minds*, *Arrow*, *Prison Break*, *The Haunting of Hill House*, *The Killing*), or already dystopian futures (*Altered Carbon*, *Snowpiercer*, *Lost in Space*, *Rise of the Planet of the Apes*, *War for the Planet of the Apes*). I found myself savouring the passivity of watching, of easy diversion from reality. But there were constant interruptions – moments when the fictional dream of a movie narrative smears, and the screen becomes more like the window of a tour bus careening through smog. I would be able to distinguish the Vancouver convention centre, the courthouse on Main Street, the shuttered mental hospital, or the diner a block from our old apartment where I watched my then-toddler son drink his first milkshake. These glimpses of the familiar quickly shift into the uncanny.

The realization that a landscape is no longer firmly affixed to a tangible world is at once exciting and distressing. In his essay "Simulacra and Simulations," Jean Baudrillard uses the example of Disneyland, a destination promoted as a delightful escape where "dreams come true." He writes, "Disneyland is presented as imaginary in order to make us believe that the rest is real, when in fact all of Los Angeles and the America surrounding it are no longer real, but of the order of the hyperreal and of simulation."[1] In Baudrillard's description of hyperreality, one's ability to differentiate between reality and a simulated reality disappears, the physical and the virtual blending together seamlessly. With my eyes fixed on the television screen, the authenticity of the world beyond the walls of the room falls into question. Once a place is shown capable of simulating any number of other spaces, the lived reality of that place – that once felt specific and solid – grows slippery. "Maybe I am living in a simulation," I say absently to the images on the screen.

Christmas in July 2021

We had been driving for a couple of hours already. Firstly, past the high-rise construction where traffic moves tightly like blood cells in a huge, restless python, then into the farmlands with the dwarfing pickup trucks, apathetic cows, and floating cranberries. There in the open, the odd hawk will draw our attention to the sky above the bowing power lines penning in our path North. Our legs are already cramped – the beginnings of sweat lines and dry mouths – and we have so much longer to drive that day. We pull the car off the highway, past the gas station, immediately slowing with the inertia of the town's main streets. Mountains lend dramatic backdrop to a strip of mostly single-storey shops and services. There is a pharmacy, a one-theatre cinema, a toy store, a hair salon, and even a Fields – one in a chain of the underwhelming clothing stores pervasive in small town British Columbia. We park between a window displaying artisan honey and frilly scarves for tourists, and our anticipated destination – the Blue Moose coffee shop conveniently located opposite the playground. It is early August, shorts weather and a time of mild delirium in the Pacific Northwest when people come to truly miss the rain. It is also the second year of a global pandemic.

"I love this place," says the six-year-old eyeing the park through the back window. "Remember the time we were here, and it was Christmas?"

I furrow my brow. The town of Hope, British Columbia, is traditionally our first stop on a two-day road trip from Vancouver to visit my in-laws in Terrace, BC. We have never made the trip in the winter, as the weather can be unpredictable, and the roads treacherous.

"It wasn't Christmas, Sweetie. . . . We've only been here in the summer, around this time."

The boy looks confused.

"But remember there was a Christmas tree and snow right there." He points towards a paved area currently displaying a rustic chainsaw carving of a bear.

I suddenly recall the time we visited when they had been filming a Hallmark movie. After researching, I learn several Hallmark Christmas-themed television movies, each centring on unsurprising heteronormative romances, have been filmed in Hope. Was it the town's name that drew them? In 2019, it was likely the set for *A Homecoming for the Holidays*, the story of a country singer returning to her hometown, destined to fall in love with a newly returned solider. A movie of tinselly Americana, it resembles a Hallmark greeting card to the masses. The set had fake snow dusting the bushes, wreaths adorning the lampposts, and teams of film people accessorized with radios and gaffer tape wheeling racks of wool coats and bins of sparkly bric-a-brac.[2]

I explain to my son, who would have been four at the time that they had been filming a movie. It was just a set decorated to look like winter. He nods pensively.

"So, it was only Christmas in a movie? It looked so nice . . ."

I offer a sympathetic smile.

"I wish it were a movie now," he says, fitting his mask to his face so we can enter the coffee shop.

***

The British Columbia Film Commission was established in 1977 by Grace McCarthy, the deputy premier and minister of recreation and travel in BC's Social Credit government. Efforts were directed at Hollywood with promotion hoping to tap into foreign capital and increase tourism. Film studios were built, and new educational programs offered industry training.[3] Already home to the popular seaside adventure series *The Beachcombers* that ran for nearly 20 years, British Columbia attracted other long-running series during this time. In *The Littlest Hobo*, a seemingly stray German Shepard dog wanders from town to town, solving people's problems. More conventional action dramas including *MacGyver*, *Neon Rider*, *Booker*, *21 Jump Street*, *The Hitchhiker*, *Danger Bay*, and *The Commish* featured crime stories ripped from the headlines. Movies like *The Changeling*, *Out of the Blue*, *Rocky IV*, *Shoot to Kill*, *Watchers*, *The Accused*, and Stephen King's horror *IT* created scenes of gritty urban streets and eerie woodlands. As a kid watching *The NeverEnding Story*, I brimmed with glee as ten-year-old Bastian chased his former bullies through Trounce Alley in Vancouver's Gastown neighbourhood from the back of the giant white luck dragon, Falkor.

In the 1990s, a tax-credit program called Film Incentive BC was introduced to boost the province's competitive edge against other Canadian and Californian locations.[4] Vancouver and the surrounding area could now offer up this wide array of geographical options at a fraction of the price. Chris Carter, creator of *The X-Files*, originally planned to shoot in Los Angeles but is said to have decided to "go where the good forests are," and so Vancouver became the set of the show's first five seasons from 1993 to 2002.[5] In the science fiction series, FBI special agents Fox Mulder and Dana Scully investigate reports of strange and unexplainable events. More sci-fi and fantasy series like *Smallville*, *Battlestar Galactica*, and *Supernatural* would keep studios busy through the aughts, with the suspenseful drama of *Riverdale* and *The Flash* starting up in the mid-2010s.

The Woods, 2021

"There's a lot of money to be made in scaring people."
– Dana Scully, The X-Files, *"Ghouli" (S11, E5, 2018)*

It is a late show, and the theatre is sparsely populated. The film *Antlers*, a supernatural horror about a boy and his family plagued by a malevolent, ancestral creature, is unsettling. As we watch, popcorn crunching under our masks, the setting takes on an eerie familiarity. It is set in an isolated Oregon town. In a street scene, the red sign of a Fields store pokes out from the corner of a shot. The unrelenting dampness of the forest, the decrepit siding of the house that acts as the boy's home, creates an atmosphere that vibrates with a particular West Coast flavour of foreboding. I slip out of the narrative, distracted. It is Hope, BC.

On the town's website, Hope promotes itself as the shooting location for *Rambo: First Blood*. The film starring Sylvester Stallone was released more than 40 years ago. And yet, the self-guided tour featured on the site still shows a map of downtown matched with stills from corresponding scenes in the film.[6] In the movie, Hope stands in for small-town Washington.

Vancouver (Figure 7.1) can be "disguised as Anywhere, USA," according to the BC Film Commission.[7] In 1996, a headline in *The New York Times* called Vancouver "The city that can sub for all of America."[8] For production companies, Vancouver is pitched as though it is like Las Vegas, a city Andrew F. Wood has described as

FIGURE 7.1 Vancouver, the port city.

Source: Courtesy of Christian Laud.

an *omnitopia* or "all-places."[9] Wood discusses *omnitopia* as a space where apparently distinct locales (New York, Paris, and Venice) converge to become part of the same experience – a place both distasteful and exhilarating. While all different genres have made use of Vancouver as "Anywhere, USA," crime dramas seem particularly enamoured. Hollywood comes here to do its killing. Already the ground zero of the province's opioid epidemic, and the very real hunting ground of serial killer Robert Pickton through the 1980s and 1990s, Vancouver can offer seedy alleyways, neglected hotels, ocean ports and bushy ravines, train yards, old barns and lonely cabins, and dark, dense forests. While the city is polished or roughed up as needed, the forests come complete with their own cinematic atmosphere.

The woods of Greater Vancouver are lush. The trees are dwarfing, limbs donning strings of a moss appropriately named witch's beard. Ferns grow as high as an adult's knees, and the forest floor is a spongy minefield of roots, rot, and slugs as big as sausages. It is beautiful, a hyperreality haunted by bodies both factual and fictitious. On film, the true danger of the woods is extended, multiplied by imagined terrors. They are made home to the supernatural, to countless gory secrets and scripted crimes. Lindsay Steenberg likens Vancouver on screen to Las Vegas' reproduction of tourist landmarks, stating, "The screen version of Vancouver offers an economical postmodern tour of the world's crime scenes."[10]

In 2010, an article in the *Vancouver Sun* reported on a woman's body found in Pacific Spirit Park. The park is a large, forested area on the Westside and part of the University of British Columbia Endowment Lands. The trails are popular with joggers, cyclists, and dog walkers. Though the body of another woman had been found in the park just the year before, several neighbours initially thought the dozen police cars on the scene were part of a film set. One nearby resident explained, "They've been filming a television crime show up the street here, so we've been seeing fake police cars around here all the time."[11] Another neighbour recounts telling her husband that all the commotion must just be a movie. But it wasn't *The Killing*, or *Motive*, or *Fringe*, or *Supernatural*, or even *Riverdale*, *Once Upon a Time*, or *Dirk Gently's Holistic Detective Agency*. The body was real.

It is the summer of 2021 and I take the kids on their bikes to Central Park in Burnaby, a large urban park with a stadium, duck ponds, a playground, and trails through the woods. We pass fluffy dogs with muddy feet and aggressive, hulking squirrels that have been overfed by misguided park-goers. This is a land of plenty. We gaze up at the canopy of trees. Around a bend a trail is taped off. A lanky man with a radio loiters at the fork directing people to take the other route. "Sorry, we're filming here," he says. We are all compliant. My kids zoom by on their little bikes. I think it might be *The Flash*. Later, I will watch *Yellowjackets* and wonder if the girls, lost in menacing woods, should be listening for the traffic sounds of Kingsway, the highway to the North, or the chimes from Patterson train station to lead them the two kilometres back to civilization.

In their essay, "Crime Film and the City," Sonia Bookman, Katelyn Mackenzie, and Austin Bernier discuss how Canadian filmmakers have depicted the city of Toronto on screen. By setting films like *Foolproof* and *Cold Blooded* which feature crime heists and gang violence in the undisguised geography of Canada's largest city, the films' fictional narratives "establish, reinforce, or rework urban place-myths and express mythologies that shape the meanings, ideas, and affects associated with the city."[12] While the majority of the films discussed in the essay presented Toronto as a cosmopolitan city, the nature of their criminal characters varied between sophisticated masterminds engaged in calculated heists, crafty individuals in a city full of creative possibilities, and criminal gangs situated in distinct ethnic enclaves. While this last group allows for representations of Toronto's cultural diversity, the neighbourhoods are shown as culturally "other," the exotic headquarters of deviant, violent behaviour. Though the lived experience of locals may vary drastically from the exciting narratives of these films, the fictional portrayals still have the potential to influence attitudes around crime, danger, and the ethnic identity of Toronto's criminals.

A handful of Vancouver-based productions – including *Da Vinci's Inquest*, a crime drama portraying the Vancouver City Coroner – have been set in Vancouver (Figure 7.2). However, most of the productions in Hollywood North continue to view the region as an avatar for elsewhere. Instead of cultivating local mythology (for better or for worse), the act of filming Vancouver is most often about erasure – obscuring evidence and constructing a false identity.

FIGURE 7.2 Downtown Vancouver through a boardroom window.

Source: Courtesy of Christian Laud.

The Motel, 2021

> *"Maybe you're just seeing what you wanted to see."*
> – *Dana Scully,* The X-Files, *"Grotesque" (S3, E14, 1996)*

"Is that where we got married?" My husband and I are sitting in our living room watching Showtime's *Yellowjackets*, a series centring around a girls' high school soccer team whose plane crash-lands in the Ontario wilderness in 1996. Ruinous conflicts took place in those woods, and years later their past actions come to complicate the lives of the now-adult women. The plot switches between timelines of the adolescent characters, and their contemporary, adult counterparts. I scrunch up my face, squinting despite the preposterous size of our new television, a purchase made after two years' worth of the pandemic spent staring at Netflix. Prior to this moment, to say I had been absorbed in the story is an understatement. I was myself a teenager in 1996. The nostalgic soundtrack alone is enough to have me hooked. In the show, Misty (played by Christina Ricci) sits with oversized glasses and poodle-style curls, spying on the conversation at another table. Indeed, she is sitting in the covered courtyard of the restaurant where I got married. Just there, we had rows of seats for our ceremony. I have pictures of our guests blowing bubbles in the air, my husband and I, childfree and ten years younger, strutting down the aisle after the nuptials, giddy. With this realization, the viewing experience changed.

FIGURE 7.3 Motel 2400 on Kingsway, East Vancouver.

Source: The author.

The first season of *Yellowjackets* was filmed in Vancouver and the surrounding area. We stare as Juliette Lewis sits in the parking lot of a motel. "That's . . . here," says my husband. And by "here," he means less than 300 metres from our living room. I pass the 2400 Court Motel (Figure 7.3) on the way to the drugstore. It has a sign boasting "75 years young," and it looks it. A traditional motor inn with individual cabins and a parking lot in the middle, on film it feels timeless, placeless, just as easily outfitted to look quaint and kitschy, or battered and seedy.

The motel's website boasts a page called "Motel 2400 On Screen."[13] Images from productions filmed onsite are showcased in a tiled mosaic. It has played set to crime dramas and horror including *Motive, The Exorcist* television series, *Halloween: Resurrection, Supernatural,* and *Bates Motel.* In 1996, scenes from "Wetwired," the 72nd episode of *The X-Files,* were filmed there. In the episode, a cable repairperson installs devices that cause people to have paranoid delusions and commit murders, killing their neighbours and loved ones when faced with hallucinations of their greatest fears.

The motel cabins can be rented by the month. They have housed Syrian refugee families new to the country. My in-laws recall renting a room for a weekend in the 1980s while on a road trip. In 2009, a guest on Tripadvisor awarded their experience of staying at the motel a single, solitary star, and wrote an unflattering review titled "Nightmare on Kingsway." It begins with "What a mess."[14]

Watching the motel on screen, I am disoriented and agitated as though I'd slept through a party at my own home. Shouldn't I have gotten some sort of notification, a Riot Grrrl bat signal that Juliette Lewis was just around the corner? The next day I walk by the motel on the way to get milk and stare at the sagging inflatable Santa strapped to the roof of the office. Since they filmed *Yellowjackets*, the motel started painting the lower half of the cabins a muddy red. I walk down the driveway towards the back of the property where the cabins haven't been painted yet. A woman sits on a white plastic lawn chair smoking. Beside her is parked a motorized wheelchair and a coffee can full of cigarette butts. I imagine Natalie, Lewis' character from the show, leaning against the neighbouring doorway, ripped fishnet tights and her own cigarette. The cabins at the back are still white with front doors thickly painted in alternating primary colours. This is how this place is supposed to look, I think. This is how I remember it . . . from the show.

The Location Scout, 2019

> *"All we can do Scully, is pull the thread and see what it unravels."*
> *– Fox Mulder,* The X-Files, *"Founder's Mutation" (S10, E2, 2016)*

I push through the door to the brewery in East Vancouver just before noon. It is November. In four months, restaurants will be shuttered in an attempt to stop the spread of the Covid-19 virus, but back then all that would have sounded like a bad science fiction movie. Christian Laud is sitting at a table, the only other customer. It has been almost 17 years since we last saw one another, but the conversation is easy. In art school, Christian had a knack for scavenging. His projects were often made up of materials he'd found while exploring abandoned sites, excavating treasures from the landscape. After years of working on film sets, it seems extraordinarily fitting that he has become a location scout. When we spoke, he had just spent his morning checking out unoccupied medical buildings for a commercial. He tells me about the day before, when he was taking pictures at Riverview, a shuttered mental hospital. Founded in 1913, the facility closed in 2012 leaving an expansive property and multiple buildings available for rent by the film industry. It has appeared as various often-sinister institutions in movies like *Halloween: Resurrection*, *Final Destination 2*, and *The Butterfly Effect*. It doubles as a sanatorium and also the rival high school in *Riverdale*, shows up as a military base in the *Watchmen*, and an animal transfer station in the quirky plot of *Dirk Gently's Holistic Detective Agency*.

Christian shows me the pictures he took of the old brick buildings and columned facades. You need permission to be on the property and there is security walking around, he tells me. Riverview attracts curious tourism as a movie set for famous fictions, and a site with a century's worth of very real and dark history of mental health services.

I tell him about the "Hollywood North Walking Tour" I attended the month before.[15] I was the only local in our small group, standing outside the Fairmont

FIGURE 7.4 The Marine Building in downtown Vancouver opened in 1930.

Source: Courtesy of Christian Laud.

Hotel Vancouver as our guide told us about a steamy scene in *50 Shades of Grey* filming in the hotel's elevator. My mind wandered to one of my first apartments, just around the block, and how the stores had gentrified since I lived there. We stopped at 1075 West Georgia Street, a monolithic concrete construction affectionately known as "the waffle building" for its square, inset windows. It played the Department of Mutant Affairs in *X-Men: Last Stand*. I have walked by it countless times. I remembered nervously waiting to sign a mortgage in the bank on the first floor.

In contrast to my perfunctory nods, the couple on the tour visiting from Florida was delighted, rapt with the comparisons between the buildings before them and the laminated movie stills supplied by our tour guide. They had seen almost every title he mentioned, endless wells of enthusiasm, quick with chortles and affirmations. I was preoccupied with my past associations with the area – the emotional phone call at that bus stop, that wet winter I saw the man in the three-piece suit slip in the snow just there. "This tour isn't working for me," I thought.

In Christian's job, he looks for spaces that can be easily costumed (Figure 7.4). He gains access to the rooftops of downtown high rises. He photographs the skyline from different angles, trying to capture how it might be framed to represent an urban environment with the fewest recognizable landmarks. He drives to the forest (Figure 7.5) to find old growth and lush greenery for a European beer advertisement. He locates a stretch of beach where, if the camera stays low, the mountains can't be seen over the dunes. They are spaces that can be isolated, cut away from

FIGURE 7.5 Golden Ears Provincial Park.

Source: Courtesy of Christian Laud.

FIGURE 7.6 Downtown Vancouver from a rooftop.

Source: Courtesy of Christian Laud.

the specific to become the general. On screen, they are made anywhere, victims of amnesia (Figure 7.6).

The tangling of fiction with lived experience is destabilizing. I watch Sandra Bullock in the movie *The Unforgivable* as her parole officer drives her character to a halfway house in Vancouver's Chinatown. They pass a grain terminal with an American flag slung over one of the silos. At the halfway house, a chaotic rabbit warren packed with agitated residents, Bullock's character examines a map of Seattle's transit system. She gets a carpentry job at a building called Chapel Arts where, years before the pandemic, I attended gallery shows, parties, and live music. I had completely forgotten about the existence of that space. The building is now immortalized in the movie, though severed from geography. In a young city, where newness is currency, little exists for long. Passing a demolished building, one scrambles on the eroding dunes of memory. "Wait, what was there before?"

The Gallery, 2018

> *"I want to remember how it was. I want to remember how it all was."*
> *– Dana Scully,* The X-Files, *"The Lost Art of Forehead Sweat" (S11, E4, 2018)*

In the spring of 2018, the Monica Reyes Gallery in Vancouver's Downtown Eastside was transformed into the basement of the J. Edgar Hoover Building in Washington, D.C. On the wall, hung a poster of a grainy horizon with the blurred form of a UFO hovering in the sky, and the words "I want to believe." A shabby, mid-century modern chair sat before a heavy desk and ominous black cabinet packed with files. It was the FBI office of Fox Mulder (played by David Duchovny), staged with original set pieces as part of the show "The Making of the X-Files."[16]

Bringing the trappings of this popular fiction into the gallery space, the show examined the history of Hollywood North. The gallery became a set. While the relationship is made explicit in "The Making of the X-Files," the development of the city's film industry has long run parallel to explorations in photography and installation in Vancouver's art scene. Since the late 1970s, artists such as Jeff Wall, Ian Wallace, Roy Arden, Vikky Alexander, and Stan Douglas among others – all working in the practice of conceptual photography described as photoconceptualism – have been grouped together under the term "The Vancouver School." Anton Lee defines photoconceptualism as "the art-historically informed understanding of the photographic surface as a representational space on a par with the pictorial and the cinematic."[17]

In Wall's photograph "In front of a Nightclub," from 2006, a line of young club-goers waits for admission to a club that appears to be in Vancouver's entertainment district. The club is a real place, in fact walking distance from where the image was actually created. However, the one shown in the photograph is a set. Due to the constraints of filming on location, Wall and his assistants spent months reconstructing the building's exterior complete with gum-spotted sidewalk in a warehouse. He studied the lighting on the real street and recreated it with precisely positioned strobes. In an interview in *The New York Times Magazine*, the artist revealed that he took hundreds of source images to inform his staging, including the many real club-goers waiting to be let in. In fact, "Wall scrutinized the photos for characters and clusterings he liked, then he hired 40 extras from a casting agency."[18] The final image is displayed in a large-scale light box, the scene illuminated like a television screen.

In her book *Engendering an Avant-Garde: The Unsettled Landscapes of Vancouver Photo-Conceptualism*, artist Leah Modigliani writes on the role of cityscapes in the work of the Vancouver School, stating that "while they stand for the material reality of Vancouver's built environment, they also remind us that these places exist everywhere in the western world."[19] Though Vancouver is featured in Wall's photography, the city is more set than character. Similarly, the photographs of artist Karin Bubaš, such as "Woman With Hollowed Tree" (2016) or "Woman In Blue Coat" (2013), show the forests and cherry blossoms that Vancouver residents identify with the area, however the images read as otherworldly, an idyllic, cinematic landscape.

Much like in television and movies, the city shown in the images of photo-conceptualism feels intangible, vanishing as it is costumed as an archetypal urban or wild environment rather than somewhere nameable, or knowable. Modigliani writes that such images "function as nodes of multiple physical, discursive, and economic sites simultaneously, reminding us that one of the most important legacies of conceptual art is its ability to be located everywhere and nowhere at once."[20]

***

In Hope, we get back in the car, sugared and caffeinated. We drive an hour and a half north to Lytton – or what is left of it. A scene from the 2001 drama *The Pledge*, starring Jack Nicholson and Robin Wright, was filmed at a ramshackle gas station in the village. In all likelihood, that building has been burnt to the ground. After setting record temperatures through the spring of 2021, the village of Lytton was consumed by a wildfire on the evening of 30 June. Though the highway reopened to general traffic the day before, vehicles are prohibited from stopping in the ravaged landscape. We drive slower than usual, staring at the still-smouldering hillsides. In a gravel stretch along the roadside, the charred frames of residents' cars and trucks have been collected. Even with the windows up, it smells like burning. It's the smell from *Mad Max: Fury Road*, from the blazing forest in *Those Who Wish Me Dead*. For kilometres, the smoke will create a screen through which we will only catch glimpses of farmhouses, mobile homes with satellite dishes, and fields with cows grazing in the noxious haze. I think of the dreaded antagonist in *The NeverEnding Story*, "The Nothing," a mysterious force that devours worlds. The highway disappears ahead. Through the windshield, my gaze searches for the horizon, but all is blank. "This," I think, "I wish this was all a movie."

Notes

1 Jean Baudrillard. "Simulacra And Simulations." in *Jean Baudrillard: Selected Writings*, 2nd ed., ed. Mark Poster (Stanford, CA: Stanford University Press, 2001), 175.
2 Jessica Peters. "Christmas Comes Early to Hope's Downtown," *Hope Standard*, October 1, 2019, www.hopestandard.com/entertainment/christmas-comes-early-to-hopes-downtown/.
3 Mike Gasher. *Hollywood North: The Feature Film Industry in British Columbia* (Vancouver: UBC Press, 2002), 69.
4 Ibid., 99.
5 Brian Lowry, Sarah Stegall, and Chris Carter. *The Truth Is Out There. The Official Guide to the X-Files. Vol. 1* (New York: HarperPrism, 1995), 17.
6 B.C. Hope (n.d.), "Rambo First Blood Filming Location Tour," accessed February 22, 2022, https://hopebc.ca/rambo-first-blood-filming-location-tour-self-guided/. This webpage is no longer accessible.
7 Gasher, *Hollywood North*, 106.
8 Justine Elias, "The City That Can Sub for All of America," *New York Times*, November 17, 1996, www.nytimes.com/1996/11/17/movies/the-city-that-can-sub-for-all-of-america.html.
9 Andrew F. Wood, "'What Happens [in Vegas]': Performing the Post-Tourist Flâneur in 'New York' and 'Paris'," *Text and Performance Quarterly* 25, no. 4 (2005): 318.
10 Lindsay Steenberg. "The Haunted Pacific: Vancouver on Crime Television," in *World Film Locations: Vancouver*, ed. Rachel Walls. World Film Locations Series (Bristol: Intellect Books, 2013), 92–93.
11 Andrea Woo, "Second Body Found in Pacific Spirit Park," *The Vancouver Sun*, June 14, 2010, www.pressreader.com/canada/vancouver-sun/20100614/283635346243982.
12 Sonia Bookman, Katelyn Mackenzie, and Austin Bernier, "Crime Film and the City: Imagining Toronto Through Entanglements of Place, Culture, and Crime," *American Review of Canadian Studies* 52, no. 1 (2022): 46–63.
13 2400 Motel. "2400 Motel On Screen," accessed February 22, 2022, https://2400motel.com/2400-motel-on-screen/.
14 Wetern, "Nightmare on Kingsway," *Tripadvisor* (blog), October 1, 2009, www.tripadvisor.ca/ShowUserReviews-g154943-d183641-r45604663-2400_Motel_Ltd-Vancouver_British_Columbia.html.
15 Fans of Vancouver, "Tours," accessed February 22, 2022, www.fansofvancouver.com/tours/.
16 Doug Sarti, "The Truth Is Out There: Fox Mulder's X-Files Office Set on Display in Vancouver," *Georgia Straight*, May 15, 2018, www.straight.com/arts/1076001/truth-out-there-fox-mulders-x-files-office-set-display-vancouver.
17 Anton Lee, "Unlearning the Vancouver School," *Canadian Art*, June 14, 2017, https://canadianart.ca/reviews/pictures-from-here-vancouver-art-gallery/.
18 Arthur Lubow, "The Luminist," *The New York Times Magazine*, February 25, 2007, www.nytimes.com/2007/02/25/magazine/25Wall.t.html.
19 Leah Modigliani, *Engendering an Avant-Garde: The Unsettled Landscapes of Vancouver Photo-Conceptualism*, 1st ed. (Manchester: Manchester University Press, 2018), 239.
20 Ibid., 240.

Bibliography

2400 Motel. "2400 Motel on Screen." Accessed February 22, 2022. https://2400motel.com/2400-motel-on-screen/.

Baudrillard, Jean. "Simulacra and Simulations." In *Jean Baudrillard: Selected Writings*, edited by Mark Poster, 2nd ed., 169–87. Stanford, CA: Stanford University Press, 2001.

Bookman, Sonia, Katelyn Mackenzie, and Austin Bernier. "Crime Film and the City: Imagining Toronto through Entanglements of Place, Culture, and Crime." *American Review of Canadian Studies* 52, no. 1 (2022): 46–63.

Elias, Justine. "The City That Can Sub for All of America." *New York Times*, November 17, 1996. www.nytimes.com/1996/11/17/movies/the-city-that-can-sub-for-all-of-america.html.

Fans of Vancouver. "Tours." Accessed February 22, 2022. www.fansofvancouver.com/tours/.

Gasher, Mike. *Hollywood North: The Feature Film Industry in British Columbia*. Vancouver: UBC Press, 2002.

Hope BC. "Rambo First Blood Filming Location Tour." Accessed February 22, 2022. https://hopebc.ca/rambo-first-blood-filming-location-tour-self-guided/.

Lee, Anton. "Unlearning the Vancouver School." *Canadian Art*, June 14, 2017. https://canadianart.ca/reviews/pictures-from-here-vancouver-art-gallery/.

Lowry, Brian, Sarah Stegall, and Chris Carter. *The Truth Is Out There: The Official Guide to the X-Files*, vol. 1. New York: HarperPrism, 1995.

Lubow, Arthur. "The Luminist." *The New York Times Magazine*, February 25, 2007. www.nytimes.com/2007/02/25/magazine/25Wall.t.html.

Modigliani, Leah. *Engendering an Avant-Garde: The Unsettled Landscapes of Vancouver Photo-Conceptualism*, 1st ed. Manchester: Manchester University Press, 2018.

Peters, Jessica. "Christmas Comes Early to Hope's Downtown." *Hope Standard*, October 1, 2019. www.hopestandard.com/entertainment/christmas-comes-early-to-hopes-downtown/.

Sarti, Doug. "The Truth Is Out There: Fox Mulder's X-Files Office Set on Display in Vancouver." *Georgia Straight*, May 15, 2018. www.straight.com/arts/1076001/truth-out-there-fox-mulders-x-files-office-set-display-vancouver.

Steenberg, Lindsay. "The Haunted Pacific: Vancouver on Crime Television." In *World Film Locations: Vancouver*, edited by Rachel Walls, 92–93. World Film Locations Series. Bristol: Intellect Books, 2013.

Wetern. "Nightmare on Kingsway." *Tripadvisor*, October 1, 2009 (Blog). www.tripadvisor.ca/ShowUserReviews-g154943-d183641-r45604663-2400_Motel_Ltd-Vancouver_British_Columbia.html.

Woo, Andrea. "Second Body Found in Pacific Spirit Park." *The Vancouver Sun*, June 14, 2010. www.pressreader.com/canada/vancouver-sun/20100614/283635346243982.

Wood, Andrew F. "'What Happens [in Vegas]': Performing the Post-Tourist Flâneur in 'New York' and 'Paris'." *Text and Performance Quarterly* 25, no. 4 (2005): 315–33.

Filmography

21 Jump Street (1987–1991), created by Patrick Hasburgh and Stephen J. Cannell, 20th Century Fox Television.

50 Shades of Grey (2015), dir. Sam Taylor-Johnson, Focus Features.

A Homecoming for the Holidays (2019), dir. Catherine Cyran, Lighthouse Pictures.

Altered Carbon (2018–2020), created by Laeta Kalogridis, Virago Productions.

Antlers (2021), dir. Scott Cooper, Searchlight Pictures.

Arrow (2012–2020), created by Greg Berlanti, Marc Guggenheim, and Andrew Kreisberg, Berlanti Productions.

Bates Motel (2013–2017), created by Carlton Cuse, Kerry Ehrin, and Anthony Cipriano, American Genre.

Battlestar Galactica (2004–2009), developed by Ronald D. Moore, NBC Universal Television Studio.

Booker (1989–1990), created by Eric Blackeney and Stephen J. Cannell, Stephen J. Cannell Productions.
Cold Blooded (2012), dir. Jason Lapeyre, Guildwood Entertainment.
Criminal Minds (2005–2020), created by Jeff Davis, Entertainment One.
Da Vinci's Inquest (1998–2006), created by Chris Haddock, Canadian Broadcasting Corporation.
Danger Bay (1984–1990), created by Paul Saltzman and Peter Dixon, Canadian Broadcasting Corporation (CBC).
Deadpool (2016), dir. Tim Miller, 20th Century Fox.
Dirk Gently's Holistic Detective Agency (2016), created by Max Landis, AMC Studios.
Final Destination 2 (2003), dir. David R. Ellis, New Line Cinema.
Foolproof (2003), dir. William Phillips, Alliance Atlantis Communications.
Fringe (2008–2013), created by J.J. Abrams, Alex Kurtzman, and Roberto Orci, Warner Bros. Television.
Halloween: Resurrection (2002), dir. Rick Rosenthal, Dimension Films.
IT (1990), dir. Tommy Lee Wallace, Green/Epstein Productions.
Lost in Space (2018–2021), created by Irwin Allen, Matt Sazama, and Burk Sharpless, Legendary Television.
MacGyver (1985–1992), created by Lee David Zlotoff, Henry Winkler/John Rich Productions.
Mad Max: Fury Road (2015), dir. George Miller, Warner Bros. Pictures.
Mission Impossible: Ghost Protocol (2011), dir. Brad Bird, Paramount Pictures.
Motive (2013–2016), created by Daniel Cerone and James Thorpe, Foundation Features.
Neon Rider (1989–1995), created by Winston Rekert and Danny Virtue, Alliance Atlantis Communications.
Once Upon a Time (2011–2018), created by Adam Horowitz and Edward Kitsis, Kitsis/Horowitz.
Out of the Blue (1980), dir. Dennis Hopper, Discovery Productions.
Prison Break (2005–2017), created by Paul T. Scheuring, Adelstein-Parouse Productions.
Rambo: First Blood (1982), dir. Ted Kotcheff, Orion Pictures, Anabasis N.V.
Rise of the Planet of the Apes (2011), dir. Rupert Wyatt, 20th Century Fox.
Riverdale (2017–), created by Roberto Aguirre-Sacasa, Berlanti Productions.
Rocky IV (1985), dir. Sylvester Stallone, United Artists.
Shoot to Kill (1988), dir. Roger Spottiswoode, Touchstone Pictures.
Smallville (2001–2011), created by Alfred Gough and Miles Millar, Tollin.
Snowpiercer (2013), dir. Bong Joon-ho, Moho Film.
Supernatural (2005–2020), created by Eric Kripke, Warner Bros. Television.
The Accused (1988), dir. Jonathan Kaplan, Paramount Pictures.
The Beachcombers (1972–2004), created by Marc Strange, Susan Strange, and Philip Keatley, Canadian Broadcasting Corporation.
The Butterfly Effect (2004), dir. Eric Bress and J. Mackye Gruber, BenderSpink.
The Changeling (1980), dir. Peter Medak, Chessman Park Productions.
The Commish (1991–1996), created by Stephen J. Cannell and Stephen Kronish, ABC Productions.
The Exorcist television series (2016–2017), created by Jeremy Slater, 20th Century Fox Television.
The Flash (2014–), developed by Greg Berlanti, Andrew Kreisberg, and Geoff Johns, Bonanza Productions.
The Haunting of Hill House (2018), dir. Mike Flanagan, Flanagan Film.

The Hitchhiker (1983–1991), created by Lewis Chesler, Riff Markowitz, and Richard Rothstein, Corazon Productions.
The Killing (2011–2014), created by Veena Sud, Fox Television Studios.
The Littlest Hobo (1983–1991), created by Dorrell McGowen, Canamac Pictures.
The NeverEnding Story (1984), dir. Wolfgang Petersen, Warner Bros. Pictures.
The Pledge (2001), dir. Sean Penn, Warner Bros. Pictures.
The Unforgivable (2021), dir. Nora Fingscheidt, Fortis Films.
The X-Files (1993–2018), created by Chris Carter, Ten Thirteen Productions.
Those Who Wish Me Dead (2021), dir. Taylor Sheridan, Warner Bros. Pictures.
War for the Planet of the Apes (2016), dir. Matt Reeves, 20th Century Fox.
Watchers (1988), dir. Jon Hess, Universal Pictures.
Watchmen (2009), dir. Zack Snyder, Warner Bros. Pictures.
X-Men: Last Stand (2006), dir. Brett Ratner, Marvel Entertainment.
Yellowjackets (2021–), created by Ashley Lyle and Bart Nickerson, Entertainment One.

8

ALWAYS THE DESERT – CREATING AFFECTIVE LANDSCAPES IN *BREAKING BAD*

Jane Lovell and Kenneth Fox

Introduction

Film, media, and transmedia tourists pursue the affective atmospheres of the fictional settings they see onscreen.[1] Tourists pursue myths and imaginaries, which are themselves composed of recurrent, familiar motifs that form the building blocks of stories.[2] For example, Lovell and Hitchmough[3] explore how the Western motifs of storytelling in TV series *Westworld* are divided into four typologies: character; action; props; and – particularly relevant for this chapter – setting. It is important to understand the tropes used in such pivotal, popular TV series, because they influence our understanding of cinematic geographies on and off the screen, feeding place myths and imaginaries. These tropes also evolve with our society and therefore charting how they are used in storytelling over time gives us an understanding of different values and perceptions.

The desert is a recurrent setting motif in the Western imaginary. Yet the desert itself can be broken down into different tropes. Beck[4] suggests that representations of the desert can be divided into spiritual experimentation and physical endurance, new capitalism, apocalypse, perceptual confusion, and a primal escape from modernity. This study offers an original analysis of Beck's tropes (some of which have existed for over two centuries) by applying them to the visual storytelling in episodes and scenes from the television series *Breaking Bad*. *Breaking Bad* (AMC, 2008–2013) is an American television series that ran for 62 episodes over five seasons. The series charts the transformation of Walter White (played by actor Bryan Cranston) from diffident high school chemistry teacher to crystal-meth-producing druglord Heisenberg, as he battles with cancer and wreaks havoc on his family, close friends and associates in the process. Joined by his confederate, Jesse Pinkman (played by Aaron Paul), the duo's descent from teacher and former pupil

DOI: 10.4324/9781003327585-8

to murderous drug manufacturers, is all set within the New Mexico landscape of Albuquerque and its desert environs.

Implicit in our chapter is the concept that the storytelling in *Breaking Bad* reinforces existing affective, mythical, touristic associations with the desert. This analysis can then be applied to other desert-set series, including space westerns such as *The Mandalorian, The Book of Boba Fett*, and *Dune*, all of which allow us to further understand how the mythical significance of desert geography in tourism imaginaries is perpetuated cinematically.

Visual storytelling

The affectiveness of the desert landscape in this chapter is conveyed through the moods created by visual storytelling methods, comprising a combination of cinematography, narrative, mise-en-scène, iconography, lighting, setting, and production design. *Breaking Bad* was shot on film as a "cinematic" piece of work.[5] At the time, the decision to shoot the series on film significantly affected the visual storytelling. Shooting on film also separated *Breaking Bad* from contemporary, digitally shot Western series such as *Justified* (2011–2015). The advent of Netflix enabled whole seasons to be streamed, so the experience of watching *Breaking Bad* heightened its sense of aesthetic difference and continuity, while the availability of HDTV further reinforced the marketing claim that the series is "cinematic."

The opening scenes of *Breaking Bad* immediately signal to the viewer that it is a Western, due to its desert location, influenced by creator Vince Gilligan's appreciation of director Sergio Leone who used desert settings in his classic 70s Westerns.[6] Gilligan establishes the desert landscape as the critical setting for his overarching vision for the series. Gilligan commented: "Gradually, after the first *Breaking Bad* episode, it started to dawn on me that we could be making a contemporary Western. So, you see scenes that are like gunfighters squaring off, like Clint Eastwood and Lee Van Cleef."[7] Gilligan's vision of *Breaking Bad* is of a "postmodern western" due to its desert and suburban settings set in and around Albuquerque, New Mexico. The desert setting shapes the overall aesthetic of the series in literal and allegorical forms and becomes the affective meta-location, as the chapter demonstrates.

The (South)Western imaginary

Deserts are classified as wilderness terrain. So, they are not simply sublimely beautiful, they are also hostile, their extremes of temperature and lack of water are often presented as malevolent onscreen, shaping the plot.[8] The desert landscapes that dominate the Southwest region of the United States form an imaginary that fuels and is fuelled by the Western myth. They include the Sonora, the Great Basin, the Mojave, and the Chihuahuan (nearest to Albuquerque, the setting for *Breaking Bad*). Buttes, mesas, canyons, and mountains shaped by wind and sand provide an affective, stark, enduring, sculpted beauty as does, as Gilligan has observed, the skyscape and the horizon. As painter and Taos resident Georgia O'Keefe remarked, "the world is so wide up there, so big . . . that sun that burns through to your bones."[9]

Gilligan was steered away from the original location of Riverside, California, and towards shooting in New Mexico by Sony because of the region's lower taxes. However, when he saw the location, he was convinced that this was the setting for the series. He persuaded the studio that New Mexico should not simply be used to emulate California, but to set the series in Albuquerque due to the cinematic geography; particularly the skyscapes, vastness, and "giantness," which attracted him. "California skies are usually blank, blue skies or a marine layer comes in and it's gray," Gilligan said, "but Albuquerque has these cumulus clouds that give you a scope of the distance out there. It's amazing. And it felt like virgin territory for cinematography."[10] As the Rough Guides guide to locations points out, the series has drawn on the qualities of the desert:

> *To venture into the bleak no-man's-land where they conduct their business, all White and sorcerer's apprentice Jesse Pinkman have to do is keep on driving when the tarmac runs out. Search on Google Earth for ABQ Studios, for example, where Breaking Bad is based, switch to Street View and face in the opposite direction, and there it is: the boundless desert, stretching away to the horizon.*[11]

According to the New Mexico Tourist Department website (New Mexico Tourism Department, n.d.), the "true Southwest," "Old West," and "Frontier Country" await you. The "true Southwest" is a form of "constructed authenticity" heavily based on the 19th-century cowboy country of popular imagination.[12] Director John Ford contributed to the Hollywood imaginary of the Southwest, by using the iconic setting for a range of his Westerns, enabling Monument Valley (the desert) to become what Kitses (2004) describes as a metonym for the Western and the West. As DeLyser observes, "the mythic West is a shifting construct: sometimes located in space at other times only in the mind; and each generation has made its own contributions to-the myth."[13]

Many Westerns, from space to realist noir, provide a combination of filmic allusions which cement the desert as hostile and otherworldly within our associations. Sam Shepard, actor, director, writer, and playwright, lived on a ranch in New Mexico for many years. His seminal work, *True West (1980)*, frames the desert as the place where Lee, the character symbolizing the Old West, disappears to lose and find himself, in contrast to the New West brother, Austin, who lives in the suburbs. Travel writings also enhance its mythological, mystic, wilderness status.[14,15]

Breaking Bad also illuminates the concept of the desert in many guises, including its classical role as a place of wilderness and survival, story and allegory. The desert is central in a trilogy of films by Italian director Michelangelo Antonioni: in metaphorical fashion in *Red Desert* (1965), and, with his critique of America at the end of the 1960s in *Zabriskie Point* (1970), representing the desire to escape from one's identity in *The Passenger* (1973). Gandy identifies the importance of the desert in Antonioni's work: "The desert is for Antonioni not only a concrete space to be conveyed in all its aesthetic complexity but also an allegorical and metaphorical realm through which we can explore different facets of human consciousness and experience."[16]

These tellings and retellings of Westerns play with desert tropes in different ways. Our contention is that *Breaking Bad*'s use of the desert has a similar heft and scope, but the series then extends and expands the Western imaginary. When we first encounter the desert in *Breaking Bad*, we are already familiar with what it signifies. For example, the opening scene of the pilot episode features desert, sky, and mesa. This draws on but then undercuts the mythological power of the Western where the iconic image of the gunfighter is replaced by Walter in his underpants, setting the revisionist tone of the show.

Methodology

The research uses a content analysis approach assessing visual narrative attributes to assess variables such as mise-en-scène,[17] cinematographic technique, and dramatic content to explore the role of the desert within *Breaking Bad*. The content analysis sources include not only the series itself but also transmedia resources, involving official show blogs, fandom sites, and a highly detailed Wiki, which supplies full location details for each episode.

We draw upon Beck's analysis of the narrative functions of the desert, which Beck sets out as five tropes,[18] which are then applied to each episode and scene involving the desert. They are addressed in this article to demonstrate the progression from emptiness and experimentation to the apocalyptic conclusion of Walt's involvement in the drugs trade:

1. Desert as a place of emptiness, experimentation physical and spiritual testing – Walter and Jesse are constantly tested against the desert backdrop;
2. Desert as a stressor upon the limits of reason and perception – supporting and resisting linearity in narrative, adding modernist design tropes and surreal effects.
3. Desert as a metaphor of apocalypse, evidence of the ultimate wasteland – the underlying relationship to Walter's cancer of the region's history of atomic testing and the landscape's resulting toxicity.
4. Desert as the primal escape from a modernity that must be resisted – the post-Anthropocene ending of the series.
5. Desert as New West capitalism – the drug trade and cartels.

Beck's tropes provide a structure for the analysis of the series. They capture the essence of the landscape and have parallels with the ways literature and film explore and represent the desert, all of which are manifested in *Breaking Bad*. Season 1 and Season 5 of *Breaking Bad* have the highest number of desert scenes as the story circles back to the opening location. The desert is a central subtext; for those scenes *not* set in the desert, the impact of events that *have* happened in the desert help shape the action of characters and knit together the multiple storylines, visual storytelling, and seriality that converge around Walter's transformation from chemistry high school teacher to mythic drug producer Heisenberg.

While examining the tropes, it is important to note that scenes could fall into more than one category. As Beck suggests, "the trope of the desert as without form and void enables the proliferation of all other tropes; the desert becomes the place of infinite metaphorical multiplicity."[19] In other words, viewers take complex meanings, positives, and negatives from stories set in the desert.

The desert as a place of emptiness, experimentation, and physical and spiritual testing

Moving on to the examination of the tropes, firstly Beck outlines the concept of desert as empty, which contains the assumption that it is free of life and placeless.[20] It is important to begin by contradicting this concept. The desertscape is far from empty or barren. Deserts are categorized as wilderness environments, which, while seemingly hostile to modern man, are full of flora and fauna which have adapted to the harsh conditions. The real Southwest is the region where the final Indian Wars of the 1890s were fought, leading to the decimation of the native tribes and the so-called end of the Frontier. A structuring absence removes signs of indigenous habitation from *Breaking Bad* for the most part, focusing on the "boonies," but the Chihuahuan desert is full of human markers and signifiers, Western and Native, from pueblos, hogans, and churches, to roads and trails, such as the Santa Fe Trail and pilgrimage routes, amalgamating to make it highly located. Also, in contrast to the concept of desert as empty place of trial are the works of Leslie Marmon Silko, the writer and poet, who was born on the Laguna Pueblo reservation 50 miles from Albuquerque. In her novel *Almanac of the Dead* (1991), she reworks the vision of the landscape as seen through the eyes of the indigenous tribes; here the desert is not a place of emptiness but a place of belonging.

In contrast, in *Breaking Bad*, it is noteworthy that the Native peoples of the area appear to have vanished. Across the run of the series, there are four appearances by indigenous peoples. In episode 101, a Native American tows Walter's RV out of the ditch. At the end of the next episode (102), two Native American children spot something behind a stone and see Walter's discarded gas mask; one of them puts it on his head, a symbol of the region's toxic past and present, a piece of evidence that will come back to haunt Walter in a later episode. The teaser in episode 306, "Sunset," sees a tribal officer encounter the "Cousins" with inevitable consequences. Finally, in episode 514, "Ozymandias," Walter buys a truck from a Native American for $10,000. While the presence of Native Americans within the series is minimal, the symbolic and narrative power of the To'Hajiille reservation, which gives its name to episode 513, is exemplified by its use as the setting for Walter's first cook for concealing his money in the pilot episode and "Buried" (episode 510), respectively.

Yet it is not simply indigenous life that is removed. A structured absence informs the artistic approach of the series, echoing Baudrillard's reference to the "aesthetics of disappearance."[21] Unreal, unpeopled places are created through editing; Roesch describes the "creative geography"[22] of the choice to leave unseen locations outside the frame. Within the mise-en-scène, a composed emptiness, stripping the desert and Albuquerque of features other than adobe architecture and expansive horizon, is a production design statement. Aside from the characteristic adobe architecture (which is red, flat-roofed, and naturalistic) and the desert, both of which signify New Mexico, the series has a studied anonymity. While selectivity must always play a part when depicting and denoting locations, in *Breaking Bad* there are no touristic shots of Albuquerque Old Town, malls, or landmarks. Many of the placeless suburban venues in *Breaking Bad*, such as the car wash or the airport, are also "deserted." They recall Baudrillard's description of the "desertification" or the "desert of the sign" in America, where symbols of the urban metropolis such as supermarkets can seem barren. In *Breaking Bad*, the studied emptiness does not simply act as a backdrop or foreground the action; it is an aesthetic catalyst for the visual storylines.

The concept of emptiness contains a narrative value and meaning; writing of the films of Antonioni, Gandy describes the desert as a "beautiful void,"[23] and Baudrillard states "the desert is no longer a landscape, it is a pure form produced by the abstraction of all others."[24] Within a discussion of the desert as a metaphor, he expounded on his perception of desertness as "the erasure of traces," the "intact silence." Emptiness as a trope of the desert is therefore perpetuated by *Breaking Bad* through its aesthetic of structured absence.

Spirituality and experimentation

The open, giant landscape does not simply foreground, but adds scope, depth, and intertextuality to *Breaking Bad*, baring emotions as a testing ground for human endurance. Addressing the aspect of spirituality in Beck's tropes, the prevailing Judaeo-Christian myth of the desert constructs it in terms of an inscrutable, hostile void where the son of God and mortals are tested spiritually, emotionally, and physically. The desert becomes the terrain of escape, renewal, and redemption. Episodes of *Breaking Bad* do not tend to contain a conscious realization by the characters that they are "emptying out" by being "at one with the desert" as described by Nicolson in his visits to the desert,[25] but there are moments of existential crisis, and the sheer scale of the landscape writ large highlights the affectiveness of small dramas of human civilization.

The desert is the site of many pivotal confrontations that occur in plain sight, in exposed yet private wilderness. It should be noted that many of the traditional stories of desert survival involve male characters proving their masculinity and the desert is presented as a gendered space in *Breaking Bad*.[26] In episode 311, "Full Measure," which serves as the finale of the third season, Walt is in full Heisenberg dress as he faces off with antagonist Gus Fring, a drug lord who fronts the restaurant Los Pollos Hermanos, in profile against the broad canvas of the desert, verbal jousting and devious plans replacing the traditional gunfight. Later, the desert is the fulcrum for emotional conflict between the pivotal characters, including Jesse, Walt, Mike (Walt's brother-in-law) and Saul (his dodgy lawyer) in episode 405, "Shotgun," when Jesse thinks Mike has asked him to dig his own grave. In "Crawlspace," episode 411, Walt is once again in the desert, fearful for his life, tasered and transported with a bag over his head as he is informed by Gus that he has been fired. In episode 501, "Live Free or Die," Walt, Jesse, and Mike meet in the desert; Mike is determined to kill Walt once he hears of Gus's death, and Jesse tests the limits of Mike's loyalty to Gus when he stands between Walt and Mike's gun.

While, as stated earlier, there are few references to indigenous habitation in *Breaking Bad*, penitent churches, pilgrimage routes, and sacred burial sites pepper the Southwest deserts, creating a sacred landscape. The indigenous spirituality of the desert is displayed in the episode 301, "No Más," which begins the third season with a striking image of a religious ceremony in a tiny Mexican village surrounded by desert scrub and mesa. Worshippers pay homage to Santa Meurte by crawling on their stomach to her shrine to make offerings. Two besuited men who look like brothers join the worshippers; they might be twins, but they become known as "The Cousins."

They emerge from a high-performance car and join the prostrated worshippers crawling towards the small grotto where offerings adorn the statue of Santa Meurte. The different colour palette signals the change of geography to Mexico at the beginning of this episode. The long shot of "The Cousins" joining the crawling villagers frames them emerging from the soil, with the sacred mountain in the background, like an altar to the Gods. The Cousins are hit men from Mexico who have Walter White/Heisenberg in their sights; they are not just doing the job as a mercenary task but as revenge for the murder of their uncle Tuco Salamanca in season 2. One provides an offering to Santa Meurte, the other pins a picture of Heisenberg to the statue. They are paying homage to the Saint of Death, invoking the old ways to take out the new threat to the cartel's interests north of the border, the mythic drug king, Heisenberg.

While this chapter examines the desert across the series of *Breaking Bad*, it also takes a deeper dive into a key episode (209) where the desert becomes a testing ground for physical endurance, spiritual reflection, and redemption. Rather than the 40 days spent by Jesus in the desert suggested by the Biblical reference, Jesse and Walt spend "4 Days Out" in the desert wilderness. Here, the action moves from water running in the clinical environment of a hospital – where Walter glimpses what he thinks is evidence that his lung cancer has spread – down a dirt-track road to the "boonies," where he and Jesse cook up another batch of crystal meth in their RV in a montage full of activity and humour. Michael Slovis's use of a filter to enhance the earth tones of the New Mexican colour palette changes it from spiritual desolation to enlightenment as they sit in their garden chairs in the brilliant sunset which illuminates their silhouettes, the wide-angle shots providing the greater scope and depth of a cinematic aesthetic which references Leone's and Antonioni's visualization of the desert.

In "4 Days Out," there is little interaction with the desert environment until they have run out of fuel and are cold and thirsty. The heat contributes to the chimera and mirage of confusion. After the engine stops for a second time, they realize they are stranded with no escape from the extreme temperatures and no way of surviving the harsh aridity. At one point Walt, a tea towel wrapped around his head against the sun in *Lawrence of Arabia*-style, is slumped in his lawn chair and Jesse fears he has died. Another scene finds Walt sitting in the scrub grass; his tea-towel-clad head inscrutable in the vast space. He is coughing blood and Jesse finally fully understands his near-death predicament. The episode finally brings the characters to moments of existential awakening, as Jesse uses contra-thinking to reflect on a choice he made not to go out with his girlfriend Jane, "Why couldn't I have just gone to Santa Fe?" Walt, meanwhile, states, "I have this coming." As Gandy argues,[27] there is a "primitivism" about the desert, which is mirrored in these emotions and life-or-death situations and subverted by the postmodern presence of the RV; an incongruous, suburban, poignant element in the scrubby terrain, windows glowing orange in the night.

The visuality of the series represents the protagonists' reflection on the prospect of death. The skyscape above the RV changes, the speeded-up movement of clouds overhead betraying their thoughts and consciousness and the burning sun and liminal sunsets, depicting the dwindling of life. Walt eventually combats their situation using science to construct a battery from nuts and bolts. The episode ends as it began, in the hospital, as Walter receives good news, with water running as a sign of life and a contrast to the arid conditions they have survived. Unusually, while a sense of existential awareness pervades this episode, the concept of redemption, an ever-present in most American stories, is absent; instead, the characters resume their behaviour when emerging from the desert to be offered a second chance. As Clark notes of Walt and Jesse, "once they leave the desert, the show introduces the conflicts that arise due to their choice to bring evidence of their violence back with them."[28] So, while the trope of existential endurance plays out in *Breaking Bad*, redemption remains elusive.

The desert alters the limits of perception

The desert also alters the limits of perception and acts as a liminal space, inverting reality through its openness, heat shimmer, and bedazzling colours and light. Baudrillard observed the primal aesthetics[29]: "The very colours – pastel blue, mauve, lilac, are the products of a slow, geological, timeless combustion. The mineral quality of the earth breaks through the surface in the crystalline flora." These pastel colours, abstraction, light, and simplicity are expressed in the paintings of New Mexican artist Agnes Martin, and, describing the sage-covered land near Taos, painter Georgia O'Keefe, stated, "The colour up there is different. It's a different kind of colour from any I'd ever seen."[30] Her early New Mexican works pair artificial flowers with the red earth hills, evoking the "dream thing" which she embraced as the surreality of the desert (ibid.).

The colourscapes and pictures of O'Keefe's buttes are referenced in the cinematography of the sunset in episode 209, "4 Days Out," where it is significant that Jesse had a choice of either going to see a Georgia O'Keefe exhibition in Santa Fe with his girlfriend Jane or cooking in the RV; the option he chose nearly caused his death. Jesse made a choice between canvases depicting a magical New Mexico or the reality of breaking down in the desert (both metaphorically and in reality). In an exasperated act of contra-thinking, he returned to this moment of choice later as a crucial moral turning-point.

Beck[31] describes the desert's sublimity and beauty, but additionally, a reaction of confusion and disorientation as a result of its vastness and light. This trope is captured by Baudrillard, who considers the desert heat metaphysical: "the spectral does not refer to phantoms or to dancing ghosts, but to the spectrum into which light disperses."[32] The expressionistic backdrop of the desert provides a surreal canvas for scenes such as those in episode 207, "Negro y Azul," where Hank and a group of DEA agents await the arrival of the informant "Tortuga." Hank moves away in panic as he spots the arrival of a tortoise with the head of Tortuga on its back with the message *Hola DEA* written on its shell. It is one of many surreal moments in the series where the audience is unsure what they are really seeing, the limits of our perception altered by the desert.

Desert as a metaphor for apocalypse: the ultimate wasteland and Walter's cancer

Beck's trope of the desert as a metaphor for apocalypse has had numerous iterations as the deserts of the Southwest have a specific link with atomic and nuclear testing at a federal level. Oppenheimer developed the Atomic bomb at Los Alamos (100 miles from Albuquerque), and the Los Alamos National Laboratories and Sandia Laboratories still conduct energy and nuclear research linking the Southwest to secret developments and radiation. This was visited and revisited in the *X-Files* (1993–2016), where *Breaking Bad*'s creator Vince Gilligan got his start in television. Treated as perceived wastelands, the reservations were used for nuclear testing and a nuclear waste dumping ground, for example, the site near Carlsbad, New Mexico (now a World Heritage Site). Walt's abandoned gas mask acts as a reference to the nuclear and other apocalyptic scenarios.

It is also important to address the apocalypse in the form of the desertification of Walter himself, as the metaphorical and emotional desert overtakes him. People often visit the desert in crisis, as Nicholson describes; when his marriage disintegrated the pull of the desert became "the geographical equivalent of my own emotional state."[33] Baudrillard conceptualizes this inner barrenness as "mental desert form."[34] Similarly, our analysis argues that the affective desert landscape within means that *Breaking Bad* also examines Walter's moral disintegration. Walter is joyless in his unfulfilled dreams, and the bland landscape of suburban Albuquerque emphasizes his lost promise. His cancer diagnosis makes him see how desperate his economic situation is in an America where health capitalism rules, so he transforms into a new identity (Heisenberg) to keep the desert from his door while all the time spreading destruction with the desertification of his secret life.

Walter's desert within (his amorality) shifts the nature and structure of his world. Walter's pathology transforms his ontology, and much of this ontological journey takes place across the landscape of Walter's body, particularly his face. *Breaking Bad*'s mise-en-scène often dwells on close-ups of Walter's shaven head, topped by what becomes Heisenberg's trademark pork-pie hat. While the desert remains unchanging and unchanged during the chemical transformation of his character, Walter's face comes to resemble the emotional desert inside him as his cancer alters his reality. As Newcomb observes, television aesthetics include developing an intimate relationship between the viewer and characters through the display of "faces, reactions, exploration of emotions" (Newcomb, 1974, pp. 245–46). His shaved head and the plains of his face become spare, stark; they are a landscape of interior struggle.

The close-ups of his face ensure that the viewer examines it like terrain, attempting to read the expression and thought processes, until, in episode 106, "Crazy Handful of Nothing," Walter morphs into Heisenberg, shaving his head as his hair is falling out due to chemo, and adopting the look he will keep as he maintains his double existence. His shaven head and moustache remain until episode 213 when Walter appears with a goatee. His granite face that of "the one who knocks," remains until episode 515 "Granite State." Once he leaves the Southwest, hiding in a log cabin in the woods of New Hampshire, his beard and hair regrow, further intertwining Walt and his environment. Allegorically, as Walter transforms, he becomes the desert and the desert (his cancer and his hubris) becomes Walter. If we consider Walter's cancer as altering the nature and structure of his world, the disease finds its ontological corollary in the desert landscape whose presence and absence help shape the structure and nature of *Breaking Bad*. Walt's journey could be termed "Southwestern Noir" in the manner in which its terrain mirrors the interior struggles of the protagonist, resembling the tropes of Nordic Noir detective series filmed in bleak locations of similar sublimity.

The desert as new capitalism

Beck's discussion of capitalism in the context of the exploitation of the desert for precious metals and petroleum places the emphasis on the New West in the next trope addressed in the paper. The capitalist aspirations in *Breaking Bad* are a key theme of episode 313, "Full Measure," which begins with a teaser flashback going back 16 years to Walt and Skyler buying their house. The house appears modest, given that Walt is working at Sandia, a prestigious laboratory. Walt's American dream is encapsulated in his statement, "We've got nowhere to go but up." In contrast, the capitalist nightmare of drug culture is a narrative strand that interweaves through the seriality of many episodes. It is encapsulated in the proximity of the border with Mexico and the involvement of a cartel as a superstructure for drug capitalism. The drug trade is the ultimate symbol of new international capitalism as identified in episode 502, "Madrigal," where the German company who own Los Pollos Hermanos (a front for drug-dealing) begin their attempt to tidy up the fall-out from Gus's death and the DEA investigation of Los Pollos. The backstories of Gustavo Fring, Hector "Tio" Salamanca, Tuco Salamanca, The Cousins, and "Tortuga" create the architecture for the series' representation of drug capitalism with the desert as its main meeting place and funeral site.

The capitalist desert is epitomized by the control of distribution, branding, and design by the cartel, and Madrigal, as the parent company of Los Pollos, follows a capitalist model of death or survival, fealty or death, profit or death. Episodes 408 and 410, "Hermanos" and "Salud," close the circle of Gus's origins as a drug distributor and his vengeance on the cartel for the death of his partner in 1989 at the hands of Tio Salamanca at cartel leader Don Eladio's house. "Salud" begins with a teaser set in a semi-arid location where Gus, Mike, and Jesse wait the arrival of a plane to carry them to Don Eladio's house in Mexico. This opening long shot, a trademark of the series, has the characters and sometimes vehicles set within a vast flat plain, confirming the Western desert aesthetic.

Once again, the desert mise-en-scène creates its own suspense in the following episode 411, "Crawl Space," which sees Walter in the desert at the mercy of Gus, who no longer needs him as meth creator; Jesse, the apprentice, is now the sorcerer. Walt's expectation for such a visit is death and a sandy, anonymous grave. Capitalism escalates as the profitability of the drugs trade overshadows the revenue of most other legitimate industries and *Breaking Bad* illustrates how the business of drugs infiltrates all aspects of society. Gus is a venture capitalist, Heisenberg (Walter) works to hold on to his blue-meth patent, Madrigal provides the distribution network for the European market, and as the profits rises so does the death toll.

The derailment of the new capitalism reaches its apex when, as Mark Zoller Seitz has identified, in episode 514 ("To'Hajiille"), the neo-Nazi gang confront Walter, Hank, and Steve Gomez, and the cinematography by Michelle MacLaren builds the tension with a series of dolly shots that move in on all the participants just before the gun battle starts. This means that the audience never loses track of where all the combatants are as the landscape becomes a killing ground, all capitalist structure breaking down as the hostile nature of the desert, a place of brutality beyond the limit of reason, returns to the archetypal Western shoot-out.

Desert as the escape from modernity and civilization

Finally, Beck's trope delineating the desert as "an elemental alternative to the rational order of 'civilized' life, or, conversely, representative of the chaos of an unordered primal 'nature' that must be resisted and expunged,"[35] seems to follow on logically as a post- or anti-capitalist phenomenon. The five tropes are permeable, and earlier this piece described how characters undergo spiritual and physical endurance tests, which take them outside time into a primal survival zone.

In addition, the desert does not simply alter the spatial but also the temporal limits of perception. Beck refers to Smithson's statement about timelessness:

"The future is always going backwards. Our future tends to be prehistoric."[36] As a serial, *Breaking Bad* cultivates familiarity in repetition, but the story plays with our concepts of time by jumping backward and forward to merge suggestions of primordial, Old West, and New West landscapes. Against this, while the desert alters perceptions, the never-changing, eternal nature of the desert offers a seamlessness ideal for narrative continuity over 62 episodes.

Episode 515, "Ozymandias," indicates how, while capitalism creates a seemingly permanent structure, the desert exerts a stronger force. The title of the episode, taken from the poem by Percy Bysshe Shelley, describes a hubristic fall from power, symbolized by the disappearance of a colossal statue into desert sands. Cranston read the poem in an AMC teaser trailer before the release of the final episodes of season five. The poem contains the lines:

> *My name is Ozymandias, King of Kings;*
> *Look on my Works, ye Mighty, and despair!*
> *Nothing beside remains. Round the decay*
> *Of that colossal Wreck, boundless and bare*
> *The lone and level sands stretch far away.*

In the teaser trailer, Cranston's voiceover becomes increasingly powerful as scenes of the desert locations from the series, empty of people, clouds racing overhead, time irrevocably passing, are shown, ending in a shot of Heisenberg's hat abandoned in the dust.

The teaser formed a critical moment for audience's understanding of the affective desert landscape. The reference to the poem in the episode title indicates how the power which Walt has built through the legend of Heisenberg is slipping away, only to be consumed by the desert. Far from emptiness, this is the annihilation of series events – and all evidence of the Anthropocene – absorbed by the sands of time. *Breaking Bad* makes it absolutely clear that the eternal desert survives any attempts at civilization; this is a subliminally vertiginous concept for the viewer to grasp, underlining the mythology of the landscape.

Conclusion

The decision to use Southwestern desert locations in *Breaking Bad* plays a fundamental part in the visual storytelling of the series and its Western aesthetic, providing an affective landscape (and skyscape) that feeds into the popular imagination. How the desert is presented onscreen eventually informs touristic impressions of the wilderness. We use Beck's tropes as a framework for our content analysis of *Breaking Bad*, providing a map of the form and function of the desert in the narrative and aesthetic. The tropes are often implicit in studies of desert tourism, such as Moufakkir and Selmi's 2017 study of tourists visiting the Sahara who "experience the perceived emptiness, closeness to God, self-renewal, refreshment and existential authenticity of the desert."[37] Making the desert tropes explicit in this study allows us to understand how storytelling shapes and consolidates the affective elements of the landscape.

The tropes can be applied to other series, for example, Space Westerns with desert settings like *The Mandalorian* and *Dune*, where motifs including spiritual and physical endurance, escape from modernity, capitalism, and apocalypse form familiar storylines. These series continue one of the aesthetic contributions of *Breaking Bad*, the postmodern Western, where the signifiers of ordinary life introduce humour to the drama of the situation: for example, the pork-pie hat that Walt assumes as part of his identity as Heisenberg, the mysterious drug dealer. But it should be noted that *Breaking Bad* has largely absented indigenous life in the desert, while the teaching and practices of desert communities are recognized as pivotal in works such as *Dune* and *The Book of Boba Fett*.

This study presents the desert as complex and contradictory: sublime, barren, picturesque, hostile, primal, capitalist, present, and absent. The exterior scenery echoes the interior desertification of the protagonist, resembling the bleak and wintry wilderness settings of Noir detective series. But ultimately, *Breaking Bad* celebrates the myth and – through its tropes – perpetuates the imaginary of the enduring, timeless desert landscape for viewers, echoing the character Saul's words: "Always, the desert."

Notes

1 Sue Beeton, Travel, Tourism and the Moving Image (Bristol, UK: Channel View, 2015); J. Lovell and S. Hitchmough, "Simulated Authenticity: Storytelling and Mythic Space on the Hyper-Frontier in Buffalo Bill's Wild West and Westworld," *Tourist Studies* 20, no.4 (2020); S. Roesch, The Experiences of Film Location Tourists (Bristol, UK: Channel View Publications, 2009).
2 Roland Barthes, *Mythologies*, trans. Annette Lavers (New York, USA: Hill, 1972).
3 Lovell and Hitchmough, "Simulated Authenticity: Storytelling and Mythic Space on the Hyper-Frontier in Buffalo Bill's Wild West and Westworld."
4 John Beck, "Without Form and Void: The American Desert as Trope and Terrain," *Nepantla: Views from South* 2, no. 1 (2001).
5 B. Mills, "What Does It Mean to Call Television Cinematic?," in *Television Aesthetics and Style*, ed. S. Peacock and J. Jacobs (London UK: Bloomsbury, 2013).
6 "*Breaking Bad'*s Camera Work Perfects the Visual Recipe, Yo," wired, updated April 7, 2012, accessed July 14, 2022, www.wired.com/2012/08/the-look-the-feel-of-breaking-bad/.
7 "Contemporary Western: An interview with Vince Gilligan," Local IQ, 2013, accessed 2017, 10 September, www.localiq.com/index.php?option=com_content&task=view&id=3019.
8 Sue Beeton, "Landscapes as Characters: Film, Tourism and a Sense of Place," Metro: Media and Education Magazine (2010); Erena Le Heron, "Placing Geographical Imagination in Film: New Zealand Filmmakers' Use of landscape," *New Zealand Geographer* 60, no. 1 (2004).
9 R. Robinson, *Georgia O'Keeffe: A life* (London, UK: Open Road Media, 2016), 237.P237.
10 Jennifer Vineyard, "Vince Gilligan at Breaking Bad Museum: That's a Bit of a Spoiler," *Vulture* (Online), *Vulture*, July 14, 2013, www.vulture.com/2013/07/vince-gilligan-breaking-bad-exhibit-tour-guide.html?mid=twitter_vulture.
11 Rough Guide Editors, "Rough Guides Exploring the Spectacular Settings of Breaking Bad," *Rough Guides* (Web post), *Rough Guides*, July 14, 2021, www.roughguides.com/article/breaking-bad-film-locations/.

12 Ning Wang, "Rethinking Authenticity in Tourism Experience," *Annals of Tourism Research* 26, no. 2 (1999), https://doi.org/10.1016/S0160-7383(98)00103-0.P349.
13 Dydia DeLyser, "Authenticity on the Ground: Engaging the Past in a California Ghost Town," *Annals of the Association of American Geographers* 89, no. 4 (1999): P610.
14 William Atkins, *The Immeasurable World: Journeys in Desert Places* (London, UK: Faber and Faber, 2018).
15 Geoff Nicholson, *Day Trips to the Desert Day Trips to the Desert: A Sort of Travel Book* (London, UK: Hodder & Stoughton Ltd, 1992).
16 Matthew Gandy, "The Cinematic Void: Desert Iconographies in Michelangelo Antonioni's Zabriskie Point," in *Landscape and Film*, ed. M. Lefebvre (Abingdon, UK: Routledge, 2007), P316.
17 G. Rose, *Visual Methodologies: An Introduction to the Interpretation of Visual Methodologies*, Reprinted Edition ed. (London, UK: Sage, 2012).
18 Beck, "Without Form and Void: the American Desert as Trope and Terrain."
19 Beck, "Without Form and Void: the American Desert as Trope and Terrain," P64.
20 E. Relph, *Place and Placelessness* (London, UK: Pion, 2008).
21 Jean Baudrillard, *In America* (London, UK: Verso, 1986), P5.
22 Roesch, *The Experiences of Film Location Tourists*, P151.
23 Gandy, "The Cinematic Void: Desert Iconographies in Michelangelo Antonioni's Zabriskie Point," P316.
24 Baudrillard, *In America*, P137.
25 Nicholson, *Day Trips to the Desert Day Trips to the Desert: A Sort of Travel Book*, P183.
26 J. J. Clark, "Breaking Bad as a Modern Western: Revising Frontier Myths of Masculinity, Savagery, and Empire" (Master of Arts (English) MA unpublished thesis, University of Colorado, 2014).
27 Gandy, "The Cinematic Void: Desert Iconographies in Michelangelo Antonioni's Zabriskie Point," P316.
28 Clark, "Breaking Bad as a Modern Western: Revising Frontier Myths of Masculinity, Savagery, and Empire," P17.
29 Baudrillard, *In America*, P137.
30 Robinson, *Georgia O'Keeffe: A life*, P327.
31 Beck, "Without Form and Void: the American Desert as Trope and Terrain," P72.
32 Baudrillard, *In America*, P30.
33 Nicholson, *Day Trips to the Desert Day Trips to the Desert: A Sort of Travel Book*. P8.
34 Baudrillard, *In America*, P5.
35 Beck, "Without Form and Void: the American Desert as Trope and Terrain," P65.
36 Beck, "Without Form and Void: the American Desert as Trope and Terrain," P63.
37 O. Moufakkir and N. Selmi, "Examining the Spirituality of Spiritual Tourists: A Sahara Desert Experience," *Annals of Tourism Research* 70 (2018), P9.

Bibliography

Atkins, William. *The Immeasurable World: Journeys in Desert Places*. London, UK: Faber and Faber, 2018.

Barthes, Roland. *Mythologies*. Translated by Annette Lavers. New York: Hill, 1972.

Baudrillard, Jean. *In America*. London, UK: Verso, 1986.

Beck, John. "Without Form and Void: The American Desert as Trope and Terrain." *Nepantla: Views from South* 2, no. 1 (2001): 63–83.

Beeton, Sue. "Landscapes as Characters: Film, Tourism and a Sense of Place." *Metro: Media and Education Magazine* (2010): 114.

———. *Travel, Tourism and the Moving Image*. Bristol, UK: Channel View, 2015.

"*Breaking Bad'*S Camera Work Perfects the Visual Recipe, Yo." *Wired*, April 6, 2012. Accessed July 14, 2022. www.wired.com/2012/08/the-look-the-feel-of-breaking-bad/.

Clark, J. J. "Breaking Bad as a Modern Western: Revising Frontier Myths of Masculinity, Savagery, and Empire." Master of Arts (English) MA unpublished thesis, University of Colorado, 2014.

"Contemporary Western: An Interview with Vince Gilligan." *Local IQ*, 2013. Accessed September 10, 2017. www.localiq.com/index.php?option=com_content&task=view&id=3019.

DeLyser, Dydia. "Authenticity on the Ground: Engaging the Past in a California Ghost Town." *Annals of the Association of American Geographers* 89, no. 4 (1999): 602–32.

Editors Rough Guide. "Rough Guides Exploring the Spectacular Settings of Breaking Bad." *Rough Guides*, July 14, 2021 (Web post). www.roughguides.com/article/breaking-bad-film-locations/.

Gandy, Matthew. "The Cinematic Void: Desert Iconographies in Michelangelo Antonioni's Zabriskie Point." In *Landscape and Film*, edited by M. Lefebvre, 340–57. Abingdon, UK: Routledge, 2007.

Heron, Erena Le. "Placing Geographical Imagination in Film: New Zealand Filmmakers' Use of Landscape." *New Zealand Geographer* 60, no. 1 (2004): 60–66.

Lovell, J., and S. Hitchmough. "Simulated Authenticity: Storytelling and Mythic Space on the Hyper-Frontier in Buffalo Bill's Wild West and Westworld." *Tourist Studies* 20, no. 4 (2020): 409–28.

Mills, B. "What Does It Mean to Call Television Cinematic?" In *Television Aesthetics and Style*, edited by S. Peacock and J. Jacobs, 57–66. London, UK: Bloomsbury, 2013.

Moufakkir, O., and N. Selmi. "Examining the Spirituality of Spiritual Tourists: A Sahara Desert Experience." *Annals of Tourism Research* 70 (2018): 108–19.

Nicholson, Geoff. *Day Trips to the Desert Day Trips to the Desert: A Sort of Travel Book*. London, UK: Hodder & Stoughton Ltd, 1992.

Relph, E. *Place and Placelessness*. London, UK: Pion, 2008.

Robinson, R. *Georgia O'Keeffe: A Life*. London, UK: Open Road Media, 2016.

Roesch, S. *The Experiences of Film Location Tourists*. Bristol, UK: Channel View Publications, 2009.

Rose, G. *Visual Methodologies: An Introduction to the Interpretation of Visual Methodologies*, reprinted ed. London, UK: Sage, 2012.

Vineyard, Jennifer. "Vince Gilligan at Breaking Bad Museum: That's a Bit of a Spoiler." *Vulture*, July 14, 2013 (Online). www.vulture.com/2013/07/vince-gilligan-breaking-bad-exhibit-tour-guide.html?mid=twitter_vulture.

Wang, Ning. "Rethinking Authenticity in Tourism Experience." *Annals of Tourism Research* 26, no. 2 (1999): 349–70. https://doi.org/10.1016/S0160-7383(98)00103-0.

9

NORDIC NOIR AND MISERABLE LANDSCAPE TOURISM

Can-Seng Ooi, Richard Ek, and Mia Larson

Introduction

Screen tourism highlights the interaction between destination development, tourism activities, and film as a pop-cultural phenomenon.[1] It is used increasingly by destinations to fabricate new spatial imageries, interwoven with pop-cultural narratives to boost and reposition the place.[2] In this context, a destination is a placement in which a screen story unfolds – and that screen experience is used to entice and attract visitors. There are many successful cases. Movies like *Crazy Rich Asians* and *Australia* showcase spectacular views of Singapore and Australia, respectively, for instance. The Singapore Tourism Board was pivotal in producing *Crazy Rich Asians*, and the movie was part of Singapore's tourism promotion strategy to market the island's sights and food cultures.[3] Another example is *Games of Thrones*, contributing to a substantial positive inflow of tourist arrivals to Dubrovnik.[4] These examples, together with other examples of screen tourism, such as *Harry Potter* and *Lord of the Rings*, are cases through which spectacular, mythical, and romantic films show beautiful landscapes to translate on-screen experiences into on-site tourist experiences.[5]

In contrast, there are also examples of films depicting depressing landscapes and these sites still attract high volumes of screen-induced tourists. For instance, in the *Twilight Saga*, Forks is portrayed as a rainy, cold, and dark place, a place Stephenie Meyer, author of the book series, characterized as necessary for her vampires and werewolves. In spite of the dreary portrayal of Forks, *Twilight* fans pilgrimage to Forks to experience and relive the story about a teenage girl falling in love with a vampire.[6] Another example is *Breaking Bad*; the successful Netflix series showing Albuquerque and the surrounding desert as the stage for drugs, crime, and violence, transforming this otherwise largely unremarkable place into a "must see"

DOI: 10.4324/9781003327585-9

destination for many.[7] Although there are a few examples of successful screen tourism showing dark and rainy scenes set in forgettable landscapes, there is a dearth of investigations into successful cases that do not celebrate beautiful places.[8]

The many nordic noir TV series are examples of screen tourism depicting the landscapes of the destinations in not an attractive way. They invoke the often cold, grey, and wet weather and dark and gloomy landscapes in Scandinavia in order to create an atmosphere of melancholia and dejection that often corresponds to the criminal narrative. These Scandinavian crime series still induce tourism, for example, *Wallander* in Ystad, *The Millennium Trilogy* in Stockholm, *Fjällbacka*, based on Camilla Läckberg's crime novels, and *Dicte*, based on Elsebeth Egholm's crime stories in Aarhus.[9]

Other examples of nordic noir movies and series generating international excitement are *The Killing*, *Borgen*, *Midnattssol*, and *Bron/Broen*. Tourists seek the nordic noir experience as they visit scenes from those series, and therefore, as Wade et al. note, there is also a potential for off-peak tourism.[10] So, the urban scenes and natural landscapes of nordic noir tourism contrast starkly against the light and cheerful images and green landscapes, typically put forward in the destination marketing of nordic destinations. These examples illustrate that the relationship between screen landscapes and tourism is more complex than selling the beautiful and positive. This chapter looks into one case of nordic noir tourism, *Bron/Broen*, to examine the relationship between screen tourism and "depressing" and "miserable" landscape. By doing so, we aim to offer three inter-related lessons.

The first lesson is on situating screen tourism depicting miserable landscapes in the wider creative tourism spectra,[11] acknowledging that attractions are not necessarily positive and beautiful – the morbid, terrifying and depressing are engaging in their own ways. By doing so, we highlight the diversity of interests and stimuli that attract visitors. Research on dark tourism, danger zone tourism, and tourists wanting to taste "scary food" have already revealed tourists' interests in death, tragedy, and the horrific, and their propensity for anxiety and adventure.[12]

While the beautiful and positive have been the focus in using movies and TV series to promote destinations, opportunities arising from the miserable and oppressive have mostly been ignored. In the context of creative tourism, Richards highlights that creativity can be understood as embedded in products (e.g. crafts and museums), generated by people (e.g. musicians and tourists engaging with public installations), found in processes (e.g. cooking and telling stories) and the environment (e.g. cultural and entertainment districts).[13]

But the interpretation of what constitutes as creative, meaningful, and worthy depends also on the tourists themselves. Tourists bring their knowledge, expectations, and preconceptions with them when they visit a destination and create their own visitor experience. While tourism and screen industries are considered part of the creative economy, movies and TV shows thrive with the beautiful and glorious, as well as the terrifying and depressing. As mentioned, dark tourism has shown that visitors may pursue the morbid and tragic, so how may miserable and depressing landscape portrayals on the screen induce tourism? Or do screen-induced tourists marginalize or ignore the depressing and miserable, and focus on other aspects of their screen experience?

The second lesson thus follows. All screen portrayals are situated in a place, and the backdrop may or may not be beautiful and spectacular. Staged sitcoms like *Friends* and *Two and a Half Men* are relatively bland but are popular because of the mere entertainment value they provide, and that induce visitors to some spots in New York and Malibu, respectively. A tourist taking the Manhattan tour connected to the show *Friends* constantly negotiate spaces of actual and virtual worlds imagined when watching the show, and in the process the everyday world of New York City is reconfigured as interesting as apartment buildings, parks, and non-celebrity chef restaurants become exhibits in an open-air museum of fictive worlds.[14] A show is multi-faceted, engaging us with sights and sounds, and are often enjoyed in comfort (whether it is at home or in a cinema), so it stimulates us emotionally and intellectually in various degrees. The idea is that the beautiful and spectacular induce travelling that privileges sight and the landscape, and neglects other aspects of the screen experience that stimulates us. These other aspects of the screen experience may also induce travelling such as scenes evoking strong emotions. An example is how *Twilight* fans travel to the outskirts of Vancouver to experience the very spot in the woods where the scene was shot when Edward broke up with Bella. Thus, supposedly unspectacular, boring, or even negative images may still stimulate the vivid imagination of viewers that encourage visitation.

The third lesson we aim to offer is a theorization of tourist experiences. Following from lessons one and two, this study locates affective imagination in a wider context, and suggests that the potential for screen-induced tourism extends beyond the visual and the beautiful, and into all other aspects of the screen experience. A more holistic and comprehensive appreciation of screen-induced tourism is needed.

Screen-induced tourism and tourism experiences

The complex nature of understanding and analysing tourism experiences is widely acknowledged.[15] With regard to screen-induced tourism, the literature has largely concentrated on the translation of the uplifting and pretty. But the movie industry has revealed that entertainment can come at different emotional levels. Comedies, serious dramas, horror flicks, and thrillers attract audiences. Different genres of screen production, including documentaries, talent shows, sitcoms, soap operas, and reality shows are popular. Not all of them portray their backdrops in spectacularly positive ways. Similarly, in tourism, ghost tours and bungee jumping are attractive for many, and to be avoided by many others. As already stated, the focus in screen-induced tourism has largely been on the beautiful, but it is important to also consider failures and unexpected screen-induced tourism projects. Nordic noir is an example of accidental film-induced success. In order to understand the complexity and diversity of screen-induced tourism experiences, this section looks at various approaches to understanding tourist experiences. There are at least six of them, all of which are interrelated (see also Figure 9.1).

Screen experiences shape the imagination

The first approach concentrates on the cognitive psychology of tourism experiences.[16] Such an approach deals with tourists' perceptions and how these affect their experiences.[17] Tourists' preconceived ideas and expectations affect how they consume, evaluate, and experience the product. This is often the most common way of understanding screen-induced tourism, as the stories and the scenery shape the imagination and supply the gaze lenses for the pursuance of visitor experiences. For instance, Platform 9¾ at King's Cross Station in London has attracted millions of *Harry Potter* movie fans, with visitors offered *Harry Potter's* London and Oxford tours.[18] There are also other unexpected aspects of movies that attract visitors, for instance, the otherwise ordinary subway grate at the corner of Lexington and 52nd Street in Manhattan have attracted visitors because it was featured in the movie *Seven Year Itch*; Marilyn Monroe was playfully pushing down her dress while air was gushing up from the grates. Tourists have layered their imagination from the movie scene onto the non-descript subway grate, so much so that it has a dedicated TripAdvisor review page.[19]

Screen portrayals contrast against the mundane

Related to the first approach on the importance of cognition in shaping tourism experiences, tourists are drawn to the contrast between the mundaneness of everyday life and what they can observe in a foreign place. John Urry highlighted that tourists experience a place by noticing things that are different from their daily life.[20] Tourists also lack the local knowledge to experience the destination in the same way as residents, which means that their tourist experiences reflect much of their own background and imagination. The screen experience transport audiences to another place that is different from their everyday existence. Screen portrayals of places and people can invent and perpetuate stereotypes and caricatures that nonetheless encourage interest, fantasy, and curiosity. For instance, without judgement here, many male visitors were not focusing on the cold and often grey weather in Copenhagen when they visited strip-joints in Denmark because they hoped to get pleasure from unrealistically stunning-looking blond girls as portrayed on screen and elsewhere.[21] In this instance, screen images introduce an imagined set of differences that contrasts to people's mundane everyday experiences that encourages visitations. Those visitors were often disappointed. Regardless, the background landscape is not central in their fantasy and imagination.

Staging and re-enacting a screen experience

Following from the first two approaches, a third perspective on situating screen-induced visitor experiences is on the staging of experiences.[22] Many visitors are inspired to recreate their screen experiences at spots where scenes were shot. Pine and Gilmore use the dramaturgical metaphor to argue that engaging experiences depend on the degree that people interact with a place. The right environment, props and cues will make people interact with the experience product at a deeper level, leading to that people feel that the product has been personalized to suit them and that they are more satisfied.[23] In this context, screen-inspirations can provide engaging plots for travellers to seek out the physical settings and props to perform and experience their stories.[24] The destination is the stage, and the screen story is the script. Bad weather and miserable landscapes can be part of the plot.

Screen-inspired personal quests

People have personal quests that sometimes are inspired through screen experiences. Therefore, they may desire certain tourist experiences that are personally beneficial, important or even considered necessary. Some of the benefits of such experiences include improving one's psychological mood and well-being, asserting one's self-identity and broadening one's horizon by learning about other places and cultures.[25] Pursuing tourism experiences can thus be a means to an end – a personal challenge – and screen experiences may inspire such pursuits. For instance, an aspiring food connoisseur will attempt to experience authentic local cuisines that can only be enjoyed at the places-of-origin.[26] Some personal emotional links may not be pleasurable but nonetheless important to pursue. For instance, many people who witnessed what they saw in the news on 11 September 2020, found a strong need to visit Ground Zero in New York to personally reflect on and memorialize that horrific day.[27] What people watch on TV and in the cinemas are tied to their interests and affinity, and the screen may nudge them further on their personal quests. Screen-induced tourist experience, in this context, is part of one's pursuit of one's own quests and realization of self and the beauty or lack of beauty of the landscape may or may not matter.

Screen-inspired optimal experiences

The fifth approach focuses on the state of mind and depth of experiential engagement.[28] Studies on experiential engagement concentrate on "optimal experiences," including the types of mental states people experience often described as special, meaningful, or out-of-the-ordinary.[29] These experiences are not only engaging but also emotionally intense, so people may "feel a sense of transcendence, as if the boundaries of the self had been expanded."[30] Tourists may feel that they are intellectually challenged, deeply attentive or that they lose their sense of time when they have optimal experiences.[31] Not all tourism experiences are optimal, however. Regardless, screen experiences may inspire some to pursue those challenges, such as base jumping in Norway or doing a meditation retreat in India, and the landscapes and locations may contribute to the experiences.

Screen experiences draw attention to selective aspects of a place

Drawing on the various dimensions in situating the tourist experience, the attention structure approach brings these dimensions together by pointing out that a tourist experience is shaped by what a person focuses on and what that person ignores or marginalizes.[32] Tourism mediators draw attention to features and stories to shape the tourist experience. Regardless, there are also distractions, such as bad weather and personal discomfort that will detract from a planned experience. As mentioned, people's own background knowledge and perceptions will influence one's experience.[33] In creating the Singapore brand experience, for instance, licensed tour guides and tourism attraction operators are given a narrative to understand multicultural Singapore and the blend of the old and new. Attention is put on constantly telling visitors how they should interpret the different things they see, such as if they see a modern building, they should also know that the building might be built on the ancient Chinese principles of geomancy.[34] The narrative draws attention to the promoted Singapore story. So, similarly in screen-induced tourism, visitors who are inspired or have internalized certain screen stories, will pursue the screen-influenced experience and pay attention to only certain things. They may concentrate on the story or a particular aspect of the show, while ignoring other aspects of the movie or at the site. With the Marilyn Monroe example given earlier, visitors ignore the fact that the original buildings in the movie have been replaced by skyscrapers and the *Seven Year Itch's* New York cityscape has largely been transformed.

Figure 9.1 summarizes the six ways of situating screen-induced tourist experiences. These different approaches highlight a range of different screen-induced tourist experiences and a plethora of factors that shape those experiences. People travel for different reasons even if the trip is screen-induced and at specific sites individuals interpret their experiences through their own screen lenses. The different approaches to understanding tourist experiences show that landscapes in screen experiences may or may not matter – the landscapes and weather conditions may even be marginalized or ignored because the story and characters are more captivating.

The different ways of situating screen-induced tourist experiences accentuate the importance of cognition and interpretation, that is, imagination and perception matter. Films, series, shows, and documentaries can provide the cognitive resources to interpret experiences. Similarly, the screen can inspire and motivate people to pursue certain tourism experiences whether they are intense, meaningful, romantic, fun, adventurous and/or scary. Destinations offer the place and the props to experience these. Tourism experiences are always mediated because visitors do not have the local knowledge, and even if they visit unprepared, they carry with them expectations and perceptions that they have garnered before coming. So, the mediation process is important in shaping tourist experiences,

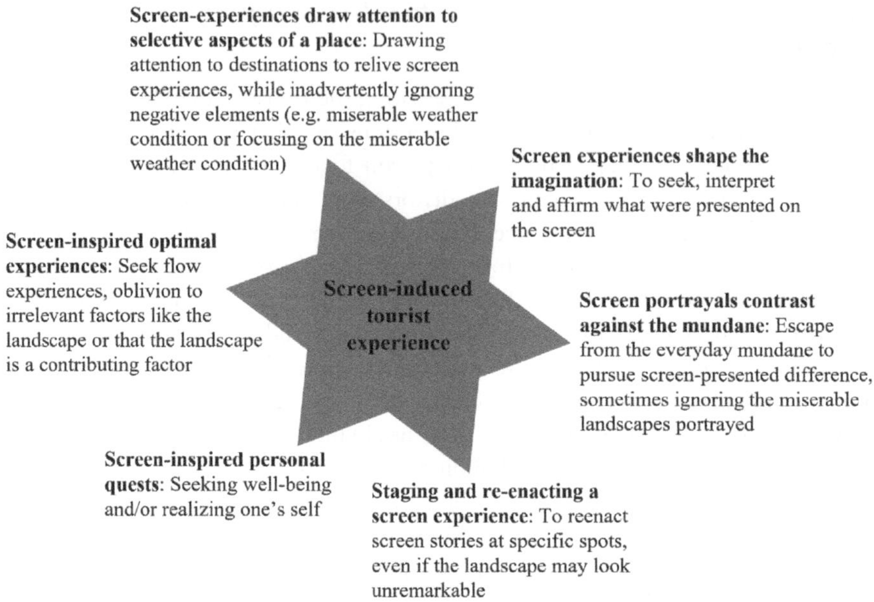

FIGURE 9.1 Six ways to situate screen-induced tourist experiences.

and therefore shows, series, films, and documentaries can inspire and draw attention to features and activities at destinations, and thus shape the tourist experience. Nevertheless, it is also important to examine what is forgotten, marginalized, and ignored when tourists experience a site since the screen-induced or screen-mediated tourist experience is selectively constructed. This is one of the points we will discuss when presenting our case – the original nordic television series *The Bridge* or *Bron/Broen*.

Miserable landscapes of *The Bridge*

For some time now, the increasingly popular nordic noir TV series have attracted pop-culture tourists to different urban and rural locations in the nordic countries. Scandinavian destination management organizations or DMO's have rapidly responded to this interest and branded specific cities and regions as well as arranged experiential tourist attractions and events, like "nordic noir tours" in Copenhagen, Denmark and Malmö, Sweden. Not an easy genre to define, nordic noir mainly refers to dramatic nordic crime fiction novels, TV series, and movies that have attracted an international audience (*Wallander*, *The Killing*, *The Girl with the Dragon Tattoo*, *Borgen*, and *Bron/Broen*). With a reference to the more established film noir, nordic noir blends fiction and reality, that is, specific cities and other surroundings in nordic countries play an active role in the unfolded fictional complex crime narratives. These places-as-character are in a destination branding context sites of potential value if the fictional work becomes successful.[35]

Bron/Broen stands out as one of the most popular and prolific recent nordic noir TV-series. It ran for four seasons (38 episodes during 2011–2018) concentrating on the police work of a Swedish female and a Danish male crime investigator. From the very beginning, it was intended to work as a typical nordic noir product, riding on the brand value of Danish (*The Killing*, *Borgen*) and Swedish (*Beck*, *Wallander*) templates but with a transnational urban agglomeration (Malmö in the South of Sweden and Copenhagen in Denmark, linked together by the Öresund Bridge) as the fictional-and-real location of the narrative.[36] The mix of well-established genres (film noir, cop genre, and serial killer genre) with some more recent dramaturgical approaches (unorthodox gender roles) packaged in a complex and meandering storyline obviously worked as the series have been sold to almost 200 countries and several adaptations have been made (*The Bridge*, taking place on the border between USA and Mexico and the French/British *The Tunnel* being the two first adaptations).[37] As the series includes and explicitly plays on a geopolitical complexity, a societal critique of increasing social polarities in the neoliberal city and identity struggles in a global contemporary milieu, it has attracted some interest of scholars in the social sciences and humanities.[38] It has, however, attracted less interest among tourism and destination branding scholars.

Bron/Broen focuses on the hunt for serial killers, one for each season. Initially the murders seem to be related to a societal mission (such as "eco-terrorism") but eventually unravels into a story about personal revenge. The central character, Saga Norén, is a skilful but socially odd police in Malmö. It is implied that she has Asperger's Syndrome,[39] but in some regards, she just acts out like the hard-boiled film noir masculine detective. Due to the particular and separate jurisdictions of the border, she is forced to work together with a male counterpart of the Copenhagen police force as the serial killer works on both sides of the Öresund – Martin Rohde (seasons one and two) and Henrik Sabroe (season three and four), both being broken and disoriented male figures even if for very different reasons (a gender reversal often found in nordic noir[40]). Besides that, the series follow the machinations of four serial killers, it also unfolds Saga's attempt to come to peace with her family-troubled past and find some sort of serenity. On a psychological plane, the series addresses issues such as responsibility, guilt, grief, betrayal, and existential confirmation. On a societal plane, the series relate to issues such as child abuse, sexual violence against women, homelessness, refugee crisis, and xenophobia, characteristic of nordic noir.

The stories of the different seasons unfold in a slow tempo, not following a straight trajectory. Action-driven sequences (hunting the serial killer, disarming bombs, etc.) intertwine with character-driven sections of the story, creating a meandering narrative constituted by several storylines that converge and diverge from each other in a complex temporal sequentiality, all juxtaposed in the conclusion of each season. Furthermore, there is intertextuality between the seasons, and incidents and the present absence of characters in the first seasons continue to influence the narratives in the later seasons, adding layers of complexity to the main characters. The past and the far away continues to, in manoeuvres of circular temporality and topological spatiality, partly chisel-out the events taking place in the current urban topography of Copenhagen and Malmö.

Essentially, *Bron/Broen* is also a tale of urban misery. Representations of the twin cities in the series communicate this misery effectively. The urban landscape is depicted in gloomy monochromatic colours, reduced to the degree that it becomes reminiscent of a black and white movie. The Öresund bridge is an uncanny melancholic architectural creature of darkness and neon that unites and separate Malmö and Copenhagen simultaneously.[41] The dominance of blue-grey cityscapes is interrupted by singular objects of colour, like Saga's yellow Porsche, that seem to be taken right out of an Edward Hopper painting, only adding an atmosphere of gloom and hopelessness into the stylized nordic design and architecture that dominates the in-door scenes.[42]

Darkness reigns in an endless Scandinavian winter, as the external environment enhances the disorientation, isolation, and apathy of the citizens of this nordic post-welfare dystopia.[43] The Scandinavian welfare state is still there, but not so much as a social safety-net for people but more like the grin of the Cheshire Cat, impossible to decipher as benevolent or malevolent.

Besides the hard-designed coldness of the affluent neoliberal city, the underbelly is repeatedly revealed – empty industrial store houses and abandoned fields, monotonous and derelict apartment building areas and multimodal traffic nodes of the restless urban landscape, inhabited by marginalized and abject individuals, acting in spite. A substantial amount of the scenes take place in motion, with the two police persons repeatedly driving over a desolated Öresund bridge at night

FIGURE 9.2 Dreary weather scene from Season 3 Episode 7 of Bron/Broen.

or at the fringes of the city, towards a trap or another victim of the serial killer.[44] A modern European urban agglomeration, based on infrastructure, restlessness, circulation, and stratified mobilities, shines out as a patchwork of non-places[45] in which even death is in motion (the halves of the first two victims of the serial killer in season one is found in the middle of the Öresund Bridge (Figure 9.2), on the very topographical border of Sweden and Denmark, juxtaposed into one Gothic Frankenstein-like creature[46]). *Bron/Broen* is truly a miserable urban tale, but nevertheless an engaging one.

Screen tourism development following *The Bridge*

Malmö Turism, the official DMO of Malmö, quickly recognized the screen tourism potential of *Bron/Broen*. During the years the show run, the series emerged into a prominent feature in the marketing of Malmö. Events were created in relation to season premiers, the website of Malmö Turism provided directions to the filming locations of the series, and *Bridge*-themed bus tours were arranged, and even a museum exhibition on a "Malmö that does not exist" were installed.[47] Visit Copenhagen, on the other hand, marketed tours in which participants could "Follow in the footsteps of Sarah Lund, Birgitte Nyborg or Saga Norén from the popular crime TV series *The Killing*, *Borgen*, and *The Bridge* around Copenhagen with nordic noir tours," and buy nordic noir tours t-shirts.[48]

The reason behind the popularity of the series is still being debated. But as we presented earlier, the series does not present Copenhagen and Malmö in a cheerful and bright manner. *Bron/Broen's* related products still attract a significant number of tourists. The miserable landscape in the series was not a hindrance to tourists enjoying their screen-induced experiences. Many tourists want to travel across the bridge, bringing their screen-induced imagination with them. They witness the places where the scenes were staged, and they could enact their own. In their experience, they are not doing the mundane – the dark and grey landscape and weather conditions may even enhance their screen-induced experiences.

Miserable landscapes in the screen-inspired tourist experience

Destinations, places, and attractions are represented in positive and beautiful ways in promotional campaigns. In the case of intentional tourism promotion through the screen, there are hardly exceptions (e.g. think of the epic landscapes in *Game of Thrones* and *Lord of the Ring*, and some nordic noir series). The British adaptation of *Wallander* includes many outdoor scenes of rolling cornfields and other pastoral panoramas of the south-eastern part of Skåne. In contrast, *Bron/Broen* presents a pseudo-real Öresund region in a miserable, gloomy, and uncanny way but still the series engaged viewers and visitors. The depiction of Malmö and Copenhagen in the series is dominated by darkness – it is a dark imaginarium of sorts that worked as a hyper-mediatized placemaking of the southern Scandinavian cross-border

Malmö-Copenhagen urban agglomeration.[49] The aesthetics of *Bron/Broen* is most of all gothic. The series outline the neoliberal urban agglomeration as a gothic space, a dark spatiality interrupted by – but also enhanced by – random colourful objects like Saga's car. The negative aesthetics of Gothicism offer a conduit for our distress about the contemporary engaging our imagination. More specifically, the darkness of *Bron/Broen* unfolds a cognitive and psychological space that invites the viewer to engage not only in a Scandinavian welfare state that cracks but also in the darkness of the human psyche. The (already more or less troubled) police officers Saga Norén, Martin Rodhe, and Henrik Sabroe become witnesses[50] to the unfolding of the seven sins (and more) in a decayed, haunted, and often deserted cityscape of two cities connected by a bridge that, as a liminal technological artefact, both unites and separates people and places symbolically and physically.[51] *Bron/Broen* (Figure 9.3) serves as an example of geopolitical television that creates a world that pulls some viewer not only to continue following the drama but also to experience the localization as tourists.[52]

In other words, tourists focus on the story and characters, making references to their own interpretation of their own situation matter. There are days in Copenhagen and Malmö that are bright, dry, and stunning. *Bron/Broen* tours and experiences will then take on a different perspective from what visitors see on television although the stories and characters remain in their imagination. In their own creative ways, visitors create their own screen-inspired experiences, regardless of the weather, whether the urban decay is obvious or not. So, within the wider creative tourism spectra, destinations do not need to be idyllic to attract people's attention and curiosity. Visitors can be creative in their imagination of these destinations in spite of the bad weather conditions and unspectacular landscape. Instead, they embrace the depressive and miserable into their imagination.

FIGURE 9.3 A featureless aerial night scene from Season 3 Episode 6 of Bron/Broen.

Miserable landscapes and melancholia – on- and off-screen experience

The unflattering screen portrayal of Öresund becomes a part of a vivified imagination that encourages tourists to physically experience Malmö and Copenhagen in the footprints of Saga Norén. The screen experience resonates with audiences, because it taps into their personal encounters with their own encounters with contemporary society and its discontents. The atmosphere of melancholia, an underestimated and intense emotion, in the series appeal to peoples' awareness and feeling that contemporary sociality can create a sense of despair and alienation. In the era of the Last Man,[53] disorientation, restlessness, and hopelessness reign materialized in a city of relentless mobility and circulation. Whether one is a citizen, a tourist, or a vagabond, one is on the move in the city of liquid modernity.[54]

The carefully commodified and packaged Scandinavian melancholia of *Bron/Broen* offers a narrative that explores liminality (and existentialism) in a borderless world still heavily stratified geographically, socially and economically.[55] For Hansen and Waade,[56] the commodified melancholic landscape is one of nordic noir's most characteristic traits, expressed and enhanced through the seasons and the topography of the nordic landscapes. In *Bron/Broen* murky and bleak weather dominates, folding the high buildings in Öresund (the Öresund Bridge and the spectacular skyscraper and landmark Turning Torso in Malmö), giving them the impression of being hazy technological artefacts in-between dimensions or worlds. Perhaps the Öresund Bridge (Figure 9.4) and the Turning Torso even can be interpreted as akin to "Cthulhuian" gods in Lovecraft's mythological universe that require to

FIGURE 9.4 Misty, grey, and melancholic view of the Øresund Bridge, scene from Season 1 Episode 6 of Bron/Broen.

be visited. In this reading the fiction tourists, like neopagan travellers,[57] search for quasi-religious meaning, and personal and cultural diagnosis by visiting the sites of the series.[58] The tourists travelling in the footsteps of Saga Norén contemplate on different location sites, depicting the bridge construction as if it was a piece of art or a miracle – taking photographs and talking about the bridge and how the view changes according to the sun and the weather conditions.[59]

These stories and personal sense of loss, depression, and anxiety draw millions to the screen. Many nordic noir series, including *Bron/Broen* tap into that sense of melancholy and darkness with the help of depressing and miserable landscapes. The miserable landscapes are part of the screen narrative and experience. In spite of the intentional presentation of a depressed place, the stories and characters come to life. Visitors embrace the story, and experience the miserable, albeit in their own comfortable, safe, and self-affirming way. They may reaffirm the screen-experiences, reaffirm their sense of melancholy, or indulge in self-pity, as they soak in the depressing atmosphere surrounded by modern technological advancements in these two cities (as seen on television). To be motivated to visit these two cities because of their screen-experience may also mean that they want to re-live that screen-induced melancholy, but in their own comfortable, safe, and even playful way.

Theorizing the experiential gap in the screen-inspired tourist experiences

Following from these last two points – visitors may ignore the miserable landscapes when they enjoy their screen-inspired tourist experiences, and visitors may embrace the miserable landscapes because it is part of the screen experience – we would like to situate the nordic noir tourist experience in a wider theoretical context. Figure 9.5 is built on Figure 9.1 and accentuates the different dimensions to the nordic noir tourist experience.

There will always be a gap between the screen experience and the visit, because the characters would not be there, the weather conditions may be significantly different, and the doctored screen views are different from the actual place. Furthermore, each person would have their own interpretation and experience of what they saw on TV or in the cinema. Figure 9.5 highlights that spectrum of nordic noir tourist experiences. The pursuit of the nordic noir experience may just be incidental and is only one experience in a larger package of experiences tourists are seeking.

The most common approaches to promoting screen-induced tourism are to present pleasant, beautiful, and glorious landscapes to attract tourists. That focus is on shaping and stimulating potential visitor imaginations, inspiring them to want to stage and enact their screen experiences on site, and to provide them with an experience that is different from their mundane everyday activities. By concentrating on the positive, the screen inspires and makes people want to travel or to have an upbeat experience at a certain site – to relive positive screen experiences.

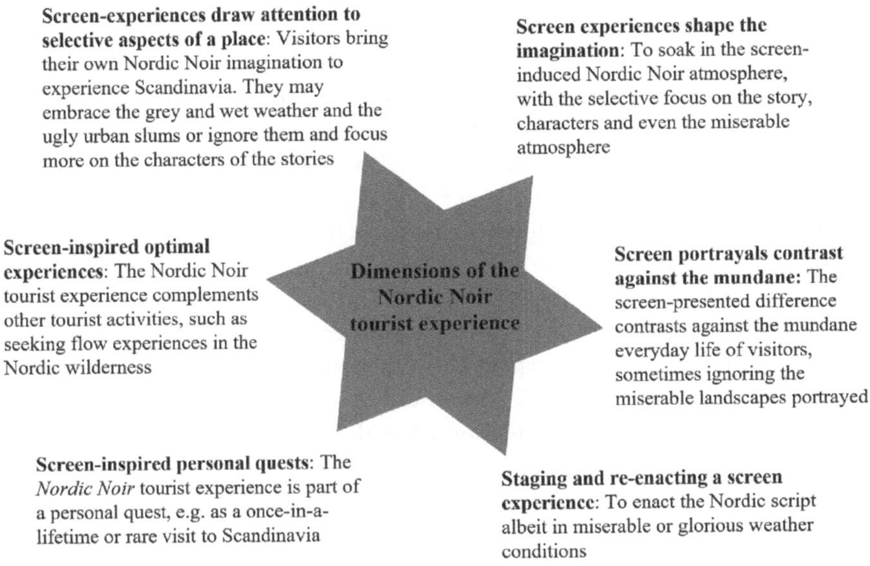

FIGURE 9.5 Situating the nordic noir tourist experience.

Our focus on miserable landscapes in nordic noir series and movies challenge that convention. Tourists fill the experiential gap between the screen portrayal and the visit, and as already discussed visitors ignore or marginalize certain dimensions when they experience a site. While there is a site for them to enact their imagination, their imagination is still selectively constructed. They can ignore elements that would otherwise detract them from what they desire, for instance negotiating with other tourists who may get in their way of their enjoyment or accept the fate that the weather conditions are not ideal for their visits.

Tourists can embrace and seek out the dark and miserable landscapes that are portrayed on the screen and want to relive that during their trip. So, their focus is different since compelling stories and strong characters are likely to draw audiences into an engaging screen universe even though it may be oppressive and tragic. Paradoxically, people seem to find pleasure in those screen experiences, and that can be translated into travel experiences. Following from dark tourism, danger zone tourism, and scary food experiences, we are suggesting that there is space for miserable landscape tourism.

Figure 9.1 also tells us that many screen-inspired experiences are part of a longer trip, and these screen-induced experiences may not be the main reason for travel. In a single trip to Scandinavia, those experiences are complemented by other experiences such as the pursuit for adventures and optimal experiences or seeking transformative encounters in the wilderness. In this context, the miserable landscapes experience may just be part of the chain of tourist experiences during a longer trip.

Conclusion

We promised three lessons, the first is to situate screen-induced tourism in the wider creative tourism economy showing that the film and tourism sectors complement each other and screen-related products and services, such as souvenirs and tours, are part of creative tourism. Similarly, enactment and story-telling are co-creative processes enhancing the screen-inspired tourist experience. The tourists themselves (in co-creation with other agents, such as guides and other tourists) are agents generating these creative processes and forming experiences. We argue that the setting, with the help of screen-inspirations, make the destination a creative environment. This lesson on creative tourism become even more evident when focusing on miserable screen experiences – how do sad, shocking, terrifying, and tragic movies and TV series sell? Can miserable weather conditions and unflattering landscapes become tourist attentions? Yes, they can, and we reiterate our arguments in the second lesson.

Just as tourism mediators select, accentuate, and aestheticize stories, artefacts and sites into tourist attractions, visitors themselves select, accentuate, and aestheticize the elements they focus upon, so as to generate their screen-inspired experiences. They may still be disappointed, as conditions may not be correct. The issue is that people selectively pay attention to elements to build their tourist experience and they may also ignore elements that may detract them from their goals. In this context, in using the attention structure approach, guides and site operators must be at hand to minimize distractions and anxiety that will befuddle the screen-inspired tourist experience. But at the same time, the less pleasant can also be weaved into the tourist experience, as miserable weather conditions and ugly urban sites become part of the story and experience. Melancholy is induced in many nordic noir series and resonates with audiences, so the screen-induced sense of melancholy is again induced by a visit to the destination.

Finally, from the first two lessons, the experiential gap between the screen experience and the screen-inspired tourist experience leaves room for manoeuvre and manipulation. The beautiful and the ugly, the cheerful and the miserable entice visitors in their own ways. But because there is a series of tourist experiences, the variety allows for visitors to recover from intense experiences. Just like a movie where there is variation of intensity and emotions, a tourism trip should also be understood in that context. Screen-induced tourist activities and experiences contribute to that variety, and therefore, there is a place for miserable landscapes tourism.

We have introduced an eclectic framework of bringing various tourist experience theories together. By doing so, we have highlighted that miserable landscape can be selectively ignored during the encounter, or that unflattering landscapes can be part of the screen experience and tourists sometimes pursue the miserable landscapes and the screen-inspired sense of misery.

Notes

1 Mia Larson, Christine Lundberg, and Maria Lexhagen, "Thirsting for Vampire Tourism: Developing Pop Culture Destinations," *Journal of Destination Marketing & Management* 2, no. 2 (June 2013): 74–84, https://doi.org/10.1016/j.jdmm.2013.03.004.
2 Szilvia Gyimóthy et al., "Popculture Tourism: A Research Manifesto," in *Tourism Research Frontiers: Beyond the Boundaries of Knowledge*, ed. Donna Chambers and Tijana Rakić (Bingley: Emerald, 2015), 13–26.
3 Nicole Tarulevicz and Can-Seng Ooi, "Food Safety and Tourism in Singapore: Between Microbial Russian Roulette and Michelin Stars," *Tourism Geographies* 23, no. 4 (July 1, 2021): 810–32, https://doi.org/10.1080/14616688.2019.1654540.
4 Marina Tkalec, Ivan Zilic, and Vedran Recher, "The Effect of Film Industry on Tourism: Game of Thrones and Dubrovnik," *International Journal of Tourism Research* 19, no. 6 (November 2017): 705–14, https://doi.org/10.1002/jtr.2142.
5 Anne Buchmann, "Planning and Development in Film Tourism: Insights Into the Experience of Lord of the Rings Film Guides," *Tourism and Hospitality Planning & Development* 7, no. 1 (February 2010): 77–84, https://doi.org/10.1080/14790530903522648; Jane Lovell, "Fairytale Authenticity: Historic City Tourism, Harry Potter, Medievalism and the Magical Gaze," *Journal of Heritage Tourism* 14, no. 5–6 (November 2, 2019): 448–65, https://doi.org/10.1080/1743873X.2019.1588282.
6 Larson, Lundberg, and Lexhagen, "Thirsting for Vampire Tourism: Developing Pop Culture Destinations."
7 Rodanthi Tzanelli and Majid Yar, "Breaking Bad, Making Good: Notes on a Televisual Tourist Industry," *Mobilities* 11, no. 2 (March 14, 2016): 188–206, https://doi.org/10.1080/17450101.2014.929256.
8 Sheela Agarwal and Gareth Shaw, *Heritage, Screen and Literary Tourism* (Bristol: Channel View Publications, 2018); Can-Seng Ooi, "Production and Consumption of Stories, Images and Fantasies: Heritage, Screen and Literary Tourism," *Annals of Leisure Research* 22, no. 3 (May 27, 2019): 408–10, https://doi.org/10.1080/11745398.2018.1506992.
9 Anne Marit Waade et al., "Nordic Landscapes in Collaborative Place-Making Interventions," in *The Nordic Wave in Place Branding*, ed. Cecilia Cassinger, Andrea Lucarelli, and Szilvia Gyimóthy (Cheltenham: Edward Elgar, 2019), 91–108.
10 Ibid.
11 Greg Richards, "Creativity and Tourism: The State of the Art," *Annals of Tourism Research* 38, no. 4 (October 2011): 1225–53, https://doi.org/10.1016/j.annals.2011.07.008.
12 Kathleen Adams, "Danger-Zone Tourism: Prospects and Problems for Tourism in Tumultuous Times," in *Interconnected Worlds: Tourism in Southeast Asia*, ed. Peggy Teo, Tou Chuang Chang, and Kong Chong Ho (London: Pergamon, 2001), 265–281; Szilvia Gyìmothy and Reidar Johan Mykletun, "Scary Food: Commodifying Culinary Heritage as Meal Adventures in Tourism," *Journal of Vacation Marketing* 15, no. 3 (July 1, 2009): 259–73, https://doi.org/10.1177/1356766709104271.
13 Richards, "Creativity and Tourism."
14 Leshu Torchin, "Location, Location, Location: The Destination of the Manhattan TV Tour," *Tourist Studies* 2, no. 3 (December 11, 2002): 247–66, https://doi.org/10.1177/14687976020023002.
15 Bong Koo Lee and C. Scott Shafer, "The Dynamic Nature of Leisure Experience: An Application of Affect Control Theory," *Journal of Leisure Research* 34, no. 3 (2002): 290–310; Richard Prentice and Vivien Andersen, "Interpreting Heritage Essentialisms: Familiarity and Felt History," *Tourism Management* 28, no. 3 (June 2007): 661–76, https://doi.org/10.1016/j.tourman.2006.02.008.
16 Can-Seng Ooi, "Tourist Experiences as Attention Products," in *Routledge Handbook of the Tourist Experience*, ed. Richard Sharpley (London: Routledge, 2022), 113–27, https://doi.org/10.4324/9781003219866-11; Lee and Shafer, "The Dynamic Nature of Leisure Experience: An Application of Affect Control Theory"; Roger C. Mannell

and Seppo E. Iso-Ahola, "Psychological Nature of Leisure and Tourism Experience," *Annals of Tourism Research* 14, no. 3 (1987): 314–31, www.sciencedirect.com/science/article/B6V7Y-46BHW9J-2B/2/aa6fc26ca237ec44ab5d5c03662aba62.

17 Gordon Waitt, "Consuming Heritage: Perceived Historical Authenticity," *Annals of Tourism Research* 27, no. 4 (2000): 835–62, www.sciencedirect.com/science/article/B6V7Y-44BMTRD-1/2/c308f2c680b6ca470da2ea78c57de346; Alison J. McIntosh and C. Prentice, "Affirming Authenticity: Consuming Cultural Heritage," *Annals of Tourism Research* 26, no. 3 (1999): 589–612, www.sciencedirect.com/science/article/B6V7Y-3X6JHDP-6/2/4280b514c2f91c9a7b04a542f664f163.

18 James Cateridge, "Deep Mapping and Screen Tourism: The Oxford of Harry Potter and Inspector Morse," *Humanities* 4, no. 3 (August 19, 2015): 320–33, https://doi.org/10.3390/h4030320; Christina Lee, "'Have Magic, Will Travel': Tourism and Harry Potter's United (Magical) Kingdom," *Tourist Studies* 12, no. 1 (April 20, 2012): 52–69, https://doi.org/10.1177/1468797612438438.

19 Tripadvisor, "Marilyn Monroe's Subway Grate," Tripadvisor, 2020, www.tripadvisor.com.au/Attraction_Review-g60763-d10533507-Reviews-Marilyn_Monroe_s_-subway_grate-New_York_City_New_York.html.

20 John Urry, *The Tourist Gaze: Leisure and Travel in Contemporary Societies* (London: Sage Publications, 1990).

21 Can-Seng Ooi, "The Danish Seduction of the China Outbound Tourism Market: New Issues for Tourism Research," in *China Business Opportunities in a Globalizing Economy* (Copenhagen: Copenhagen Business School Press, 2008), 171–91.

22 B. Joseph Pine and James H. Gilmore, *The Experience Economy: Work Is Theatre and Every Business a Stage* (Boston: Harvard Business School Press, 1999).

23 Ibid.

24 Sangkyun Kim, "Extraordinary Experience: Re-Enacting and Photographing at Screen Tourism Locations," *Tourism and Hospitality Planning & Development* 7, no. 1 (February 2010): 59–75, https://doi.org/10.1080/14790530903522630.

25 Lee and Shafer, "The Dynamic Nature of Leisure Experience: An Application of Affect Control Theory"; Richard C. Prentice, Stephen F. Witt, and Claire Hamer, "Tourism as Experience: The Case of Heritage Parks," *Annals of Tourism Research* 25, no. 1 (1998): 1–24, https://doi.org/10.1016/S0160-7383(98)00084-X.

26 Can-Seng Ooi and Jesper Strandgaard Pedersen, "In Search of Nordicity: How New Nordic Cuisine Shaped Destination Branding in Copenhagen," *Journal of Gastronomy and Tourism* 2, no. 4 (October 17, 2017): 217–31, https://doi.org/10.3727/216929717X15046207899375.

27 Can-Seng Ooi and Ana María Munar, "Digital Social Construction of a Tourist Site," in *Tourism Social Media: Transformations in Identity, Community and Culture*, ed. Ana María Munar, Szilvia Gyimóthy, and Liping A. Cai, Vol. 18 (Bingley: Emerald, 2013), 159–75, https://doi.org/10.1108/S1571-5043(2013)0000018011.

28 Gary D. Ellis, Catherine Morris, and Judith E. Voelkl, "Measurement and Analysis Issues With Explanation of Variance in Daily Experience Using the Flow Model," *Journal of Leisure Research* 26, no. 4 (1994): 337–56.

29 Gordon J. H. Walker, R. Bruce Hull, and Joseph W. Roggenbuck, "On-Site Optimal Experiences and Their Relationship to Off-Site Benefits," *Journal of Leisure Research* (National Recreation & Park Association, 1998).

30 Mihalyi Csikszentmihalyi, *Flow: The Psychology of Optimal Experience* (New York: Harper & Row, 1990).)

31 Ellis, Morris, and Voelkl, "Measurement and Analysis Issues With Explanation of Variance in Daily Experience Using the Flow Model"; Walker, Hull, and Roggenbuck, "On-Site Optimal Experiences and Their Relationship to Off-Site Benefits."

32 Thomas H. Davenport and John C. Beck, *The Attention Economy, Understanding the New Currency of Business* (Boston: Harvard Business School Press, 2001); Ooi, "Tourist Experiences as Attention Products."

33 Yue Ma, Can-Seng Ooi, and Anne Hardy, "Chinese Travelling Overseas and Their Anxieties," in *Asian Cultures and Contemporary Tourism. Perspectives on Asian Tourism*, ed. Elaine Yang and Catheryn Khoo-Lattimore (Singapore: Springer, 2018), 201–20, https://doi.org/10.1007/978-981-10-7980-1_11; Ooi, "Tourist Experiences as Attention Products."

34 Can-Seng Ooi, "Branding and the Accreditation Approach: Singapore," in *Destination Brands: Managing Place Reputation*, ed. N. J. Morgan, A. Pritchard, and Roger Pride (Oxford: Elsevier, 2011), 185–96.

35 Kim Toft Hansen and Anne Marit Waade, *Locating Nordic Noir: From Beck to The Bridge* (Basingstoke: Palgrave Macmillan, 2017).

36 Toft Hansen and Waade, *Locating Nordic Noir: From Beck to The Bridge*, 269–271; Annette Hill, "Push – Pull Dynamics: Producer and Audience Practices for Television Drama Format The Bridge," *Television & New Media* 17, no. 8 (December 1, 2016): 754–68, https://doi.org/10.1177/1527476416658131.

37 Tobias Steiner, "Bron/Broen, the Pilot as Space between Cultures, and (Re)Negotiations of Nordic noir," in *The Scandinavian Invasion: The Nordic noir Phenomenon and Beyond*, ed. Richard McCulloch and William Proctor (Bern: Peter Lang, forthcoming).

38 Tina Askanius, "Engaging with The Bridge : Cultural Citizenship, Cross-Border Identities and Audiences as 'Regionauts,'" *European Journal of Cultural Studies* 22, no. 3 (June 8, 2019): 271–90, https://doi.org/10.1177/1367549417722093; Isadora Garcia Avis, "Adapting Landscape and Place in Transcultural Remakes: The Case of Bron | Broen, The Bridge and The Tunnel," *Series: International Journal of TV Serial Narratives* 1, no. 2 (2015): 127–38; Caius Dobrescu, "The Limit as Centre: Some Considerations on the Political Imagination of the In-Between, Starting from the Central Symbol of the Crime Series Bron/Broen – The Bridge," *Acta Universitatis Sapientiae, Philologica* 11, no. 1 (November 1, 2019): 7–20, https://doi.org/10.2478/ausp-2019-0001; Emily Gray, "In/between Places: Connection and Isolation in The Bridge," *Aeternum: The Journal of Contemporary Gothic Studies* 1, no. 1 (2014): 73–85. Anders Marklund, "Foreign Influences on Nordic (noir) Borderlands," *European Journal of Scandinavian Studies* 49, no. 1 (April 24, 2019): 178–96, https://doi.org/10.1515/ejss-2019-0011; Robert A. Saunders, "A Dark Imaginarium: The Bridge, Malmö and the Making of a 'Non-Existent' Place," *Journal of Urban Cultural Studies* 4, no. 3 (September 1, 2017): 361–85, https://doi.org/10.1386/jucs.4.3.361_1; Robert A. Saunders, "Geopolitical Television at the (b)order: Liminality, Global Politics, and World-Building in The Bridge," *Social & Cultural Geography* 20, no. 7 (September 2, 2019): 981–1003, https://doi.org/10.1080/14649365.2017.1404122; Robert A. Saunders, "Small Screen IR: A Tentative Typology of Geopolitical Television," *Geopolitics* 24, no. 3 (May 27, 2019): 691–727, https://doi.org/10.1080/14650045.2017.1389719.

39 Avis, Isadora, "Adapting Landscape and Place in Transcultural Remakes: The Case of Bron | Broen, The Bridge and The Tunnel," 133.

40 Gray, "In/between Places: Connection and Isolation in The Bridge," 77.

41 Saunders, "A Dark Imaginarium: The Bridge, Malmö and the Making of a 'Non-Existent' Place," 368–70.

42 Toft Hansen and Waade, *Locating Nordic Noir: From Beck to The Bridge*, 275.

43 Askanius, "Engaging With The Bridge: Cultural Citizenship, Cross-Border Identities and Audiences as 'Regionauts,'" 277–80; Gray, "In/between Places: Connection and Isolation in The Bridge."

44 Isadora Avis, "Adapting Landscape and Place in Transcultural Remakes: The Case of Bron | Broen, The Bridge and The Tunnel," 133.

45 Marc Augé, *Non-Places: Introduction To Anthropology Of Supermodernity* (London: Verso, 1995).

46 Gray, "In/between Places: Connection and Isolation in The Bridge," 80.

47 Saunders, "A Dark Imaginarium: The Bridge, Malmö and the Making of a 'Non-Existent' Place," 272–77.

48 Visit Copenhagen, "Nordic noir tours," accessed May 28, 2020, www.visitcopenhagen.com/copenhagen/planning/nordic-noir-tours-gdk918519.
49 Saunders, "A Dark Imaginarium: The Bridge, Malmö and the Making of a 'Non-Existent' Place," 361.
50 Toft Hansen and Waade, *Locating Nordic Noir: From Beck to The Bridge*, 85.
51 Gray, "In/between Places: Connection and Isolation in The Bridge."
52 Saunders, "Geopolitical Television at the (b)order: Liminality, Global Politics, and World-Building in The Bridge", 985.
53 Francis Fukujama, *The End of History and the Last Man* (New York: Free Press, 1992).
54 Zygmunt Bauman, *Globalization: The Human Consequences* (London: Polity Press, 1998); Zygmunt Bauman, *Liquid Modernity* (Cambridge: Polity Press, 2000); Zygmunt Bauman, *Liquid Times. Living in an Age of Uncertainty* (Cambridge: Polity Press, 2007).
55 Gray, 2014.
56 Toft Hansen and Waade, "*Locating Nordic Noir: From Beck to The Bridge*", 87–90.
57 Robert A. Saunders, "Pagan Places: Towards a Religiogeography of Neopaganism," *Progress in Human Geography* 37, no. 6 (2012): 786–810.
58 Saunders, "A Dark Imaginarium: The Bridge, Malmö and the Making of a 'Non-Existent' Place", 378; Toft Hansen and Waade, "*Locating Nordic Noir: From Beck to The Bridge*", 8.
59 Toft Hansen and Waade, "*Locating Nordic Noir: From Beck to The Bridge*", 8.

Bibliography

Adams, Kathleen. *Danger-Zone Tourism: Prospects and Problems for Tourism in Tumultuous Times*. Edited by Peggy Teo, T. C. Chang, and K. C. Ho. London: Pergamon, 2001.

Agarwal, Sheela, and Gareth Shaw. *Heritage, Screen and Literary Tourism*. Bristol: Channel View Publications, 2018.

Askanius, Tina. "Engaging with The Bridge : Cultural Citizenship, Cross-Border Identities and Audiences as 'Regionauts'." *European Journal of Cultural Studies* 22, no. 3 (June 8, 2019): 271–90. https://doi.org/10.1177/1367549417722093.

Augé, Marc. *Non-Places: Introduction to Anthropology of Supermodernity*. London: Verso, 1995.

Avis, Isadora, Garcia. "Adapting Landscape and Place in Transcultural Remakes: The Case of Bron | Broen, the Bridge and The Tunnel." *Series: International Journal of TV Serial Narratives* 1, no. 2 (2015): 127–38.

Buchmann, Anne. "Planning and Development in Film Tourism: Insights Into the Experience of Lord of the Rings Film Guides." *Tourism and Hospitality Planning & Development* 7, no. 1 (February 2010): 77–84. https://doi.org/10.1080/14790530903522648.

Cateridge, James. "Deep Mapping and Screen Tourism: The Oxford of Harry Potter and Inspector Morse." *Humanities* 4, no. 3 (August19, 2015): 320–33. https://doi.org/10.3390/h4030320.

Csikszentmihalyi, Mihalyi. *Flow: The Psychology of Optimal Experience*. New York: Harper & Row, 1990.

Davenport, Thomas H., and John C. Beck. *The Attention Economy, Understanding the New Currency of Business*. Boston: Harvard Business School Press, 2001.

Dobrescu, Caius. "The Limit as Centre: Some Considerations on the Political Imagination of the In-Between, Starting from the Central Symbol of the Crime Series Bron/Broen – The Bridge." *Acta Universitatis Sapientiae, Philologica* 11, no. 1 (November 1, 2019): 7–20. https://doi.org/10.2478/ausp-2019-0001.

Ellis, Gary D., Catherine Morris, and Judith E. Voelkl. "Measurement and Analysis Issues With Explanation of Variance in Daily Experience Using the Flow Model." *Journal of Leisure Research* 26, no. 4 (1994): 337–56.

Gray, Emily. "In/between Places: Connection and Isolation in the Bridge." *Aeternum: The Journal of Contemporary Gothic Studies* 1, no. 1 (2014): 73–85.

Gyimóthy, Szilvia, Christine Lundberg, Maria Lexhagen, and Mia Larson. "'Popculture Tourism: A Research Manifesto'." In *Tourism Research Frontiers: Beyond the Boundaries of Knowledge*, edited by Donna Chambers and Tijana Rakić, 13–26. Bingley: Emerald, 2015.

Gyìmothy, Szilvia, and Reidar Johan Mykletun. "Scary Food: Commodifying Culinary Heritage as Meal Adventures in Tourism." *Journal of Vacation Marketing* 15, no. 3 (July 1, 2009): 259–73. https://doi.org/10.1177/1356766709104271.

Hill, Annette. "Push – Pull Dynamics: Producer and Audience Practices for Television Drama Format the Bridge." *Television & New Media* 17, no. 8 (December 1, 2016): 754–68. https://doi.org/10.1177/1527476416658131.

Kim, Sangkyun. "Extraordinary Experience: Re-Enacting and Photographing at Screen Tourism Locations." *Tourism and Hospitality Planning & Development* 7, no. 1 (February 2010): 59–75. https://doi.org/10.1080/14790530903522630.

Larson, Mia, Christine Lundberg, and Maria Lexhagen. "Thirsting for Vampire Tourism: Developing Pop Culture Destinations." *Journal of Destination Marketing & Management* 2, no. 2 (June 2013): 74–84. https://doi.org/10.1016/j.jdmm.2013.03.004.

Lee, Christina. "'Have Magic, Will Travel': Tourism and Harry Potter's United (Magical) Kingdom." *Tourist Studies* 12, no. 1 (April 20, 2012): 52–69. https://doi.org/10.1177/1468797612438438.

Lee, BongKoo, and C. Scott Shafer. "The Dynamic Nature of Leisure Experience: An Application of Affect Control Theory." *Journal of Leisure Research* 34, no. 3 (2002): 290–310.

Lovell, Jane. "Fairytale Authenticity: Historic City Tourism, Harry Potter, Medievalism and the Magical Gaze." *Journal of Heritage Tourism* 14, no. 5–6 (November 2, 2019): 448–65. https://doi.org/10.1080/1743873X.2019.1588282.

Ma, Yue, Can-Seng Ooi, and Anne Hardy. "Chinese Travelling Overseas and Their Anxieties." In *Asian Cultures and Contemporary Tourism. Perspectives on Asian Tourism*, edited by Elaine Yang and Catheryn Khoo-Lattimore, 201–20. Singapore: Springer, 2018. https://doi.org/10.1007/978-981-10-7980-1_11.

Mannell, Roger C., and Seppo E. Iso-Ahola. "Psychological Nature of Leisure and Tourism Experience." *Annals of Tourism Research* 14, no. 3 (1987): 314–31. www.sciencedirect.com/science/article/B6V7Y-46BHW9J-2B/2/aa6fc26ca237ec44ab5d5c03662aba62.

Marklund, Anders. "Foreign Influences on Nordic (noir) Borderlands." *European Journal of Scandinavian Studies* 49, no. 1 (April 24, 2019): 178–96. https://doi.org/10.1515/ejss-2019-0011.

McIntosh, Alison J., and Richard C. Prentice. "Affirming Authenticity: Consuming Cultural Heritage." *Annals of Tourism Research* 26, no. 3 (1999): 589–612. www.sciencedirect.com/science/article/B6V7Y-3X6JHDP-6/2/4280b514c2f91c9a7b04a542f664f163.

Ooi, Can-Seng. "Branding and the Accreditation Approach: Singapore." In *Destination Brands: Managing Place Reputation*, edited by N. J. Morgan, A. Pritchard, and Roger Pride, 185–96. Oxford: Elsevier, 2011.

———. "The Danish Seduction of the China Outbound Tourism Market: New Issues for Tourism Research." In *China Business Opportunities in a Globalizing Economy*, 171–91. Copenhagen: Copenhagen Business School Press, 2008.

———. "Production and Consumption of Stories, Images and Fantasies: Heritage, Screen and Literary Tourism." *Annals of Leisure Research* 22, no. 3 (May 27, 2019): 408–10. https://doi.org/10.1080/11745398.2018.1506992.

———. "Tourist Experiences as Attention Products." In *Routledge Handbook of the Tourist Experience*, edited by Richard Sharpley, 113–27. London: Routledge, 2022. https://doi.org/10.4324/9781003219866-11.

Ooi, Can-Seng, and Ana María Munar. "Digital Social Construction of a Tourist Site." In *Tourism Social Media: Transformations in Identity, Community and Culture*, edited by Ana María Munar, Szilvia Gyimóthy, and Liping A. Cai, vol. 18, 159–75. Bingley: Emerald, 2013. https://doi.org/10.1108/S1571-5043(2013)0000018011.

Ooi, Can-Seng, and Jesper Strandgaard Pedersen. "In Search of Nordicity: How New Nordic Cuisine Shaped Destination Branding in Copenhagen." *Journal of Gastronomy and Tourism* 2, no. 4 (October 17, 2017): 217–31. https://doi.org/10.3727/2169297 17X15046207899375.

Pine, B. Joseph, and James H. Gilmore. *The Experience Economy: Work Is Theatre and Every Business a Stage*. Boston: Harvard Business School Press, 1999.

Prentice, Richard, and Vivien Andersen. "Interpreting Heritage Essentialisms: Familiarity and Felt History." *Tourism Management* 28, no. 3 (June 2007): 661–76. https://doi.org/10.1016/j.tourman.2006.02.008.

Prentice, Richard C., Stephen F. Witt, and Claire Hamer. "Tourism as Experience : The Case of Heritage Parks." *Annals of Tourism Research* 25, no. 1 (1998): 1–24. https://doi.org/10.1016/S0160-7383(98)00084-X.

Richards, Greg. "Creativity and Tourism: The State of the Art." *Annals of Tourism Research* 38, no. 4 (October 2011): 1225–53. https://doi.org/10.1016/j.annals.2011.07.008.

Saunders, Robert A. "A Dark Imaginarium: The Bridge, Malmö and the Making of a 'Non-Existent' Place." *Journal of Urban Cultural Studies* 4, no. 3 (September 1, 2017): 361–85. https://doi.org/10.1386/jucs.4.3.361_1.

———. "Small Screen IR: A Tentative Typology of Geopolitical Television." *Geopolitics* 24, no. 3 (May 27, 2019): 691–727. https://doi.org/10.1080/14650045.2017.1389719.

———. "Geopolitical Television at the (b)Order: Liminality, Global Politics, and World-Building in The Bridge." *Social & Cultural Geography* 20, no. 7 (September 2, 2019): 981–1003. https://doi.org/10.1080/14649365.2017.1404122.

Steiner, Tobias. "Bron/Broen, the Pilot as Space Between Cultures, and (Re)Negotiations of nordic noir." In *The Scandinavian Invasion: The nordic noir Phenomenon and Beyond*, edited by Richard McCulloch and William Proctor. Bern: Peter Lang, n.d.

Tarulevicz, Nicole, and Can-Seng Ooi. "Food Safety and Tourism in Singapore: Between Microbial Russian Roulette and Michelin Stars." *Tourism Geographies* 23, no. 4 (July 1, 2021): 810–32. https://doi.org/10.1080/14616688.2019.1654540.

Tkalec, Marina, Ivan Zilic, and Vedran Recher. "The Effect of Film Industry on Tourism: Game of Thrones and Dubrovnik." *International Journal of Tourism Research* 19, no. 6 (November 2017): 705–14. https://doi.org/10.1002/jtr.2142.

Toft Hansen, Kim, and Anne Marit Waade. *Locating nordic noir: From Beck to the Bridge*. Basingstoke: Palgrave Macmillan, 2017.

Torchin, Leshu. "Location, Location, Location: The Destination of the Manhattan TV Tour." *Tourist Studies* 2, no. 3 (December 11, 2002): 247–66. https://doi.org/10.1177/14687976020023002.

Tripadvisor. "Marilyn Monroe's Subway Grate." *Tripadvisor*, 2020. www.tripadvisor.com.au/Attraction_Review-g60763-d10533507-Reviews-Marilyn_Monroe_s_subway_grate-New_York_City_New_York.html.

Tzanelli, Rodanthi, and Majid Yar. "Breaking Bad, Making Good: Notes on a Televisual Tourist Industry." *Mobilities* 11, no. 2 (March 14, 2016): 188–206. https://doi.org/10.10 80/17450101.2014.929256.

Urry, John. *The Tourist Gaze: Leisure and Travel in Contemporary Societies*. London: Sage Publications, 1990.

Visit Copenhagen. "Nordic noir Tours." Accessed May 28, 2020. www.visitcopenhagen.com/copenhagen/planning/nordic-noir-tours-gdk918519.

Waade, A. M., J. C. Pasgaard, M. Meldgaard, and T. Nielsen. "Nordic Landscapes in Collaborative Place-Making Interventions." In *The Nordic Wave in Place Branding*, edited by C. Cassinger, Andrea Lucarelli, and Szilvia Gyimóthy, 91–108. Cheltenham: Edward Elgar, 2019.

Waitt, Gordon. "Consuming Heritage: Perceived Historical Authenticity." *Annals of Tourism Research* 27, no. 4 (2000): 835–62. www.sciencedirect.com/science/article/B6V7Y-44BMTRD-1/2/c308f2c680b6ca470da2ea78c57de346.

Walker, Gordon J. H., R. Bruce Hull, and Joseph W. Roggenbuck. "On-Site Optimal Experiences and Their Relationship to Off-Site Benefits." *Journal of Leisure Research*, 30. National Recreation & Park Association, 1998.

10

SERIAL KILLER CINEMA AND DARK TOURISM

The Affective Contours of Genre and Place

Jane Stadler and Lesley Hawkes

Introduction

In 1993, the concept of dark tourism debuted simultaneously in research by sociologist Chris Rojek and in a star-studded feature film, *Kalifornia*.[1] *Kalifornia* is a psycho-thriller road movie in which David Duchovny plays an erudite writer researching a book about serial killers and the locations where their heinous crimes took place. He and his fashionable, artistic girlfriend (Michelle Forbes) plan a road trip in their matte black 1963 Lincoln Continental convertible across the United States from Louisville to Los Angeles, documenting the sites of serial killings. When they advertise for passengers to share petrol costs, an ill-educated pair played by Brad Pitt and Juliette Lewis leave their trailer home and come along for the ride. In true Hollywood fashion, Pitt's character turns out to be a serial killer himself and the two couples end up in mortal conflict at a remote desert location – the historic Nevada Nuclear Test site, which now offers tours to those fascinated by sites of death and destruction. With its attractive movie stars and troubling locations, *Kalifornia* thus became one of the first cinematic representations of the practice of dark tourism and also a catalyst for screen-induced serial killer tourism.

This chapter focuses on screen-induced dark tourism related to serial killings, analysing the affective qualities of both the screen experience and tourists' visits to locations related to the crimes or the screen productions. The phenomenon of dark tourism has gained traction in relation to media-induced tourism since Rojek[2] first introduced the idea of the commercialization of "black spots" or graves that attract visitors fascinated by celebrity deaths and sites of violence. Soon after, Malcolm Foley and John Lennon[3] focused on historical locations that drew tourists to them because of the traumatic events that occurred at locations such as concentration camps, battle grounds, and assassination sites. Dark tourism has since been linked to modern mass travel and global communication technologies in that the media

DOI: 10.4324/9781003327585-10

"are inherent in both the events which are associated with a dark tourism product and are present in the representation of the events for visitors at the site itself."[4] This dark shade of tourism breaks with dominant conceptions of tourism (and, indeed, media consumption and moviegoing) pertaining to leisure activities typically associated with entertainment, positive emotions, and fun-filled events and destinations.

By 2018, dark tourism and its close link to media had penetrated contemporary culture to the extent that it became the subject of a popular Netflix documentary series, *Dark Tourist*, in which journalist David Ferrier investigates morbid locations and the practices of tourists who are attracted to sites of catastrophe and mass murder. Indeed, tourism related to notorious serial killers such as Jeffrey Dahmer and Charles Manson features prominently in episodes of the *Dark Tourist* Netflix series and has been explored in academic research.[5] This phenomenon indicates that although tourism and media consumption are leisure and entertainment activities, they also have disquieting dimensions including the experience of negative emotions for tourists fascinated by suffering and death and for fans of media genres that trade on dark emotions and unpleasurable feelings.

In this chapter, we investigate the role that genre conventions and the representation of hunting grounds and gravesites in serial killer film and television play in the construction of audience engagement and touristic behaviour. Methodologically, we draw together critical tourism studies, genre studies, and theories of affective responses to examine how understandings of tourists' feelings and practices relate to audience engagement with media narratives. We consider how screen media represent and create an interest in troubling events, locations, and experiences, exploring how the emotional appeal of dark tourism sites is mediated by film and television associated with serial killers. In particular, we develop and apply theories of mediated affect and embodied experiences of place to two case studies from different genres, *Snowtown* and *Wolf Creek*.[6] We argue that the genre conventions through which murder sites are represented on screen and the degree of isolation, memorialization, or commodification present during site visits are defining features of the affective experience of dark tourism.

Snowtown is a biographical social realist feature film that documents and dramatizes how John Bunting and his accomplices killed 12 people in the 1990s and hid their bodies in barrels in a disused bank vault in rural Snowtown, South Australia. In stark stylistic contrast, *Wolf Creek* is the first film in a psycho-thriller road movie franchise influenced by *Kalifornia*. Director Greg McLean adapted the American genre to the dangerous isolation of the vast Australian landscape and traded on the fascination with charismatic serial killers. The media franchise McLean created is based loosely on actual tourist murder cases, with direct reference to Ivan Milat's notorious "backpacker murders," in which the bodies of seven slain travellers were found in 1992 and 1993. Using these disturbing destinations and media genres to examine the cultural practices associated with dark tourism, we analyse the extent to which such tourism may involve the search for extreme experiences to recreate the affective thrill of the encounter with death and establish a sense of spatial and emotional proximity with the killer and/or their victims, much as serial killers revisit the scene of the crime, take souvenirs, and re-enact the experience.

The Dark Side of Screen Tourism: Mediated Affect and Screen Landscapes

Film and tourism have been interconnected since early silent cinema's scenic travelogue films,[7] but when *Wolf Creek* became a surprise international sensation in 2005, there were concerns that "*Wolf Creek* would do for backpacking what *Jaws* did for swimming" and that it would harm Australia's reputation as a tourist destination, because it featured the gruesome demise of backpackers on a road trip in the outback. However, as film scholar Gemma Blackwood points out, "tourism research shows that frightening or negatively themed films have been able to create travel to film locations, and not scare people away."[8] By the time *Wolf Creek II* was released in 2013, tourism operators reported that young, international travellers were flocking to the eponymous Wolfe Creek meteorite crater and other sites *because* of the horror films, their spectacular shooting locations, and the morbid thrill of retracing the journeys of the murdered travellers whose real misfortunes inspired the franchise.[9]

When we consider the serial killings in *Wolf Creek* and *Snowtown*, it is worth noting that the figure of the outsider (the explorer, the escaped convict, the city slicker and, by definition, the tourist) has long played a central role in Australian landscape cinema and genre fiction.[10] In *Wolf Creek*, the killer Mick Taylor (John Jarratt) epitomizes the lone wolf bushman who shuns urban life and hunts city-dwellers and foreigners who intrude in the outback. In the bland, suburban, and semi-rural locale of *Snowtown*, however, the outsider is the social outcast – Bunting, his accomplices, and his victims are all socioeconomically vulnerable, mentally unwell, non-heteronormative, or drug-dependent persons living on the margins of society. In what follows, we demonstrate that the narrative strategies and location choices evident in each of these case studies interpellates the audience in very different ways in relation to both the sites of murder and the emotional connection to the people and places involved.

The films themselves, along with the act of spectatorship and visiting locations related to the murders shown on screen, can to some extent be conceptualized via practices of witnessing and memorialization. As Philip Stone states in his introduction to *The Palgrave Handbook of Dark Tourism Studies*, "Contemporary memorialisation is played out at the interface of dark tourism, where consumer experiences can catalyse sympathy for the victims or revulsion at the context."[11] Building on Stone's work, our research shows that the forms of memorialization during tourist site visits give rise to very different affective responses, depending on the location, the context, and the way the grisly story is framed by genre. The style, nature, and extent of media coverage and cinematic representations can augment or deter tourism and, as Richard Sharpley and Daniel Wright point out, people may be "drawn to places of death, disaster, and tragedy, whether recent or historical, to view the site or representations/memorials of the event."[12]

Given that the negative sentiment of serial killer media might reasonably be expected to deter tourism, this chapter analyses what aesthetic and rhetorical strategies are used to engage the affective sensibilities of audiences and dark tourists in ethically confronting cases such as those of the gruesome murders committed by Milat and Bunting. Media representations of serial killings range from interviews, news reports, and court footage of the murderers and their trials through television documentary, docudrama, and re-enactment, to low-budget realist films and feature-length horror films. The *Wolf Creek* franchise draws on the crimes perpetrated on tourists in isolated areas of Australia, principally Milat's aforementioned abduction and slaughter of seven hitchhikers in the 1990s and the murder by Bradley Murdoch of British tourist Peter Falconio and the narrow escape of Falconio's partner Joanna Lees while on a road trip in 2001. In *Wolf Creek* Milat's fictional counterpart, Mick Taylor, lives at a disused mine site called Navitalim Mining, which is an anagram of Ivan Milat's name. As Blackwood notes,

> The presence of the tourist as naïve victim in many of Australia's most notorious crimes in recent decades – from the disappearance of Azaria Chamberlain at an Uluru campsite in 1980 to the Port Arthur massacre in Tasmania in 1996 – lends extra legitimacy to [*Wolf Creek*'s] narrative.[13]

Milat's predatory behaviour is also documented in the true crime book *Sins of the Brother* and a television docudrama miniseries based on that book, *Catching Milat*,[14] and numerous documentaries. Murdoch's story is dramatized in the made for TV movie *Murder in the Outback* and documented in Lees' book *No Turning Back*.[15]

The titles of the two films, *Wolf Creek* and *Snowtown*, are both place names ostensibly designating murder sites. Murdoch committed his crimes on a remote stretch of road near Barrow Creek in the Northern Territory desert and the Milat backpacker murders took place deep in Belanglo State Forest in New South Wales; however, in *Wolf Creek*, these killings were relocated to the vicinity of Wolfe Creek meteorite park in Western Australia and other footage was filmed in the Flinders Ranges, South Australia. This complex film geography showcases some of Australia's most dramatic landscapes, yet it also dislocates the murder sites from the filming locations so that audience members become armchair tourists whose affective journey is orchestrated by the commercial logic of genre filmmaking.

Snowtown was shot on location where the events occurred. Snowtown itself is the location where the bodies of Bunting's victims were hidden, after being killed in or near an unremarkable, downtrodden suburb of Adelaide, Salisbury North. As with Milat, Bunting's serial killings feature in literary, cinematic, and televisual narratives and Snowtown's notorious "bodies in barrels" even found their way to Miami, Florida, in the fictional crime TV series *Dexter*, inspiring the "Barrel Girl

Murders" storyline of season five.[16] In each of these cases, despite the sensational media coverage of the killings and the film and television programs that recreate the crime scenes and relocate them to more dramatic, touristic sites such as Miami and Wolfe Creek meteorite crater, the actual places where the bodies were discovered – Belanglo Forest and Snowtown – retain a grim fascination for dark tourists.

Affective Subject Positions on the Continuum From Fact to Fiction

We argue that films positioned at varied points on the aesthetic continuum from factual to fictional representations of serial killings make different affective appeals to the audience. Where genre films offer a visceral emotional charge often combining horror with disturbing closeness to the killer and their victims, social realism and true crime provide socio-cultural context and more often invite audience alignment with an investigative or evaluative role akin to that of a police detective or juror. Genre and style also influence whether moral emotions such as compassion and contempt are elicited. As TV scholar Anita Biressi points out, by attempting to represent the unrepresentable – trauma, memory, and fear – true crime "renders private trauma knowable via public narratives."[17] The biographical and forensic work of true crime media thereby not only destabilizes the boundaries separating the public from the private and fact from fiction, but it also troubles subject positions such as investigator or juror that the audience is invited to occupy and it problematizes how we acquire and internalize understanding by cognitively and emotionally processing evidence and responding to affective and aesthetic cues.

Drama, comedy, romance, and the powerful charge of fear and suspense can augment emotional associations with the places depicted on screen and enhance the affective appeal of the landscapes featured in film and television. The term "affect" relates to bodily feelings and sensory experience, the degree of excitation involved, and whether those feelings are positively or negatively valenced.[18] Affect is a component of the broader category of emotion, which also involves complex cognitive and behavioural triggers and responses. For this reason, visiting the site of a terrifying, tragic event like a murder (or visiting the location where it was restaged and filmed) can potentially create a pleasurable (positively valenced) feeling of excitement (affective arousal) rather than a negative emotion like fear or a moral emotion such as compassion for the victims. Similarly, watching a horror movie or thriller is a leisure activity that strongly engages bodily affect and can make the audience fearful in a safe and exciting way that phenomenologist Julian Hanich describes as oscillating between dreadful fascination and pleasurable fear.[19] For example, the isolation and inhospitable terrain of *Wolf Creek*'s arid landscape serves to augment the victimized characters' feelings of helplessness and terror while also amplifying the negatively charged excitation that the audience experiences. For tourists visiting a site or spectators watching an atrocity on screen, pleasure and fascination are positively valenced components of the affective experience that are made possible by spatio-temporal distance from and experiential proximity to the killings.

Regarding affective responses to serial killers, media representations of their crimes, and tourists' reactions to visiting crime sites, moral emotions such as compassion, shame, and guilt are feelings that function simultaneously as ethical judgements about the self in relation to society. A shocked, nauseated response to atrocity is one way emotions place the audience on moral high ground because "gut reactions" are taken to express an authentic form of ethical evaluation – a visceral, somatic alignment with victims produces a literal embodiment of the judgement that serial killings are repulsive moral violations. However, the ghoulish fascination that fuels horror fandom and dark tourism associated with commercial "Helter Skelter Dearly Departed" Manson location tours, as shown in the Netflix *Dark Tourism* series, cannot be considered moral emotions. In contrast with the negative affective thrill of horror movies, watching social realist films, documentaries, and true crime is associated with the investigative impulse and the collection of evidence (similar to visiting crime scenes, searching for clues, and taking photos) is connected with deliberation and judgement via alignment or identification with police investigators, journalists, and jury members. Dark tourists' macabre fascination with serial killers evinces a voyeuristic desire to see inside the serial killer, which is generally disassociated from feelings such as empathy for victims.

The ethics of representation and genre film production are called into question when serial murder is stylized, the fact-fiction continuum is warped, and the suffering of victims who died within living memory becomes entertainment. For instance, when Mick Taylor tells his victim that he will stab her spine to paralyse her while he rapes and tortures her in *Wolf Creek*, the script incorporates Milat's actual words as reported in the non-fiction book *Sins of the Brother*: "This little procedure is called making a head on a stick. Once your spine's severed, that's what you are."[20] This intrusion of horrifying reality into genre fiction created for leisure and entertainment is problematic because it capitalizes on and sensationalizes death and cruelty, arguably trivializing it by embedding it in a film genre that offers cheap thrills.

In contrast with genre films intended to thrill and entertain, Tanja Horeck states that true crime stories

> with their strong affective pull on audiences, open up a set of vital questions about the politics of empathy, and about the functions of emotionality in a contemporary culture in which the lines between the private and the public, the individual and the collective, are in the process of being dramatically reconfigured.[21]

At one end of the spectrum, news footage circumvents the moral imagination to some extent by providing an avenue for judgement, rather than a rich mixture of evidence and context. In the middle of the spectrum, educational docudramas like *Catching Milat* and other realist narratives map out the social milieu and the context in which identity and behaviours are formed. The realist storytelling strategies often make a stronger appeal to the audience's moral emotions, imagination, and empathic insight. For instance, through dramatic reconfiguration, *Snowtown*'s naturalistic style and the use of biographical detail and socio-economic contextualization gives coherence, emotional impact, and aesthetic form to the conflicting and fragmented narratives and images around the tragic discovery of bodies in the Snowtown bank vault and the abusive and abused life narrative of Bunting, the serial killer. On the fictionalized end of the continuum, a horror film like *Wolf Creek* co-opts sensationalized genre tropes that may commodify suffering and foster fascination with celebrity criminals. A similar continuum of ethical concerns about commodification, affective responses, and leisure or entertainment activities and practices occurs with respect to dark tourism site visits and the marketing of such destinations.

Visitations and Motivations: Placing the Tourist in the Crime Scene

Tourism operators use the media to tap into emotional motivations and establish scripted roles for dark tourists visiting serial killer sites. Annaclaudia Martini and Dorina Maria Buda examine dark tourism as an affective socio-spatial encounter, arguing that places and affective responses can be manipulated by the tourism industry to incite certain reactions from visitors and these experiences are framed by political and ethical values.[22] As Sangkyun Kim and Gareth Butler state in their article about community perspectives on dark tourism, "the media coverage of the Snowtown murders created a fascination and emotion amongst the general population, who would otherwise not have any desire to visit Snowtown as a destination."[23] The same is true of Belanglo Forest's bland pine plantations and scrubby trails, which bear no resemblance to the gothic desert landscapes that feature in the *Wolf Creek* franchise. Although Belanglo has few amenities or attractions for tourists other than its dark history,[24] the occasional curious traveller, journalist, or dark tourist documents walking through the forest in search of the places where people are known to have been killed.

In Belanglo (Figure 10.1), the unprepossessing place where the bodies were found is in an area known as Executioner's Drop on Longacre Fire Trail, marked by a modest plaque commemorating the memory of the seven backpackers whose remains were found there. Although the *Wolf Creek* franchise favours spectacular locations in Western Australia and South Australia, the docudrama *Catching Milat* was filmed in and around Sydney in the areas where Milat lived, hunted, and worked, incorporating archival footage and aligning viewers with the investigative role of the detective. One harrowing scene featuring a memory sequence even purports to take Paul Onions, a British hitchhiker who narrowly escaped abduction by Milat, back to Belanglo to revisit his escape and help identify his attacker. In this way, the use of experiential footage and real locations on screen helps to evoke empathy for Milat's victims and the trauma they experienced.

FIGURE 10.1 Belanglo State Forest Sign.

Source: Photograph courtesy of Brian Yap, Creative Commons CC BY-NC 2.0, www.flickr.com/photos/yewenyi/3413383281.

One dark tourist, Julian Morgans, reflects on his visit to Belanglo in search of the plaque commemorating the victims,

> It stands at the end of a long 4WD track, and you couldn't find it unless you looked. It seemed a painfully inadequate tribute to seven lives. . . . I kept thinking about these kids who came from so far away, only to die in this shitty forest.[25]

The memorial plaque for the victims may be inadequate, but the fact that the site is not marketed as a tourist destination seems ethically appropriate and it is more likely to elicit moral emotions and empathy than visiting the dramatic film location of the Wolfe Creek meteorite crater. As Morgans writes:

> I don't think I'm alone when I say I find serial killers interesting. It's the mystery of psychopathy that does it. . . . So I wanted to go out to Belanglo as a sort of ultimate Wikipedia excursion. I thought I'd head out with a friend and have a big weird, spooky day. But no one wanted to come. And it wasn't weird, just incredibly sad. . . . I'd just found a small shrine to murdered kids and then I'd felt awful.[26]

In stark contrast with this commemoration of a tragic serial murder site, in 2015 Goulburn Ghost Tours advertised midnight "extreme terror tours" of Belanglo's "killing fields," charging dark tourists AUD$150 per person to "Come with us to Belanglo where Ivan Milat buried the bodies of his victims! Once you enter Belanglo state forest you may never come out." On the tour's website (since removed) the marketing pitch shamelessly stated, "Who else is out here? Is there another victim waiting to be found?" This seeks to create a horror movie thrill, playing on the theory that Milat did not act alone and one of his relatives may still be hunting tourists, but since it is addressing prospective tourists, it also establishes a connection to the subject position of Milat's victims with the phrase, "Are we being followed . . . watched?" In this way, the tour simultaneously traded on the fascination with Milat himself while ostensibly proffering a sympathetic alignment with his victims. However, it primarily situated tourists in an investigative role within a detective experience, mixing the genre tropes of true crime and horror to attract dark tourists based on the hypothesis that Milat had an accomplice and murdered more victims than those that have been found:

> Belanglo State Forrest (sic) is the old stomping ground of Ivan Milat, convicted serial murderer. Learn about his crimes and use paranormal techniques to help solve the baffling murder of Angel, believed to be murdered AFTER Ivan Milat was jailed![27]

These "terror tours" were shut down after complaints that they were capitalizing on the suffering of Milat's innocent victims and their grief-stricken families. This underscores the immorality of sensationalizing and making money from murder sites: commercialization is insensitive to victims and locals and it tends to elicit inappropriate responses such as trivialization and black humour.

A further aspect of dark tourism that is morally problematic is the cult-like fandom that forms around celebrity criminals. For instance, as compere David Farrier observes in the episode of *Dark Tourist* featuring sites frequented by the Manson family, "Manson was an outsider who attracted fans, followers and misfits, and that hasn't changed."[28] The same can be said for Milat, a rebellious, roguish figure who garnered quite a following including copycat killers during his time behind bars and a "Milat Letters" Facebook group that has over 8,000 followers created by Milat's nephew Alistair Shipsey who claimed Milat was framed to generate tourism revenue.[29] Earlier we noted the strong appeal of the outsider figure in Australian serial killer films and the affinity with outsiders that is felt by tourists who want to get off the beaten path and distinguish themselves from mainstream tastes and practices. Yet, in some respects, tourists may visit particular sites to feel a sense of fitting into a group, even if that group is an oppositional subculture or counter fandom that frequents extreme terror tours or celebrity criminal fansites. Regarding our culture's interest in death, psychopathy, and criminal acts and artefacts, the emergence of "dark fandoms" is to be expected: "These communities allow those with a shared interest in those who have committed heinous acts, like mass or serial murderers, to collectively share their interests."[30] Participants in dark fandom often come to feel a sense of identity and community and engage in practices similar to more conventional fandoms, albeit more "deviant and morbid."[31] Indeed, the current owners of Milat's house even joke about renting rooms out to backpackers as a tourist attraction.[32]

Within this complex context, tourism scholars Avital Biran and Dorina Maria Buda find that an "important aspect of dark tourism – that of human suffering – has made it difficult to apply traditional tourism motivational theories."[33] However, as we have detailed and as Dirk Gibson's tourism research demonstrates, serial killings often attract tourists to travel to crime sites for "sightseeing, entertainment, dining, refreshments, lodging, photo-opportunities, and the acquisition of souvenirs."[34] There are varied motivations for visiting murder locations, including wanting to be present at the site after seeing a cinematic or televisual retelling of the crime. Motivations may differ depending on the genre. The desire to feel close to the victims and to pay respect may arise in the case of visiting graves after watching a documentary, the desire to experience the horror and terror of the event may be evoked after watching a genre film, the desire to solve the crime may occur after viewing a true crime series, and a sense of connection with the murderer may even form in the case of fans of celebrity criminals. In many cases, there is also the desire to feel a connection with the landscape or location itself. Dark tourists want to visit these sites in the hope that the locations will somehow release emotional traces of horrific occurrences and they will feel the affective thrill of the experiences and events that occurred there. In tourism inspired by film and television, the sites represented on screen may have been shot in several different locations. However, tourists are still led to believe the locations may offer up an insider's understanding of the story and such visitations are often linked to the desire to confront something taboo or forbidden and make it less mysterious or threatening.[35]

We argue that both the stylized representation of murder sites on screen and actual site visits themselves add important dimensions to the affective experience. As Sue Beeton's film-induced tourism research suggests, tourists may be motivated to visit film locations partly to reignite a fantasy, an emotion, or an experience represented on screen, or to feel a connection to celebrity, whether an actor or their screen persona.[36] Beeton asserts "attachment to the emotion engendered in a movie or TV series is more important than simply visiting a site."[37] Dark tourists may be motivated by the belief that the actual site holds memories and sensations that can be reignited. They may seek to reactivate the same excitement and fear they experienced while watching the film. Daniel Olsen and Max Korstanje argue that dark tourism involves the act of being witness to traumascapes that haunt the imagination and that dark tourism journeys take on a "kind of metempsychosis" where the tourist has the sense that they are in the presence of another.[38] For example, the Milat backpacker murders haunt Australian society and visiting the locations associated with these crimes may be a way of working through these historic traumas. Taking a related approach, Dorota Golanska looks at dark tourism with a focus on "affective aesthetics" and the way the body responds to its surroundings and experiences a sense of becoming "other."[39] The socially expected critical and resistant attitude is dispelled while at these sites, giving rise to a bodily reaction that could involve pleasure, excitement, sadness, or a complex affective combination. These instances, Golanska argues, are moments when the mind needs to listen to or bear witness to the body.[40]

Snowtown: Commodification and the Horror of the Everyday

Although many dark tourists may be motivated by the desire to experience the affective thrill of a potential encounter with danger, there may be times when the tourist does not feel any dramatic emotional gravity at the site. Dark tourism locations often turn out to be rather ordinary and banal. Warwick Frost finds that many film directors choose almost "anonymous scenery" to emphasize the moral wasteland and a sense of alienation or entrapment.[41] The apparent normalcy of these sites and the horror of the everyday may explain why dark tourists venture to quite unremarkable destinations that are far from conventional tourist locations. In *Affective Landscapes in Literature, Art and Everyday Life*, Christine Berberich, Neil Campbell, and Robert Hudson contend that strong feelings often emerge from encounters with the small scale, the ordinary, and the everyday because it is these encounters that are more intimate and personal.[42]

Snowtown's locations appear unremarkable. Snowtown itself is a small rural farming town about a two-hour drive north of Adelaide. It is known in contemporary popular culture as the town where the bodies in the barrels were found. When the film of these murders was released, locals were quick to emphasize that the murders did not take place in the town and the murderers had no connection with the town. For dark tourists, however, this is what makes the town appealing. It is the chosen site for keeping the bodies – a named location that can be found on a map and reached on a daytrip from Adelaide. Previously, the town based its identity on the rural idyll and notions of safety, security, and old-fashioned values. The town's official tourist page gives its history as one of wool, wheat, and lamb breeding. The dark tourists are not interested in any of these industries; rather, they want to visit the barrels in the bank.

Visiting the bank provides unmediated evidence of the crimes, going beyond photographs and footage seen in the news or on film and allowing tourists to witness artefacts and locations associated with death.[43] The bank, which is colonial in appearance and harkens back to an earlier period in Australian history, provides the link the tourist needs to feel the thrill of being in close proximity with the crime, the murderers, and the victims. The tourists are not interested in the contextual background of the murders nor that the bodies in barrels were moved to the empty Snowtown bank (Figure 10.2); instead, they are interested in the fact that the barrels are the vessels where the bodies were located. They are identifiable artefacts of the murders. The barrels and the empty bank become a way for tourists to reconcile the ordinary and the horrific. The *Snowtown* movie trailer calls the story, "a web of evil with no escape" and the shuttered bank emphasizes this prison-like atmosphere. A lot of the film consists of interior shots of low-cost housing and

FIGURE 10.2 Snowtown Bank.

Source: Photograph courtesy of Jane Stadler.

remnants of waste from consumer society (discarded shopping trolleys, junk from packaging, and car bodies). Everyday items take on a darker tone and a visit to the town reinforces this unease.

Jeffrey Podoshen finds that the tourist is "stimulated by physical encounters with relics" and being in the actual location "gives rise to further thoughts and emotions as part of a constant feedback-oriented iterative process."[44] The creation of a feedback loop is also the case with a visit to Snowtown. The media emphasized the gruesome disposal of the victims' bodies and dark tourists are stimulated by being in the location where the barrels were found. Each site visitor plays their role in keeping the narrative alive, adding to and reigniting interest in the location and the story. The more recent the event, the more heightened the affective impact will be for the tourist because temporal distance dilutes the affective charge. However, there are ethical considerations when it comes to the time between the crime and the visit to the location. Kim and Butler indicate in their research on Snowtown that even though "locals were generally supportive of the idea of dark tourism development, they believed the timing of the development to be a crucial factor," because the traumatized community needs time to heal after the crimes.[45] Tensions develop because many dark tourists seek to visit as soon as possible after the event to feel the heightened affective impact and to discover whether the landscape will reveal its secrets. As Philip Stone writes, "dark tourism offers visual signifiers and a multiplicity of meanings within touristic landscapes, as global visitor sites function as retrospective witnesses to acts of atrocity or tragedy."[46] When crimes are recent, the act of witnessing becomes too close and takes on more of a voyeuristic tone that can come across as being insensitive to the victims, their families, and locals.

Literary and cultural scholar Brian Jarvis's work on serial killers and consumer culture looks at how "the commodification of violence in popular culture is structurally integrated with the violence of commodification itself," particularly in genre fiction and in relation to the sale and purchase of serial killer tours and merchandise or "murderabilia."[47] Just as criminals collect mementos of their crimes and movie fans re-watch favourite scenes, the tourist can collect location mementos and souvenirs that create a tangible, affective connection between the tourist, the location, the crime, and the victim or the criminal. Within this context, as Jarvis shows, violent crime is commodified as a spectacle available for consumption.[48] For instance, tourists visiting Snowtown can buy cheap fridge magnets showing skeletons jumping out of barrels, proclaiming "I visited Snowtown and survived!" Prior to going out of business, the proprietor of the local curio shop stated, "I'm doing something that can bring money into Snowtown. You can't tell me the people who made the *Snowtown* movie didn't cash in on the town?"[49]

The locals of Snowtown express conflicted ethical perspectives regarding the financial gain that dark tourists could bring to the region. Some locals think it is disrespectful to the victims and the town, to the extent that Snowtown itself has "become a victim of the publicity, of the murders," while others think they should use the association between the crimes and the town to benefit the community financially.[50] Tom Michael, a local farmer, understands that tourism could bring much-needed funds to the town but he struggles with the ethical concerns: "But the thing is, how do I talk to my little kids about profiting off it?" Tom's father Neville favours a plaque in memory of the victims rather than establishing a museum to the dark event. Both men would prefer the town not become involved in the promotion of dark tourism.[51] However, tourists come regardless of whether they are welcome or not. Perth resident John Dawson, for instance, detoured from his South Australian road trip to take some photos of Snowtown to "check this place out just to have a look" and to take a quick photo of the bank.[52]

Even taking photos of locations can become an act of mapping and collecting evidence or proof of connection, as the many photos of tourists at the bank and by the large barrel-shaped tables at the Snowtown Hotel across the road from the bank attest. It is the sharing of these photos or mapped locations that enable dark tourists to feel connected to a wider community and to relive and record this connection. Site visits and the practice of collecting photographs, souvenirs, and murderabilia may enable the tourist to feel a sense of connection with something bigger than themselves. As Lindsay Steenberg's insightful examination of crime tourism suggests,

> When a crime becomes famous (or infamous), it no longer functions exclusively as an event or an act – it becomes a place to visit and to map, and provides tourists with tangible souvenirs, in an uncanny echo of the serial killer's propensity to collect.[53]

This desire to connect with the serial killer and their dark shade of celebrity through the locations they frequented may also help explain the desire for dark tourists to visit the film locations in *Wolf Creek*.

Isolation, Awe, and Trauma in the Landscapes of *Wolf Creek*

The locations chosen for the film *Wolf Creek* (Figure 10.3), just like the events in the film itself, are hybrid in nature. The Australian outback is vast, and it encompasses many different landscapes: red deserts, mountain ranges, rock formations, thick mulga scrub, and grassy plains. The commonality among these landscapes is a fearsome sense of isolation and awe. As discussed, the outback has been the setting for Australian films in a number of different genres, ranging from romantic blockbusters such as *Australia* through comedies like *Crocodile Dundee* to a number of horror films. In horror films, the mythic outback retains its majestic and sublime effect; however, the meaning of its vastness is subverted and the outback setting becomes a place of menace. For example, both the desolate landscape scenes of *Wolf Creek* and the lonely, untrammelled bush of Belanglo evoke the alienation and isolation associated with the Milat crimes. While Belanglo Forest is only two hours from Sydney, there are unlikely to be passers-by to help someone in trouble and that is even more true for those who venture out of cell phone

FIGURE 10.3 Wolfe Creek Meteorite Crater

Source: Photograph courtesy of Jane Stadler.

range along the corrugated dirt road approximately 150 km (nearly 100 miles) from the nearest mechanic or water source to the Wolfe Creek Meteorite. For even the best-equipped travellers, the sense of risk is as palpable as it is during the eerie scene in the movie when the ill-fated backpackers' car won't start and they are stranded at the crater.

The harsh landscapes surrounding the crime scenes (both the real location and its cinematic substitute) echo the horrific nature of the actual crimes and present a stark reminder of the dangers of travelling in remote regions of Australia. It is this affective echo of danger and crime that tourists may connect with when they visit. When present at the site, the tourist may come to know and feel the horror of the encounter as the affective response to the landscape triggers alertness and heightened awareness of danger. The Wolfe Creek crater is a spectacular site of awe and wonder that has been manufactured as a location of horror in the genre film. The massive meteorite crater evokes a strong sense of the otherworldly, creating the powerful effect of an unknowable landscape.

In many Australian horror films, the outback is the place where outsiders vanish: when they misunderstand the dangers inherent in these locations, they pay the ultimate penalty. *Wolf Creek* continues this tradition. When Mick Taylor first comes across the tourists stranded at the meteorite in *Wolf Creek* he says, "What the bloody hell are you doing out here?" The tourists reply with an uneasy laugh. Dark tourists desire this same sense of unease, even if they are unfamiliar with the "Where the bloody hell are you" Tourism Australia campaign. In a sense, tourists are tempting fate when they visit these outback sites to experience the awe and wonder of the location and experience the element of risk. Traditionally, "death, conflict, and atrocity sites which elicit sadness, distress, and an inherent sense of danger have been predominantly considered deterring factors for tourists,"[54] but in the case of *Wolf Creek*, these factors become motivations for visiting the sites. Tourists want to feel the anxiety of not knowing whether a Mick Taylor figure may be around the corner. As Mick says in the film, "You will never know where I might pop up." If they survive the trip, their photographs and souvenirs become evidence that they are not a "naïve tourist." Dark tourists who visit Wolfe Creek and Belanglo Forest take Mick's words as a dare, an invitation to take their chances off the beaten track and test their skills in the outback, flaunting warnings that unwary travellers deserve everything they get if they disregard cautions about venturing into dangerous, isolated territory.

The *Wolf Creek* franchise also implies that the landscape is somehow implicated in the crimes. While *Snowtown* emphasizes the lived reality of social outsiders who are pushed to the margins of society because of poverty and neglect, *Wolf Creek* conveys the menace that still exists in the Australian landscape (Figure 10.4). Podoshen found in his research that dark tourists attempt to "reconcile comparisons between imagined landscapes and topographical reality."[55] In the case of visits to the outback by dark tourists, these comparisons can be reconciled as

FIGURE 10.4 *Wolf Creek* Flinders Ranges film location.

Source: Photograph courtesy of Jane Stadler.

Australia's topography lives up to the awe-inspiring landscapes of the films. The outback becomes a prime example of a traumascape and dark tourists expect to feel heightened emotions when they visit. They desire the sense of unease they felt while watching the films and, in the case of the Wolfe Creek meteorite, the harsh landscape far-removed from cellphone towers, petrol stations, police, or passers-by creates sensations of vulnerability and isolation. Tourists visiting the actual crime sites in Belanglo Forest may have a different experience. They may feel uneasy but this unease may result from the recognition that the site bears little resemblance to the stylized and spectacular film locations. The reality of the crime scene brings a stark reminder that they are standing in a location of extreme suffering and the lack of recognition makes them feel like an intruder.

The locations chosen for the *Wolf Creek* and *Snowtown* films have vast differences. One is awe-inspiring and sublime, while the other is bland and ordinary. Yet these locations reveal the diverse motivations driving dark tourists to make these trips. As Berberich and colleagues write, "affectual relations with landscapes are always political for they scale humanity to the world, draw us intimately to place and time and change so that we might sense our own vulnerability as susceptible parts of the greater whole."[56]

Conclusion

In crime tourism involving tours or reconstructions of crime scenes and in film-induced dark tourism, the serial killer's infamy intersects with the discourse of celebrity and incites fan behaviours such as taking photos and recording footage at key sites, keeping clippings of interviews and images, and collecting memorabilia and autographs.[57] We have argued that the inclination to visit the dark sites represented on screen and to collect photographs or murderabilia is in some ways akin to the killer's compulsion to revisit crime scenes and keep souvenirs and that this distinguishes dark tourists' behavioural and affective contours from those who are merely fans of horror or true crime. Nevertheless, the affective experience of visiting murder sites is not disconnected from movie fandom or emotional engagement with the characters, events, and locations seen on screen. Visiting the scene of the crime when engaging in dark tourism has a perverse thrill that is comparable to the interplay of pleasurable fear, revulsion, and fascination involved in watching horror movies, whereas true crime documentaries appeal more to the "*mortality-related* motives – including curiosities to understand how atrocities happened, social responsibility, identity construction, empathetic identification, and remembrance and commemoration."[58] It is film and photography that connect these different media genres and the practices of dark tourists who visit the film locations and murder sites. The pervasive sense that the camera can capture documentary evidence provides a tangible connection to real places and past events. As Lennon states:

> Visualisation would appear to be critical to appreciation of the divided emotions of horror and fascination. . . . The nature of viewing such dark images and indeed visiting such sites is linked to the viewers/visitors' often-inexpressible experience. The compulsion to visit, to stare fixedly at content and to record feelings and visual images is typical of the behavioural response of the visitor.[59]

These touristic practices and the feelings that dark tourists may experience are informed by both genre and place. At each site, the feelings and images that visitors record differ in accordance with the media narratives that inspired the site visit and the characteristics of the location. Screen-induced dark tourism linked to sensational genre tropes often foregrounds tourists' personal fears and fascination with the killer. On the other hand, documentary and social realism cultivate affective responses that tend to elicit moral emotions for the victims. This chapter has demonstrated how *Snowtown* and *Catching Milat* both use realism to elicit other-oriented emotions and an evaluative, investigative stance; however, the difference between commodifying death and suffering or memorializing victims involves more than representational strategies. Location-specific features also influence the affective experiences and behavioural responses of visitors, such as the fearsome isolation of an ancient landform, the presence of a souvenir shop, or a simple memorial for those who lost their lives.

Notes

1 Chris Rojek, *Ways of Escape: Modern Transformations in Leisure and Travel* (London: Palgrave Macmillan, 1993); *Kalifornia*, dir. Dominic Sena (USA, Polygram, 1993).
2 Rojek, *Ways of Escape*, 136, 138.
3 Malcolm Foley and John Lennon, "JFK and Dark Tourism: A Fascination with Assassination," *International Journal of Heritage Studies* 2, no. 4 (1996): 198–211.
4 Malcolm Foley and John Lennon, *Dark Tourism: The Attraction of Death and Disaster* (London: Continuum, 2000), 16.
5 James A. Fox and Jack Levin, *Overkill: Mass Murder and Serial Killing Exposed* (New York: Plenum Press, 1994); Dirk C. Gibson, "The Relationship Between Serial Murder and the American Tourism Industry," *Journal of Travel & Tourism Marketing* 20, no. 1 (2006): 45–60.
6 *Snowtown*, dir. Justin Kurzel (Australia, Screen Australia, 2011); *Wolf Creek*, dir. Greg McLean (Australia, Film Finance Corporation, 2005). The *Wolf Creek* franchise created by McLean includes *Wolf Creek II* (2013) and two seasons of a spin-off television series (Australia, Stan, 2016–2017).
7 Jennifer Lynn Peterson, *Education in the School of Dreams: Travelogues and Early Nonfiction Film* (London-Durham: Duke University Press, 2013); Jeffrey Ruoff, ed., *Virtual Voyages: Cinema and Travel* (London-Durham: Duke University Press, 2006).
8 Gemma Blackwood, "*Wolf Creek*: an UnAustralian Story?," *Continuum: Journal of Media & Cultural Studies* 21, no. 4 (2007): 491; Sue Beeton, *Film-Induced Tourism* (Clevedon: Channel View Publications, 2005), 24; Roger Riley, Dwayne Baker, and Carlton S. Van Doren, "Movie Induced Tourism," *Annals of Tourism Research* 25, no. 4 (1998): 932.
9 John Scott and Dean Biron, "*Wolf Creek*, Rurality and the Australian Gothic," *Continuum* 24, no. 2 (2010): 307–22; Tromp, Belinda, "Horror Film *Wolf Creek* Brings Tourists to the Outback," *ABC*, February 17, 2014, <www.abc.net.au/radionational/programs/archived/bushtelegraph/horror-film-wolf-creek-brings-tourists-to-the-outback/5265056>
10 Jane Stadler, "*Atopian* Landscapes: Gothic Tropes in Australian Cinema," in *A Companion to Australian Cinema*, ed. Felicity Collins, Jane Landman, and Susan Bye (Redmond, Vic.: Wiley-Blackwell, 2019), 336–354; Warwick Frost, "Life-changing Experiences: Film and Tourists in the Australian Outback," *Annals of Tourism Research* 37, no. 3 (2010): 707–726.
11 Philip Stone, "Introduction" in *The Palgrave Handbook of Dark Tourism Studies*, ed. Philip Stone, Rudi Hartmann, Tony Seaton, Richard Sharpley, and Leanne White (London: Palgrave Macmillan, 2018), viii, https://link.springer.com/content/pdf/10.1057/978-1-137-47566-4.pdf.
12 Richard Sharpley and Daniel Wright, "Disasters and Disaster Tourism: The Role of the Media," in *The Palgrave Handbook of Dark Tourism Studies*, ed. Philip Stone, Rudi Hartmann, Tony Seaton, Richard Sharpley, and Leanne White (London: Palgrave Macmillan, 2018), 377.
13 Blackwood, "*Wolf Creek*," 491.
14 Mark Whittaker and Les Kennedy, *Sins of the Brother: The Definitive Story of Ivan Milat and the Backpacker Murders* (Sydney: Pan Macmillan, 2001); *Catching Milat*, dir. Peter Andrikidis (Australia, Shine, 2015).
15 *Murder in the Outback*, dir. Tony Tilse (Australia, Channel 10, 2007); Joanne Lees, *No Turning Back* (Hodder & Stoughton, 2006).
16 *Dexter*, created by James Manos Jr (USA, Showtime, 2006–2013).
17 Anita Biressi, "Inside/out: Private Trauma and Public Knowledge in True Crime Documentary," *Screen* 45, no. 4 (2004): 405.
18 Julian Hanich, *Cinematic Emotion in Horror Film and Thrillers: The Aesthetic Paradox of Pleasurable Fear* (New York: Routledge, 2010), 183.
19 Hanich, *Cinematic Emotion*, 186.

20 Whittaker and Kennedy, *Sins of the Brother*, 150.
21 Tanya Horeck, "'A Film That Will Rock You to Your Core': Emotion and Affect in *Dear Zachary* and the Real Crime Documentary," *Crime, Media, Culture* 10, no. 2 (2014): 165.
22 Annaclaudia Martini and Dorina Maria Buda, "Dark Tourism and Affect: Framing Places of Death and Disaster," *Current Issues in Tourism* 23, no. 6 (2020): 679–692.
23 Sangkyun Kim and Gareth Butler, "Local Community Perspectives Towards Dark Tourism Development: The Case of Snowtown, South Australia," *Journal of Tourism and Cultural Change* 13, no.1 (2015): 78–89, 81. Kim and Butler's otherwise thought-provoking examination of tourism in Snowtown incorrectly describes the genre of *Snowtown* as a "crime horror-style production" and erroneously states it is directed by Katrina Sedgwick rather than Kurzel.
24 In addition to the hitchhikers slain by Ivan Milat, the forest is also the site where Milat's (then) 17-year-old great-nephew, Matthew Milat, and his accomplice Cohen Klein brutally murdered their school friend David Auchterlonie in 2010, according to Michael McGowan, "Ivan Milat's Chilling Serial Backpacker Murders Still Haunt Australia," *The Guardian Australia*, October 27, 2019, www.theguardian.com/australia-news/2019/oct/27/ivan-milat-chilling-serial-murders-haunt-australia-after-death.
25 Julian Morgans, "A Sad Walk Through Australia's Serial Killer Forest," *Vice*, June 30, 2016, www.vice.com/en_au/article/wdb5v4/a-very-sad-walk-in-australias-serial-killer-forest.
26 Morgans, "A Sad Walk."
27 Daisy Melwani, "Backpacker Murder Tour Banned," *Travel Weekly AAP*, July 15, 2015, www.travelweekly.com.au/article/backpacker-murder-tour-banned/.
28 *Dark Tourist* S1E8, "Back in the USA" (Netflix, 2018).
29 Alistair Shipsey, "The MILAT Letters," *Facebook*, June 12, 2020, www.facebook.com/pages/category/Author/The-MILAT-Letters-378352079013220/.
30 Ryan Broll, "Dark Fandoms: An Introduction and Case Study," *Deviant Behavior* 41, no. 6 (2019): 11.
31 Broll, "Dark Fandoms," 11.
32 Gavin Butler, "Meet the Lovely Couple who Moved into Ivan Milat's House," *Vice*, August 20, 2018, www.vice.com/en_au/article/9kmv3p/meet-the-lovely-couple-who-moved-into-ivan-milats-house.
33 Avital Biran and Dorina Maria Buda, "Unravelling Fear of Death Motives in Dark Tourism," in *The Palgrave Handbook of Dark Tourism Studies*, ed. Philip Stone, Rudi Hartmann, Tony Seaton, Richard Sharpley, and Leanne White (London: Palgrave Macmillan, 2018), 515.
34 Gibson, "Serial Murder and the American Tourism," 45.
35 Biran and Buda, "Unravelling Fear," 517.
36 Sue Beeton, *Film-Induced Tourism* (Clevedon: Channel View Publications, 2005).
37 Sue Beeton, "The Advance of Film Tourism," *Tourism and Hospitality Planning & Development* 7, no. 1 (2010): 4.
38 Daniel Olsen and Maximiliano E. Korstanje, *Dark Tourism and Pilgrimage* (Oxfordshire: CAB International, 2020), 16.
39 Dorota Golanska, "Affective Spaces, Sensuous Engagements: In Quest of a Synaesthetic Approach to 'Dark Memorials'," *International Journal of Heritage Studies* 21, no. 8 (2015): 773–790.
40 Golanska, "Affective Spaces," 787; see also Martini and Buda, "Dark Tourism and Affect," 681.
41 Frost, "Life-changing Experiences," 720–722.
42 Christine Berberich, Neil Campbell, and Robert Hudson, *Affective Landscapes in Literature, Art and Everyday Life* (Ashgate: Routledge, 2015).
43 See John Lennon's introduction to Ambroise Tezenas', *I was Here: Dark Tourism* (London: Dewi Lewis Publishing, 2014).

44 Jeffery Podoshen, "Dark Tourism Motivations: Simulation, Emotional Contagion and Topographic Comparison," *Tourism Management* 35 (2013): 266.
45 Kim and Butler, "Local Community Perspectives," 86.
46 Stone, "Introduction," viii.
47 Brian Jarvis, "Monsters Inc.: Serial Killers and Consumer Culture," *Crime, Media, Culture* 3, no. 3 (2007): 326–7.
48 Jarvis, "Monsters Inc.," 328.
49 Heather Kennet, "Sick Snowtown Souvenirs are Making a Killing," *The Sunday Telegraph*, July 15, 2012, www.dailytelegraph.com.au/news/nsw/sick-snowtown-souvenirs-are-making-a-killing/news-story/f0058ee8ee42e329c4f27943fe6a82d6.
50 Daniel Keane and Patrick Martin, "Life After Death: Dark Tourism and the Future of Snowtown," *ABC* May 20, 2019, www.abc.net.au/news/2019-05-20/can-snowtown-ever-shake-off-its-dark-past/11082778?nw=0&r=HtmlFragment.
51 Keane and Martin, "Life After Death."
52 Keane and Martin, "Life After Death."
53 Lindsay Steenberg, "Crime and Celebrity," *Oxford Research Encyclopedia of Criminology*, (2017): 14, https://oxfordre.com/criminology/view/10.1093/acrefore/9780190264079.001.0001/acrefore-9780190264079-e-26.
54 Biran and Buda, "Unravelling Fear," 515.
55 Podoshen, "Dark Tourism Motivations," 269.
56 Berberich, Campbell, and Hudson, *Affective Landscapes*, 16.
57 Steenberg, "Crime and Celebrity."
58 Biran and Buda, 2018, 518.
59 John Lennon, "Dark Tourism Visualisation: Some Reflections on the Role of Photography," in *The Palgrave Handbook of Dark Tourism Studies*, ed. Philip Stone, Rudi Hartmann, Tony Seaton, Richard Sharpley, and Leanne White (London: Palgrave Macmillan, 2018), 585–602, 586.

Bibliography

Beeton, Sue. *Film-Induced Tourism*. Clevedon: Channel View Publications, 2005.

———. "The Advance of Film Tourism." *Tourism and Hospitality Planning & Development* 7, no. 1 (2010): 1–6. https://doi.org/10.1080/14790530903522572.

Berberich, Christine, Neil Campbell, and Robert Hudson. *Affective Landscapes in Literature, Art and Everyday Life*. Ashgate: Routledge, 2015.

Biran, Avital, and Dorina Maria Buda. "Unravelling Fear of Death Motives in Dark Tourism." In *The Palgrave Handbook of Dark Tourism Studies*, edited by Philip Stone, Rudi Hartmann, Tony Seaton, Richard Sharpley, and Leanne White, 515–532. London: Palgrave Macmillan, 2018.

Biressi, Anita. "Inside/Out: Private Trauma and Public Knowledge in True Crime Documentary." *Screen* 45, no. 4 (2004): 401–12.

Blackwood, Gemma. "*Wolf Creek*: An UnAustralian Story?" *Continuum: Journal of Media & Cultural Studies* 21, no. 4 (2007): 489–97.

Broll, Ryan. "Dark Fandoms: An Introduction and Case Study." *Deviant Behavior* 41, no. 6 (2019): 1–13. https://doi.org/10.1080/01639625.2019.1596453.

Busby, Graham, and Julia Klug. "Movie-induced Tourism: The Challenge of Measurement and Other Issues." *Journal of Vacation Marketing* 7, no. 4 (2001): 316–32.

Butler, Gavin. "Meet the Lovely Couple Who Moved Into Ivan Milat's House." *Vice*, August 20, 2018. www.vice.com/en_au/article/9kmv3p/meet-the-lovely-couple-who-moved-into-ivan-milats-house.

Cohen, Erik. "Thanatourism: A Comparative Approach." In *The Palgrave Handbook of Dark Tourism Studies*, edited by Philip Stone, Rudi Hartmann, Tony Seaton, Richard Sharpley, and Leanne White, 157–172. London: Palgrave Macmillan, 2018.

Crawford, Robert. "Learning to Say G'Day to the World: The Development of Australia's Marketable Image in the 1980s." *Consumption Markets and Culture* 13, no. 1 (2010): 43–59.

Crofts, Stephen. "Re-imaging Australia: *Crocodile Dundee* Overseas." *Continuum: The Australian Journal of Media and Culture* 2, no. 2 (1989): 129–42.

Foley, Malcolm, and John Lennon. "JFK and Dark Tourism: A Fascination with Assassination." *International Journal of Heritage Studies* 2, no. 4 (1996): 198–211.

Foley, Malcolm, and John Lennon. *Dark Tourism: The Attraction of Death and Disaster.* London: Continuum, 2000.

Fox, James A., and Jack Levin. *Overkill: Mass Murder and Serial Killing Exposed.* New York: Plenum Press, 1994.

Frost, Warwick. "Life-changing Experiences: Film and Tourists in the Australian Outback." *Annals of Tourism Research* 37, no. 3 (2010): 707–26.

Gibson, Dirk C. "The Relationship Between Serial Murder and the American Tourism Industry." *Journal of Travel & Tourism Marketing* 20, no. 1 (2006): 45–60.

Golanska, Dorota. "Affective Spaces, Sensuous Engagements: In Quest of a Synaesthetic Approach to 'Dark Memorials'." *International Journal of Heritage Studies* 21, no. 8 (2015): 773–90.

Hanich, Julian. *Cinematic Emotion in Horror Film and Thrillers: The Aesthetic Paradox of Pleasurable Fear.* New York: Routledge, 2010.

Heather Kennet, "Sick Snowtown Souvenirs Are Making a Killing." *The Sunday Telegraph*, July 15, 2012, www.dailytelegraph.com.au/news/nsw/sick-snowtownsouvenirs-are-making-a-killing/news-story/f0058ee8ee42e329c4f27943fe6a82d6.

Hooper, Glenn, and John J. Lennon, eds. *Dark Tourism: Practice and Interpretation.* London: Routledge, 2017.

Horeck, Tanya. "'A Film That Will Rock You to Your Core': Emotion and Affect in *Dear Zachary* and the Real Crime Documentary." *Crime, Media, Culture* 10, no. 2 (August 2014): 151–67. https://doi.org/10.1177/1741659014540293.

Jarvis, Brian. "Monsters Inc.: Serial Killers and Consumer Culture." *Crime, Media, Culture* 3, no. 3 (2007): 326–44. https://doi.org/10.1177/1741659007082469.

Keane, Daniel, and Patrick Martin. "Life After Death: Dark Tourism and the Future of Snowtown." *ABC News*, May 20, 2019. www.abc.net.au/news/2019-05-20/can-snowtown-ever-shake-off-its-dark-past/11082778.

Kim, Sangkyun, and Gareth Butler. "Local Community Perspectives Towards Dark Tourism Development: The Case of Snowtown, South Australia." *Journal of Tourism and Cultural Change* 13, no. 1 (2015): 78–89. https://doi.org/10.1080/14766825.2014.918621.

Lennon, John. "Dark Tourism Visualisation: Some Reflections on the Role of Photography." In *The Palgrave Handbook of Dark Tourism Studies*, edited by Philip Stone, Rudi Hartmann, Tony Seaton, Richard Sharpley, and Leanne White, 585–602. London: Palgrave Macmillan, 2018.

Lennon, John, and Malcolm Foley. *Dark Tourism.* London: Continuum and Thompson Learning, 2000.

Martini, Annaclaudia, and Dorina Maria Buda. "Dark Tourism and Affect: Framing Places of Death and Disaster." *Current Issues in Tourism* 23, no. 6 (2020): 679–692. https://doi.org/10.1080/13683500.2018.1518972.

McGowan, Michael. "Ivan Milat's Chilling Serial Backpacker Murders Still Haunt Australia." *The Guardian Australia*, October 27, 2019. www.theguardian.com/australia-news/2019/oct/27/ivan-milat-chilling-serial-murders-haunt-australia-after-death.

Melwani, Daisy. "Backpacker Murder Tour Banned." *Travel Weekly AAP*, July 15, 2015. www.travelweekly.com.au/article/backpacker-murder-tour-banned/.

The Mercury. "Tasmania Invites You to Share Her Charms!" *The Mercury*, September 9, 1926. http://nla.gov.au/nla.news-article29458084.

Morgans, Julian. "A Sad Walk Through Australia's Serial Killer Forest." *Vice*, June 30, 2016. www.vice.com/en_au/article/wdb5v4/a-very-sad-walk-in-australias-serial-killer-forest.

Olsen, Daniel, and Maximiliano E. Korstanje. *Dark Tourism and Pilgrimage*. Oxfordshire: CAB International, 2020.

Peterson, Jennifer Lynn. *Education in the School of Dreams: Travelogues and Early Nonfiction Film*. London and Durham: Duke University Press, 2013.

Podoshen, Jeffery. "Dark Tourism Motivations: Simulation, Emotional Contagion and Topographic Comparison." *Tourism Management* 35 (2013): 263–71.

Riley, Roger, Dwayne Baker, and Carlton S. Van Doren. "Movie Induced Tourism." *Annals of Tourism Research* 25, no. 4 (1998): 919–35.

Rojek, Chris. *Ways of Escape: Modern Transformations in Leisure and Travel*. London: Palgrave Macmillan, 1993.

Rosenbaum, Adam T., Rudi Hartmann, John Lennon, Daniel P. Reynolds, Alan Rice, Adam T. Rosenbaum, and Philip R. Stone. "Discussion: The History of Dark Tourism." *Journal of Tourism History* 10, no. 3 (2018): 269–95. https://doi.org/10.1080/1755182X.2018.1545394.

Ruoff, Jeffrey, ed. *Virtual Voyages: Cinema and Travel*. London and Durham: Duke University Press, 2006.

Scott, John, and Dean Biron. "*Wolf Creek*, Rurality and the Australian Gothic." *Continuum* 24, no. 2 (2010): 307–22.

Seaton, Anthony. "Guided by the Dark: From Thanatopsis to Thanatourism." *International Journal of Heritage Studies* 2, no. 4 (1996): 240, 243.

Sharpley, Richard, and Daniel Wright. "Disasters and Disaster Tourism: The Role of the Media." In *The Palgrave Handbook of Dark Tourism Studies*, edited by Philip Stone, Rudi Hartmann, Tony Seaton, Richard Sharpley, and Leanne White, 335–354. London: Palgrave Macmillan, 2018.

Sharpley, Richard, and Philip R. Stone, eds. *The Darker Side of Travel: The Theory and Practice of Dark Tourism*. Bristol: Channel View Publications, 2009.

Steenberg, Lindsay. "Crime and Celebrity." In *Oxford Research Encyclopedia of Criminology*, 1–26, March 29, 2017. https://oxfordre.com/criminology/view/10.1093/acrefore/9780190264079.001.0001/acrefore-9780190264079-e-26.

Stone, Philip, Rudi Hartmann, Tony Seaton, Richard Sharpley, and Leanne White, eds. *The Palgrave Handbook of Dark Tourism Studies*. London: Palgrave Macmillan, 2018. https://link.springer.com/content/pdf/10.1057/978-1-137-47566-4.pdf.

Tarlow, Peter. "Dark Tourism: The Appealing 'Dark' Side of Tourism and More." *Niche Tourism: Contemporary Issues, Trends and Cases*, edited by Marina Novelli, 47–58. London: Routledge, 2005.

Tromp, Belinda. "Horror Film *Wolf Creek* Brings Tourists to the Outback." *ABC*, February 17, 2014. www.abc.net.au/radionational/programs/archived/bushtelegraph/horror-film-wolf-creek-brings-tourists-to-the-outback/5265056.

Whittaker, Mark, and Les Kennedy. *Sins of the Brother: The Definitive Story of Ivan Milat and the Backpacker Murders*. Sydney: Pan Macmillan, 2001.

11

DOWN THE RABBIT HOLE

Disneyland Gangs, Affective Spaces, and Covid-19

Stephanie Williams-Turkowski

Introduction

Walking down Main Street USA isn't just a song cast members sing as you enter Disneyland's gates, it is a mantra for many fans of Disney theme parks who enter an evocative and immersive space, a place where narrative, symbolism, and emotional attachment merge into a tangible text. This heterotopia is described perfectly by the plaques above the tunnel entrances into Disneyland: "Here you leave today and enter the world of yesterday, tomorrow, and fantasy." The scenery, the aromas, and the sounds are unforgettable, an intense cultural experience that encompasses visitors as soon as they enter. Time does not exist in theme parks, as it does outside of them. Instead, nostalgic ideas, dreams, and futuristic mindsets all come together to provide a narrative escape for visitors. To avid Disney fans, the pilgrimage to the resort parks is a celebration of identity, a symbolic escape, and a realm in which fantasy becomes reality. But for one fan group in particular, the Disney parks have become a collaborative stomping ground for chaos, festivity, and acceptance.

This chapter considers how the theming and affective landscapes of Disneyland Resort and California Adventure generate fandom identity and ethos, specifically the Social Clubs of Disneyland, who have publicly become known as the "Disneyland Gangs." This study, undertaken over several weeks of onsite interviews and observation at the theme parks in Anaheim, CA (pre-Covid), and through virtual interviews conducted after the Disneyland Resort reopened to guests, examines how the Disney Gangs/Social Clubs use the park's landscape to provide a sense of belonging, transportation, and identity before the pandemic and how new regulations have changed several dynamics of their fandom.

DOI: 10.4324/9781003327585-11

Disneyland as a Heterotopic Space

The aesthetic qualities of Disney theme parks allow visitors to step into their favourite childhood movies and become a character within them. Theme parks, like Disneyland, invite visitors to enact their imaginations and become a part of the culture they love. With the variety of themed spaces, sounds, smells, and the interaction with characters, visitors are immersed in the space. For instance, Disneyland and Disney World are both divided into different lands. These lands are distinguished from each other by colour, smell, architectural styles, landscaping, shops, music, and landmarks.[1] These lands allow nostalgic ideas, dreams, and futuristic mindsets to all come together and provide a narrative escape for visitors. "The theme park can be defined as: A social artwork designed as a four-dimensional symbolic landscape, evoking impressions of places and times, real and imaginary."[2]

Many who visit theme parks are looking for authentic experiences, and this authenticity comes from the multisensory, multi-experiential space that theme parks, like Disneyland, provide. Simply visiting the park is not enough, the Walt Disney Company has put decades of research into creating a barrier from the outside world of "symbolic landscapes of cultural narratives."[3]

By providing an immersive space that brings everything to life, these affective spaces provide freedom to guests where they can "meet" characters, live out stories, and become a part of the narrative in a special way. Whether performing as a fan or experiencing it as a casual visitor, theme parks allow guests to create authenticity with time, attention, and detail with visceral involvements. Those who visit the parks give them authority, which allows them to invest themselves in the narrative, "letting the object of such investments speak for him or herself."[4] There is an architectural form that depicts the cultural context in which we live by becoming icons and symbols, such as Disney's Castle, that communicates what is important to those societies.[5] "Disneyland's power, shared with other compelling landscapes, is rooted in culture and shaped as a matrix of cues that point to our most passionate and deeply felt ideas about who we are and how we think as people."[6] Through these ideas, Disney encompasses a close relationship between identity and caring, a correlation that signifies the affective quality of the brand and parks.[7]

Part of the appeal of Disney is being able to interact as an active audience member instead of a passive one at home. It is about going beyond the basic need for entertainment and escape and into consumption of the brand through immersive experiences. By doing this, guests must be engulfed in the spaces around them. Fans actively constitute places through the organization of affective investments by making certain things and practices matter. By doing so, fans authorize the affective places and items to become spokespersons for their identity.[8]

Disneyland, in comparison to other Disney parks in Florida or outside the United States, encourages this investment because it is the original of all the parks; the park that Walt Disney himself planned and walked through. There you will find that characters roam more freely throughout the park (pre-Covid) for spontaneous interactions, bringing the interactive element of theme parks front and centre compared to the other parks. Disneyland is also significantly smaller than its counterpart in Florida: Disneyland houses only 500 acres of land compared to Walt Disney World, which boasts 27,520 acres. There is also the difference in crowds. Walt Disney World may be bigger, but Disneyland has more annual passholders visiting the park each day, creating a different atmosphere for those visiting for vacation trying to maneuverer around those who are popping into the park to eat or catch a show, creating very different experiences for fans when they visit.

It is because each park has their own unique personality that they become a medium of study, a text in which consumers are transported to different lands with characters. "A text can only mean something in the context of the experience and situation of its particular audience."[9] Disney itself provides these experiences through strategic tactics and simulations that can be found in their landscape and ride theming, character meet-and-greets, merchandizing, and signage.

The Disney parks provide one of the biggest reprieves from reality not only as a staple in the theme park industry but as a cultural icon. The Disney parks aren't just a material realm of, but a state of mind.[10] "The structure is cinematic. Visitors are placed in the middle of scenes that unfold in a carefully choreographed manner as they move through them."[11] This translates to Disney being a cultural brand that anyone and everyone can relate too, a community memory of sorts. But this community memory only becomes significant when the text, or theme park in this matter, serves a function, and that function can be different for everyone, including the Disney Land Social Clubs.

Heigh Ho, Heigh Ho, It's Off to Disneyland I Go

In the summer of 2019, I spent a total of 28 days visiting Disneyland and California Adventure from 19 June through 9 July, and then returned for Dapper Day Weekend 31 October through 3 November 2019. During that time, I spent an average of four to eight hours at the park interacting in the Disney Gang/Social Club (SC) scene. I first heard of this unique fandom from a colleague who was from Southern California and frequented the parks on several occasions. It was her description that enticed me to investigate why adults would want to change the aesthetic feeling of a Disney Park, usually associated with innocence and magic, by mixing Mickey with the iconography of the FX show *Sons of Anarchy*.

Known for their rowdy behaviour and controversial attire (Figure 11.1) while visiting the parks, club members use the parks to escape, socialize, and provide unique experiences separate from those visiting the park as casual fans. Additionally, their compelling narratives often describe how they view the park and use it as a form of expression and identity. This provides an understanding of affective landscapes through the lens of fandom culture, for a "text can only mean something in the context of the experience and situation of its particular audience."[12] The SC community, which prefers their "social club" title over "gang," is made up of several clubs, each presenting a discernible hierarchy of voted-in members who are

FIGURE 11.1 Members of various Disneyland Social Clubs take a photo during a ride take over.

seeking other non-traditional Disney fans with whom to celebrate their fandom. Similar to the Lost Boys of Neverland, these fans are cloaked in denim vests with club insignia patches on the back and roam parks for a purpose:

> *[F]or members to feel like they belong somewhere. Too often, those who don't fit the mould – the artists, the bikers, the self-proclaimed freaks – find themselves ostracized and alienated from society at large. The clubs give them a place to commune with others like them – who also happen to love Disney.*[13]

Starting with The Neverlanders in 2013, the SCs have grown from a few clubs made up of annual passholders to more than 150, with many more looking to establish themselves. Disneyland SCs revolve around Disney characters, narratives, rides, and even Walt Disney himself, with names like The White Rabbits, The Flynn Riders, Walt's Most Wanted, and many more, encompassing anything and everything Disney. Some clubs are smaller with less than ten members, while others boast (about) large memberships of 200.[14]

Although the clubs claim that they are not gangs, you can see many of them in the parks behaving in ways that might seem aggressive: having a drink, aiming to take over rides and areas, and creating possible access issues for the general visiting public. Many of the group members say that their aim is to simply celebrate Disney and participate in charitable giving[15] but that does not stop other fans and park-goers from judging them:

> *Not only do Disneyland social club members take their fandom to the next level, but the way they present themselves to the world – and the way they're perceived by other park guests – tells a complex story. Nonetheless, the outsider aesthetic causes a lot of problems for the social clubs. Numerous Reddit comments gripe about the social clubs' habit of "ride takeovers" (basically, ensuring that all members make it onto a ride vehicle at the same time); taking up too much space in congested areas; or perhaps most amusingly, shouting obscure Disney trivia at each other while waiting in line.*[16]

But to the clubs, their presence in the parks isn't meant to cause controversy, but rather, they are there to enjoy the parks just like any other fan.

Although the clubs state that they are not gangs, a lawsuit has been filed between two clubs regarding protection money connected to a charity walk, causing a negative narrative towards all clubs.[17] Deemed as a "dark undercurrent" to this Disney fandom, this topic is a sore spot for several in the community who say that this episode does not represent them or their fandom in the best light.

What's in a Vest?

The White Rabbits' reputation came with many warnings before I ever approached one of the members. Park-goers told me of their unruly behaviour and gang-like mentality. Other SCs described them as old school compared to their "new school" club approach, warning me that they would not want me talking to them. And of course, the media narratives surrounding *any* Disney SC all but assured me that conversations with these groups would be impossible. It was not until I had been turned away as a researcher by several SCs for asking too many questions that I became brave enough to seek out a vest with a white rabbit emblazoned on the back. It just so happened that interviewing Nick, a prominent member of the White Rabbits, would break down the walls I needed to make connections. From there, I was welcomed into the SC community with open arms and six months later, I would become a White Rabbit.

With over 100 SCs visiting the parks every week for Sunday Funday, it comes as no surprise that guests feel their presence; there is a change of authority and identity when the clubs are around. This change in affective investments can be attributed to how the clubs identify with the parks compared to a casual visitor. The Disneyland resort is not just a theme park for the SCs, but a second home, a place where being an adult who loves Disney is never questioned. But the identity that the media and other park goers have given the SCs is not one they enjoy or want to adopt. To them, they are simply Disney fans enjoying the park.

> *I think it's the total escapism. I think that's what sets it apart from other theme parks, is that is especially Disneyland or Magic Kingdom. They try their absolute best to*

> *engulf you in their brand of magic and it works completely. They are trying to get you to step out of your life and have this amazing experience, so, it is totally overwhelming but it's, it's a good overwhelming most of the time. They provide a like 360 degrees, all senses experience on every level.*[18]

But as they enjoy the park and total affective immersion it provides, many SC members wonder what is so off-putting about them. Is it their aesthetic which is not always the squeaky-clean persona that many Disney fans have, but instead a vibrant masterpiece of tattoos and brightly coloured hair? What about their vulgar, yet hysterical theatrics when together? Or is it simply the stigma that comes with the three-piece patch set attached to a sleeveless denim vest, a symbol of many biker organizations? Jack, the president and founder of the White Rabbits, reflects upon the impression made upon others in the park environment by SC members and the dichotomy it provides:

> *They're always going to talk shit. Do you think if we would have been wearing, like, collegiate cardigans, we wouldn't have got the same shit as wearing the denim biker vest? Probably not. I guess it would be no different than a large group of adults wearing the same t-shirts.*[19]

Jack's sentiment regarding the attire is just one opinion among the SC community. Some agree with him, while others honour the tradition and sentiment that comes with the vest, tagging them as outsiders. The deep-rooted oppositional nature of the biker vest causes a push and pull on the SCs. The association of a patched vest is inherent in the biker culture, especially within the Harley Davidson sub-culture indicating a "hard-core and outlaw" ideology that is associated with acting out and being wild in the general public[20] Significant resonance associated with such iconography carries symbolic weight in many cultures, and even with further integration among other groups or fandoms, these associations can be hard to shake.[21] To some, the idea of the vest and its ideology and past is an interesting dichotomy within the parks. All in all, the SC scene has made a mark in the Disney fandom and certainly within the walls of Disneyland. Daniel, a member of Tomorrowland's Ravengers, says, "We're nothing but Disney nerds in vests"[22] but, in reality, SCs represent far more. By being outsiders in a place known around the world, members of different SCs have bonded together to show that being a Disney fan can be represented in a lot of different ways (Figure 11.2).

However, some of the most critical judgements have come from others in the overall Disney fandom. Jack states:

> *The social club community have actually had to battle a bad rap amongst the Disney fan subcultures, too. In fact, some of the biggest haters of the social club community come from within other Disney subcultures and it's mainly people that have never hung out . . . they never really talked to them.*[23]

FIGURE 11.2 Members of The White Rabbits and the Bangerrang Babes pose for a "Sunday Funday" photo at Disneyland.

Source: Photo provided by a member of the Bangerrang Babes.

Because affective experiences are so tightly related to emotion,[24] anything outside of the normal Mickey Mouse aesthetic is usually seen in the parks as dissimilar or negative. But the SC affiliation to Disney isn't necessarily wrong, it just creates a different dynamic. The parks become not only platforms for fans to tangibly interact with the narratives but also a place to represent their fandom without judgement.

King[25] describes theme parks as "social artwork(s) designed as a four-dimensional symbolic landscape, evoking impressions of places and times, real and imaginary." Roxanne, the President of the Bangarang Babes, described going to Disney as magic, a place that transports you to a different world:

> *[A]s long as you are willing to just be a kid and go with it. But like, you can't be too cool, you can't be stuffy, to be any of that, you have to just leave all that at the gate and go in and like you're seven years old, and have the time of your life, and if you do, then all of a sudden, there's no strings on anything . . . and you can just let it take you away.*[26]

Theme parks also establish an immersive space that brings everything to life providing a sense of connection, loyalty, and freedom to guests within the park.[27] The aesthetic qualities of a theme park allow fans to live out experiences that they have read or seen through attention to narrative detail. By interacting and sharing the thematic spaces of the Disneyland Resort, the social club fandom has created their own habitus.

With so many different clubs, what sets each one of them apart is not so much the vest, but the patch on the back. The characters, rides, and names that each club wear represents who they are and what they stand for. But that is just the back of the vest. The fronts of the vest are weighed down with Disney pins, non-Disney pins, supporter patches, and Disney-related patches. "The entire vest is steeped in enough symbolism to rival the most medal-heavy military uniform, combining the romance of group membership with the thrill of self-expression."

Inclusion and Exclusion as a Club Member

Wearing a vest in the park was an interesting experience. I became a member of the White Rabbits before I returned to Disneyland for the second part of my data collection. Unlike many of the SCs, the White Rabbits do not have an initiation process. If the President, Jack, meets you and thinks you fit in, you are good to go, but you get three strikes to "fuck up" and after that, you are out. Many of the other clubs make their new members go through several months of tasks to join, getting them acquainted with the park and other clubs. But no matter the process to get in, once you are a part of the SC, you are treated like family – by those with vests on, at least. The cast members and other park-goers are a different story.

It was interesting to see how cast members treated me, especially when going through security. I was often stopped and talked to longer by cast members while checking my bag. They would ask what I was up to in the parks that day or ask me about my pins to ensure I was there slightly longer than when I did not wear my vest. I was also not called "princess" by cast members while wearing my vest (Figure 11.3), a common signifier they use for women of all ages when interacting with them in the park. I do not want to say that I was being monitored or singled out, but it felt as if I was being watched more than the average guest. When one wears a

FIGURE 11.3 Stephanie Williams-Turkowski poses for a "back patch" photo during her first visit as a White Rabbit in the parks.

vest in the park, even if they are not part of a group, they are noticed. The vest and patch unite the members, but it also singles them out as outsiders in comparison to others in the park that may represent a cleaner and family-friendly aesthetic:

> *Disneyland and its Magic Kingdom clone lead visitors through scenes that depict happiness, innocence, industriousness, and resourcefulness – all watered-down versions of morality play virtues but nonetheless appropriate to a "have a nice day" America. These mini virtues are found in a quasiutopian context of cleanliness, friendliness, safeness and relative efficiency. These positive characteristics – the Disney version of family entertainment – triumph both narratively and literally over meanness, anger, rudeness, idleness, slovenliness, and critical thinking.*[28]

So how does wearing a symbol of violence in American culture change the experience? Well, personally, it changed the way in which I performed within the park and how I was treated by others. I wasn't a young adult enjoying rides and the theming of my favourite films; I felt as if I was a part of something more. I was symbolically tagged and used the theme park not as a playground, but as an association with an authority. With affect comes different forms of authenticity[29] and while wearing a vest, my new affective investments in the park weren't always accepted by others. Sandvoss[30] states

> Fandom is facilitated by a meaningful and affective bond between the self and the object world, and the deep connection one shares with a community is manifested by central aspects of the fan's identity and the values on which one reflects.

Focused on creative practices that take a text far beyond casual enjoyment, fans are often seen as active audiences that closely identify with, interpret, and sometimes rework texts through various interactions, both individually and with a fandom community. I wasn't simply in a place to have fun; I felt as if I owned a piece of the parks. I was emotionally tied to more than just the characters and media, and I was socially entwined into a community that used the thematic elements and areas as their headquarters.

Known for taking over areas of the park for Sunday Funday Meetups and trying to fit as many SC members as possible in line for a ride at a time, their love for all things Disney does not go unnoticed. Take, for instance, Nick the first White Rabbit I spoke to. He is a very tall male and stands apart from the crowd because of his height. Add a biker vest, his tattoos, and piercings, and he could be taken as a menacing individual. Therefore, he takes as much time as needed to talk to anyone that asks about the SC scene, the White Rabbits, or just his love for Disney, because he wants to shed the negative association with the community. He goes on to explain that because he wears a vest, he, along with other members of the SC community, take it upon themselves to be a governing force:

> *For me, it's different than any other fan, because I have a vest on my back and social clubs hold each other to a very high standard – like, beyond any standard a normal guest would have. Because if I'm a jerk and I'm walking around without my vest on, they just say, "oh, it's some tall dude with a black hat on backwards." But if I'm a jerk with my vest on, then they're like, "Oh, it's a tall dude with a black hat AND a White Rabbit vest on." It's a lot easier to identify.*[31]

Prior to being patched in as a White Rabbit, if I turned up to a Sunday Funday hangout, I was questioned by many members about my intentions and who invited me. Their reputation has been so tainted by media, rumours, and a few individuals' actions that they wanted to make sure anyone who came around was authentic to the cause. When I returned in November with a patch on my back, the atmosphere changed significantly. I was no longer the outsider asking questions and intruding on their time at the parks. I was one of them. I was welcomed with smiles and hugs rather than cautious looks and muted greetings. It was just accepted that I was going on rides with members and a part of the conversations, which changed from guarded answers to hilarious antics. For instance, when I returned, I was immediately put in line with several clubs to ride Pirates of the Caribbean. I had ridden rides with the SC community before, but they all seemed to be on their best behaviour and talked normally among each other. This time it was different. We got into our boat on the ride and immediately the chastising of others on the ride began. Inside of Pirates is a very expensive restaurant where guests are eating and can see others riding the ride and vice versa. Before the vest, they would have enjoyed the ride together, but with a vest, I was witness to them yelling at restaurant-goers asking "those bougie-ass people to throw them some rolls." To be honest, I laughed along with them and realized that they were just being themselves, despite the many families aboard with us. Seeing the parks in a different light also shed insight into just how different my experience is when the affective collectively shifts from one group to another.

Park Rituals

The change in atmosphere and interaction allowed me to see a different side to the SC community, one that represented them so much better than the guarded persona I had encountered before. However rough their external shell, they were simply kids at heart enjoying the parks in a more adult and humorous way. But to the SC community, it is not just about visiting the park and seeing each other. To them, being a part of a family atmosphere that is so huge is about giving back to each other and those outside of the community, especially children, because that is so inherent in Disney's culture.

Many clubs raise money for charity by selling support patches for their club. These patches dictate a friendship among the hundreds of clubs and allow members of different SCs to support one another. This support is often seen when new members are patched in at "patching ceremonies" which are usually held somewhere in the park. Members from all over the community gather to say welcome and congratulate newly established members. But the support does not end there.

One ritual among the SC community is using social media to "check in" to the parks by posting a selfie or back-patch picture to indicate that they are in the parks and ready to hang out. Each club has a particular place in the park that has been claimed as their check-in spot and often, many other clubs will post selfies at each other's check-ins to show support. Their appropriation and symbolic ownership of park spaces interweaves their narrative into the park even more. Again, they are not simply park goers, but act as if they own the park. By claiming certain areas as their own, they are using the close relationship between identity and caring to create their own emotional significance and intuitive space with the already established narrative. Most importantly, however, "it is a search for an unmediated experience, of putting oneself, literally, in the object of fandom and the self that goes beyond mere consumption and fantasy."[32] By doing this, the SC members are using the space in a unique way that it unravels around them, engulfing them into the narrative.[33] As Koren-Kuik[34] states:

> *Disney encourages participatory fandom . . . the jewels in the crown of this participatory fandom construct are Disney's theme parks because they offer the consumer/fan a spatial platform which brings all the elements together in the experience of walking or riding through narrativized space. The land of Disney relies on two elements to ensure its success: a particular kind of utopia and a special type of fandom.*

This "utopia" is provided by the affective spaces that allow visitors to relive nostalgic experiences. The SC experience uses their rituals to provide a utopia for those highly involved in the community. Theme parks also establish an immersive space that brings everything to life providing a sense of connection, loyalty, and freedom to guests within the park.[35] So, what happens when the park is no longer available?

412 Days Without Magic

"It felt like part of the magic had been taken away from the world" was a common sentiment among the SC community after Disneyland was shut down due to Covid-19. Their second home, their security, their fellowship, and their escape had been taken from them during a time when many were looking for an escape. The entire Disneyland fandom went without access to the parks for a total of 412 days. Although the parks partially opened for shopping and a small food festival, the vast majority of the Disney experience was inaccessible for over a year after closing down in March 2020.

As with other sectors of life in the time of Covid-19, some social club activities were pursued virtually. Many clubs met through Zoom with Disneyland-themed backgrounds or held get-togethers with safe distancing once restrictions started to soften, but many still felt the pull of normal club activities within the park. To fill that void, some SC members began exploring other options in the Disney fandom. Roxanne, for example, turned to *Disney+*, the Disney streaming service that launched the November before the pandemic:

> *Covid hit, and they close the parks, and I kind of had to think, am I as big a Disney fan, as I thought I was without the parks? It made me realize that my particular brand of Disney fandom centres around the Disney parks . . . but that changed when I was at home without the ability to go to the parks.*[36]

The constant changes and updates to park policies and experiences, however, had some members wondering if this could be the end to the Disney aspect of the SC experience:

> *If anything, I think that Covid has proven that social clubs don't need Disneyland, as much as social clubs want Disneyland. I don't think it's a necessity in order for the social clubs to exist. . . . I think social clubs will adapt. I just. . . . I would love to see social clubs get back to the way they were, but with the limitations that Disney has put on annual passes right now, it's not going to happen anytime in the foreseeable future.*[37]

However, even after the parks opened back up to the public on 30 April 2021, the atmosphere wasn't quite the same. Many members of the SC community, as well as the general public, believed that although it was great for the safety of everyone, requiring masks and displaying hand-washing stations throughout the park took away from the escapism. It brought the real world into the parks, creating a dichotomy in guest experiences. As Roxanne suggests, "you can leave your bills at the gate [of the park], and you can leave your job and the argument you got in with your co-worker at the gate, but something like Covid-19 . . . you can't."[38]

Restrictions tied to the re-opening of the park included limiting capacity to 25 percent, temperature checks, requiring masks indoors and outdoors, social distancing, mobile food ordering, spaced-out ride lines, no shows, no parades, and infrequent character sightings at extreme distances. While many fans loved the smaller crowds for once, the overall ambience seemed to be missing an important affective quality: magic. Brandon, a member of the 33rd Alley Cats, believes that this will have the biggest impact on clubs returning.

> *Many of the changes that have been put into place have tremendously impacted the community. Disney was once a place of location that pulled us together. The Magic is gone and never returned from the opening. Some days it is very difficult to fully enjoy the Disney experience we once had before the pandemic closure.*[39]

Within fandom, affect plays a crucial role in how fans organize their social life and for the SC community, the new restrictions placed on the parks, the clubs found themselves in a constant limbo in achieving stabilization of their fan identity. Fandom is also a measure of self-sustained engagement and fan groups that are more resilient often thrive, while those that aren't have difficulty creating a robust community.[40] Although the Social Clubs have seen their fair share of drama within the community, the parks closing measured their true dedication to the patch. Without the parks or the ability to attend on a regular basis with their annual passes, their group cohesion and activities were often challenged. Gone were weekly rituals, meetings, patch-ins, and ride takeovers. Without these rituals, or their hypothetical glue that binds the SC community together, the community began to see they must adapt with the parks new rules or die out. Older clubs seemed to adapt better than clubs formed in the last few years, showing their dedication meant more than visiting the theme parks.

As the clubs adapted, so did the parks. Disneyland executives are continuing to change the experience, causing a volatile shift in the experience. For one, the annual passes system was dissolved and instead turned into a reservation system, creating an even bigger lull in the club's Sunday Funday adventures. Without annual passes and reservations, it became almost impossible to plan a Disney day with your club. Instead, individual members were going with their families, some wearing their vests, and others deciding that it wasn't worth it if there wasn't a strong presence. Many members, on the other hand, gave up the idea of going altogether, deciding the cost of a ticket wasn't worth the lacklustre experience. Others simply couldn't afford a ticket, and many were unsure of venturing into public spaces.

The ease of being able to visit the parks whenever an annual pass holder is a significant part of the SC scene and without it, coming together as a community after the shutdown became even more difficult. Trying to make reservations for large groups on the same day was virtually impossible, especially when you could only book nine visits within 90-days. The overall Disney community was not happy with the reservation system, but recently, as if Disney executives heard these wishes, annual passes were re-branded and marketed as Magic Keys on 25 August 2021, four months after the re-opening.[41] Magic Key holders, who still have three reservations at a time, are placed in a separate pool within the reservation system than the regular public and are disciplined if they miss a reservation. This new venture allowed members of the SC community to see a light at the end of the metaphoric Disneyland entrance tunnel. But others aren't concerned about the Covid restrictions, but more about how these changes will impact the park experience going forward. Nick said the experience will never be what it was before, stating it won't be freeing, but more about Disney having more control over fans through continual ticket price increases, reservation restrictions, and the ongoing push for visitors to spend more money. Some of the blame goes to the pandemic, but many of the SC Community, as well as various other parts of the Disney fandom, blame the new leadership. Brandon said the greed, in particular, is what caused the biggest issues with the reopening, stating, "Disneyland's number one slogan is 'Bringing the Magic Back,' but the greed never left, prices are still soaring and leaving people to think if it's worth the Disney 'quality' anymore."[42]

But almost a year after the reopening, the parks are back to full capacity causing not only crowded parks with limited spacing in an ongoing pandemic but also limited access to the park, even with a Magic Key. The reservation system still plagues many members of the community, but they adapt and have begun to create a new normal. Yet, as soon as they feel as if things are back to normal, the parks shift again. Similar to what Nick and Brandon described, Disney has started to introduce new ideas that require more money from visitors. Not only do the Magic Keys cost more than the previous annual pass did with restricted access, but merchandise and food prices have also increased. But one of the biggest changes comes with ride rituals. Now guests are seeing the introduction to Genie Plus and lightning lanes, the new Fastpass system that used to be free, is now a significant up charge. In the past, a group such as the SC could all grab fast passes for a ride at a certain time, now they must wait for what seems like hours to overtake rides or hang out in certain spaces within the park, interrupting rituals that have been in place for years.

However, more and more clubs are returning to the parks and the ongoing love for the community is being re-established, as well as small details that they say bring the magic to the parks, such as getting to visit and hug characters without social distancing and more exclusive Disney events being held for pass holders. It will be interesting to see how changes in the next few years will impact the community as the park continues to adapt, change, and reconstruct to account for Covid-19, economic changes, and fan tolerance.

Different Patch, Same Love – Different Times, Same Clubs

The overall relation between fans and their object, such as that of the Social Club community to the parks, relies on their interaction and the subculture and identities they have created to function within both spaces. What once was a place where you could leave the real world and its troubles outside the entrance gates is now inundated with rules and regulations on how you must participate in themed worlds. Even with Covid-19 restrictions being lifted every day and "normal" operations returning to the park, the overall experience has changed not only the physical theming but also those that visit it.

Many Social Clubs have tried to pick up where they left off before Covid, while others have become disjointed, and many new clubs are being formed by newcomers to the community and Disneyland fandom. How the community and the parks will co-exist will continue to change, just as the rest of the world continues to adapt. However, after speaking to so many members of this community and becoming a member myself, I believe that "Gangs of Disneyland" will continue to thrive, as they are very self-sustaining group of individuals who, despite being

FIGURE 11.4 Crowd in Front of Castle in Dropbox. The Social Clubs of Disneyland pose for their first annual Castle Photo post-Covid-19 to show off their patches and celebrate being back in the parks.

Source: Photo provided by the Bangerrang Babes Social Club.

different from others in their fandom, have found each other (Figure 11.4). Similarly, they will adapt to a new era in the Social Club scene that reflects who they are within a new Disneyland Park and how it feels on a Sunday afternoon.

Notes

1 Arnold Berleant, "The Critical Aesthetics of Disney World," *Journal of Applied Philosophy* 11, no. 2 (1994).
2 Margaret J. King, "The New American Muse: Notes on the Amusement/Theme Park," *The Journal of Popular Culture* 15, no. 1 (1981): 56–62.
3 Margaret J. King, "The Theme Park: Aspects of Experience in a Four-Dimensional Landscape," *Material Culture* 34, no. 2 (2002): 1–15.
4 Lawrence Grossberg, "Is There a Fan in the House?: The Affective Sensibility of Fandom," in *The Adoring Audience* (Routledge, 1992), 50–65.
5 Umberto Eco, "2. Function and Sign: Semiotics of Architecture," in *The City and the Sign* (Columbia University Press, 1986), 55–86.
6 Margaret J. King, "The Theme Park: Aspects of Experience in a Four-Dimensional Landscape," *Material Culture* 34, no. 2 (2002): 1–15.
7 Lawrence Grossberg, "The Space of Culture, the Power of Space," in *The Post-Colonial Question* (Routledge, 2002), 169–186.
8 Scott A. Lukas, *The Immersive Worlds Handbook: Designing Theme Parks and Consumer Spaces* (London: Routledge, 2012).
9 Lawrence Grossberg, *The Power of Space: The Postcolonial Question: Common Skies, Divided Horizons* (London and New York: Routledge, 1993).
10 Scott A. Lukas, in *Theme Park* (London: Reaktion, 2008).
11 Stephen M. Fjellman, Vinyl Leaves: Walt Disney World and America (Routledge, 1992).
12 Grossberg, Lawrence. "The Space of Culture, the Power of Space," in *The Post-Colonial Question* (Routledge, 2002), 53

13 Donn Saylor, "Things You Never Knew About Disney Social Clubs," Ranker (Ranker, November 12, 2019), www.ranker.com/list/facts-about-disney-groups/donn-saylor.
14 "Social Clubs of Disneyland," socialclubs, accessed 2020, www.socialclubsofdisney.com/.
15 Michelle Kaminsky, "The First Rule of Disneyland Social Clubs: Don't Demand Protection Money From Another Social Club," (Forbes Magazine, February 20, 2018), www.forbes.com/sites/michellefabio/2018/02/12/the-first-rule-of-disneyland-social-clubs-dont-demand-protection-money-from-another-social-club/?sh=52e03eb32403.
16 E. J. Dickson, "Guys in 'Disney Social Clubs' Are Wilding out at the Happiest Place on Earth. How Long Will the Park Allow It?," (MEL Magazine, January 28, 2019), https://melmagazine.com/en-us/story/disney-social-clubs?fbclid=IwAR0-bJmcClANZ0.
17 Martin Hugo, "Defendants in Disneyland Social Club Lawsuit Are Evading Process Servers, Lawyer Says" (Los Angeles Times, March 14, 2018), www.latimes.com/business/la-fi-disneyland-social-club-service-20180314-story.html.
18 President of The Bangarang Babes, Roxanne. Personal, August 17, 2021.
19 President of the White Rabbits, Jack. Personal, July 9, 2019.
20 Hans Ouwersloot and Gaby Odekerken-Schröder, "Who's Who in Brand Communities – and Why?," *European Journal of Marketing* 42, no. 5/6 (2008): 571–585.
21 Cornel Sandvoss, *Fans: The Mirror of Consumption* (Malden, MA: Polity, 2005).
22 Member of Tomorrowland's Ravengers, Daniel. Personal, June 20, 2019.
23 President of the White Rabbits, Jack. Personal, July 9, 2019.
24 Brian Massumi, *Parables for the Virtual: Movement, Affect, Sensation* (Durham: Duke University Press, 2021).
25 Margaret J. King, "The Theme Park: Aspects of Experience in a Four-Dimensional Landscape," *Material Culture* 34, no. 2 (2002): 387–389.
26 President of The Bangarang Babes, Roxanne. Personal, August 17, 2021.
27 Scott A. Lukas, *The Immersive Worlds Handbook: Designing Theme Parks and Consumer Spaces* (London: Routledge, 2019).
28 Stephen M. Fjellman, *Vinyl Leaves: Walt Disney World and America* (Routledge, 1992), 317.
29 Lawrence Grossberg, "Is There a Fan in the House?: The Affective Sensibility of Fandom," in *The Adoring Audience* (Routledge, 2002), 50–65.
30 Cornel Sandvoss, "Toward an Understanding of Political Enthusiasm as Media Fandom: Blogging, Fan Productivity and Affect in American Politics," *Participations* 10, no. 1 (2013).
31 Member of the White Rabbits, Nick. Personal, June 20, 2019.
32 Sandvoss, Cornel. *Fans: The Mirror of Consumption* (Malden, MA: Polity, 2005), 61.
33 Scott A. Lukas, in *Theme Park* (London: Reaktion, 2008).
34 Meyrav Koren-Kuik, "Desiring the Tangible: Disneyland, Fandom and Spatial Immersion," *Fan CULTure: Essays on Participatory Fandom in the 21st Century* (2013).
35 Scott A. Lukas, *The Immersive Worlds Handbook: Designing Theme Parks and Consumer Spaces* (London: Routledge, 2012).
36 President of The Bangarang Babes, Roxanne. Personal, August 17, 2021.
37 Member of the White Rabbits, Nick. Personal, August 8, 2021.
38 President of The Bangarang Babes, Roxanne. Personal, August 17, 2021.
39 President of the 33rd Alley Cats, Brandon. Personal, December 21, 2021.
40 Zoe Fraade-Blanar and Aaron M. Glazer, *Superfandom How Our Obsessions Are Changing What We Buy and Who We Are* (New York, NY: W.W. Norton & Company, 2017).
41 Disneyland.disney.go.com, accessed 2021, https://disneyland.disney.go.com/magic-key-terms-conditions/.
42 President of the 33rd Alley Cats, Brandon. Personal, December 21, 2021.

Bibliography

Berleant, Arnold. "The Critical Aesthetics of Disney World." *Journal of Applied Philosophy* 11, no. 2 (1994): 171–80.

Brandon, interview by PhD Stephanie Williams-Turkowski. *President of the 33rd Alley Cats*, December 21, 2021.

Daniel, interview by PhD Stephanie Williams-Turkowski. *Member of Tomorrowland's Ravengers*, June 20, 2019.

Dickson, E. J. "Guys in 'Disney Social Clubs' Are Wilding Out at the Happiest Place on Earth. How Long Will the Park Allow It?" January 28 2019. https://melmagazine.com/en-us/story/disney-social-clubs?fbclid=IwAR0-bJmcClANZ0.

Eco, Umberto. *The Semiotics of Architecture: Signs, Symbols and Architecture*. New York: Columbia University Press, 1986.

Fjellman, S. M. *Vinyl Leaves: Walt Disney World and America*. Boulder: Westview Press, 1992.

Fraade-Blazer, Zoe, and Aaron M. Glazer. *Superfandom: How Our Obsessions Are Changing What We Buy and Who We Are*. New York: Profile Books, 2017.

Gary, interview by PhD Stephanie Williams-Turkowski. *Member of the White Rabbits*, August 8, 2021.

Grossberg, Lawrence. "Is There a Fan in the House? The Affective Sensibility of Fandom." In *The Adoring Audience: Fan Culture and Popular Media*, 50–65. London: Routledge, 1992.

———. *The Power of Space: The Postcolonial Question: Common Skies, Divided Horizons*. London and New York: Routledge, 2002.

Jack, interview by PhD Stephanie Williams-Turkowski. *President of the White Rabbits*, July 9, 2019.

Kaminsky, Michelle. "The First Rule of Disneyland Social Clubs: Don't Demand Protection Money from Another Social Club." February 12, 2018. www.forbes.com/sites/michellefabio/2018/02/12/the-first-rule-of-disneyland-social-clubs-dont-demand-protection-money-from-another-social-club/?sh=52e03eb32403.

King, Margaret J. "Introduction: The New American Muse: Notes on the Amusement/Theme Park." *Journal of Popular Culture* 15, no. 1 (1981): 56.

———. "The Theme Park: Aspects of Experience in a Four-Dimensional Landscape." *Material Culture* 34 (2002): 1–15.

Koren-Kuik, Meyrav. "Desiring the Tangible: Disneyland, Fandom and Spatial Immersion." In *In Fan Culture: Essays on Participatory Fandom in the 21st Century*, edited by Kristin Barton and Jonathan Lampley, 146–58. Jefferson, NC: McFarland, 2013.

Lukas, Scott A. *Theme Park*. London: Reaktion Books, 2008.

———. *The Immersive Worlds Handbook: Designing Theme Parks and Consumer Spaces*. London: Routledge, 2012.

Martin, Hugo. "Defendants in Disneyland Social Club Lawsuit Are Evading Process Servers, Lawyer Says." March 14, 2018. Accessed October 16, 2018. latimes.com/business/la-fi-disneyland-social-club-service-20180314-story.html.

Massumi, Brian. *Parables for the Virtual*. Durham: Duke University Press, 2002.

Nick, interview by PhD Stephanie Williams-Turkowski. *Member of the White Rabbits*, June 30, 2019.

Nick, interview by PhD Stephanie Williams-Turkowski. *Member of the White Rabbits*, August 8, 2021.

Ouwersloot, H. and G. Odekerken-Schröder. "Who's who in brand communities – and why?" *European Journal of Marketing* 42, no. 5–6 (2008): 571–85.

Resort, Disneyland. "Disneyland Resort Magic Key Terms and Conditions." n.d. Accessed 2021. https://disneyland.disney.go.com/magic-key-terms-conditions/.

Roxanne, interview by PhD Stephanie Williams-Turkowski. *President of the Bangerang Babes*, August 17, 2021.

Sandvoss, C. *Fans: The Mirror of Consumption*. Cambridge: Polity, 2005.

———. "Toward an Understanding of Political Enthusiasm as Media Fandom: Blogging, Fan Productivity and Affect in American Politics." *Journal of Audience and Reception Studies* 10 (2013): 252–96.

Sayor, Don. "Things You Never Knew About Disney Social Clubs." November 12, 2019. Accessed January 1, 2020. www.ranker.com/list/facts-about-disney-groups/donn-saylor.

Social Clubs of Disneyland. "Social Clubs of Disneyland." 2020. www.disneylandsocialclubs.com/about.

12

IMMERSIVE WORLDS AND SITES OF PARTICIPATORY CULTURE

The Evolution of Screen Tourism and Theme Parks

Carissa Baker, Ray Eddy, and David Bailey

Introduction

While many may feel that screen tourism is defined by enthusiastic fans seeking out famous or obscure locations where popular film and television stories were created or referenced, we argue that studios and theme parks generate their own kind of screen tourism. From the earliest studio tours to the most recent, technologically advanced theme park attractions, a kinship has always existed between film/television and attraction-based visitation. This association has been mutually beneficial, a symbiotic relationship that not only builds a fan's appreciation for a story franchise that can encourage them back to a movie theatre or their television, but also drives repeat attendance in theme park spaces.[1] This fact alone makes an analysis of the relationship between film tourism and theme parks worthwhile.

In addition, this relationship has continued to evolve over time, both interactively and technologically. Starting with the original studio tours that grew into theme park rides, these experiences have become fantastically ornate and complex, more interactive, and deeply immersive, expanding upon existing storylines and even creating new ones. This chapter explores this evolution, starting with how theme park attractions and shows based on film and television franchises evolved from studio tours into various iterations of the "making a movie" style of show. It then examines how these shows have become much more technologically advanced and immersive. And finally, it will demonstrate that some theme park shows, and rides have even become source material for films and television shows, reversing the normal trend in the industry where theme parks expand based on successful story franchises.

DOI: 10.4324/9781003327585-12

Screen Tourism as a Third-Person Experience

While amusement parks have been in existence since the 1500s,[2] their evolution into theme parks is widely considered to have taken place with the opening of Disneyland in 1955.[3] This also marks a time when parks were directly connected to one particular film studio's works. Hollywood certainly took notice of Disneyland's success in driving the Disney brand, and in the 1960s, a number of film studios (Universal, MGM, Fox Studios, Warner Brothers, NBC, and Columbia) created tours of their film and television studios for the same purpose.[4] The effect of this type of tour was to make the viewer feel as though they were given a "sneak peek" into the inner workings of movie-making magic. For Universal, their original tours began way back in 1915,[5] but by 1964 had grown into Universal Studios Hollywood,[6] which evolved into a full-fledged theme park. Disney followed this behind-the-scenes trend by creating their own version of a studio-based park with Disney's MGM Studios (now called Disney's Hollywood Studios) in 1989.[7] Over the years, several other movie-based parks have opened, with recent examples including Cinecittá World in Italy (2014), Bollywood Parks Dubai in the United Arab Emirates (2016), and Huayi Brothers Movie World in China (2018).

While fans of particular films may always want to see remote locations where iconic scenes are shot, such as the mountains of New Zealand from the *Lord of the Rings* franchise or London's Platform 9 ¾ from *Harry Potter*, the motivation behind studio tour and theme park experiences was to create a screen tourism destination. Fans of various films – or simply fans of filmmaking itself – would be motivated to visit these locations, and the fact that the films being highlighted would change over time could drive repeat visits. One cannot ignore the fact that, love of films and filmmaking notwithstanding, production companies only exist if they continue to make a profit.

Backlot tour experiences evolved into more dynamic, active attractions over time. For example, Disney's version (the Backlot Tour) included a scene where randomly selected guests would be placed into the presentation, such as standing at the steering wheel of a boat (wearing some thoughtfully provided raingear) as it was doused in a deluge of water. Another sequence provided the experience of fire, flooding water, and even a (fake) 18-wheel semi-truck sliding down an embankment towards the guests. The culmination of this was a tram ride around to the back of the system, showing how the props reset for the next show, and all of the water and electrical systems that controlled the mock disaster they had just witnessed. This was no longer a view into the making of an actual film or television show, but instead a recreation of how filmmaking works in general.

Universal took this philosophy a step further by modifying their "tours" into fully immersive experiences, albeit less of a "behind-the-scenes" version. Their Earthquake and Twister attractions allowed guests to see and (somewhat) experience tremors and whirlwinds, respectively. Their Jaws attraction went even further, taking guests on a pontoon boat for a tour, during which some playful acting by their "tour guide" would reveal that they were actually being stalked by Jaws himself. An animatronic great white shark would appear lunging towards the boat, creating some terrific jump scares, but of course the boat would always *barely* escape, with the tour guide asking passengers not to reveal to the next party that anything had gone wrong.

The final iteration of the "how movies are made" experience in a theme park was to provide an attraction where the audience is told that they are *actually* on the set where an *actual* film was being created. This was never true, of course, but the illusion was maintained by a mock film crew holding (usually fake) cameras. A primary difference between these "making a movie" shows and a more traditional theme park show is that the former is essentially telling two stories: one is the story being told by the film itself, and the other is the story of how the film is being made. For this reason, this type of show essentially consisted of two separate storylines being told by two separate casts. One cast was the performers in the (pretend) movie being shot, and the other was the (pretend) film crew explaining to the audience exactly what they were doing and why. While a traditional show would simply tell a standard story from beginning to end, a behind-the-scenes show was necessarily more disjointed.

A prime example of this kind of show was *The Indiana Jones Epic Stunt Spectacular* (see Figure 12.1) at Disney's Hollywood Studios (1989), where one of the authors was a cast member for eight years. This show started with an actor portraying a Casting Director who revealed to the audience that they were about to watch a movie being made. Volunteer extras were then recruited from the audience to be in a future scene. After that point, the performance switched into "action" mode and proceeded as a traditional show would, with the Indiana Jones character appearing on set and recreating a scene from the film *Raiders of the Lost Ark*. At the end of the scene another actor portraying the Director of the film would yell "Cut!" and interact with the stunt actor playing Indiana Jones *as an actor*. That is, the stunt actor was introduced as himself, rather than Indiana Jones. This sequence continued through the remainder of the show, switching between these two modes of performance.

The brilliance of this kind of show design is that the interstitial scenes ("here's what we just filmed, and here's what we're doing next") between the action scenes (recreations from *Raiders of the Lost Ark*) created the space for set pieces to move around, actors to recover and reset for the next scene, and props to be placed. In addition, it provided a logical reason why the scenes being re-enacted could jump from one time and location to another: that's just how films are made! But from an audience perspective, this provided what seemed to be an authentic filmmaking experience. From the clap of a clapboard to "action" and "cut," to cameras moving around the edge of the scene, to a tangible experience of rumbling music, piercing sound effects, and heat from real flames, the audience truly felt as though they were part of the show. Various performers even described what they were doing on the set, from stunt performers explaining their process to camera operators noting that they will shoot "36 frames per second instead of the usual 24 so we can get that shot in really smooth slow motion."

FIGURE 12.1 A scene from the Indiana Jones Epic Stunt Spectacular.

Source: Photograph by the author.

It is undeniable that this kind of show format created an experience where audience members felt as though they learned something about moviemaking. Besides selling tickets to the parks themselves, a secondary desire was to create a stronger affinity for the properties in the shows themselves. Building up fan loyalty and a desire to engage with the films outside the shows in the parks would create future customers. This is an often-seen phenomenon in the *Star Wars* universe, as older generations introduce their children to the film franchise, who then pursue it on their own terms.

One counterpoint, however, was that the film depicted in the Indiana Jones stunt show had already been made, so there was always a bit of non-reality about it. This aspect was rectified by Disney's next iteration in this style that started as Moteurs . . . Action! at Disneyland Paris and was copied (with some modifications) at Disney's Hollywood Studios as Lights, Motors, Action! in 2005 (see Figure 12.2). (Note: the same author was also a cast member in this show.) The stunts in this production were primarily vehicular; cars, motorcycles, and jet skis performed intricate choreography amidst simulated gunfire and real explosions. While the original intent was for this show to be based on the character of James Bond, licensing issues prevented that from coming to fruition, so the show simply became "the filming of an upcoming spy action movie." What this allowed, though, was the ability to pretend that the actors were filming an actual movie that was being made for release. (Adult members of the audience would realize, of course, that shooting

FIGURE 12.2 A scene from the Lights, Motors, Action! Stunt show.

Source: Photograph by the author.

the same sequences in every show every day eliminated this leap in logic, but some kids were actually convinced that it was a real movie being filmed.)

The counterpoint to *this* narrative is that because there was no *actual* film connected to Lights, Motors, Action!, there was no film-specific screen tourism connected to it. The demographic for this show was not fans of a particular film, unlike with the Indiana Jones production, where fans might have been attracted to the characters in that show. Rather, a screen tourist in this audience would be an individual who loved watching live action sequences, or others who were purely fans of moviemaking.

Screen tourism from this perspective has run the gamut from riding a tram around a studio lot looking at sets and props, to active experiences with performers on a simulated tour, to mock film sets created to give the audience a feeling of being in the Director's chair. In most cases, this experience was connected to a certain film (e.g. *Raiders of the Lost Ark*, *Twister*) or a particular studio (e.g. Universal Studios), while at other times, it was an original creation designed simply for that specific experience (e.g. Lights, Motors, Action!, and Earthquake). The intent was to bring the audience into the filmmaking world rather than to create a pure storytelling experience. With the evolution in theme parks towards more fully immersive experiences, a la *Harry Potter* and *Star Wars*, however, the trend now is that theme park show design seems to be moving rapidly away from "you are watching an actual film as it is being made" and towards "you are IN the movie." The latter phenomenon will be discussed in the next section.

Immersive Worlds of Themed Screen Tourism

Several platforms have evolved to incorporate interactivity. A simple, well-known literary example would be Choose-Your-Own-Adventure books; an early example from cinema is Raduz Cincera's use of live narration between two scene choices in the 1967 film *Kinoautomat*. Other interactive developments have involved technological advances, such as Charlie Brooker's *Black Mirror: Bandersnatch*, released on Netflix in 2018. That episode allows for real-time participant interactivity throughout a viewing, leading to significantly different outcomes. Attractions within theme parks are following the same path, allowing park guests to fully embrace screen interactives surrounding fantasy. In a way, they are merging film viewing and theme park guest experiences with an infusion of digital technologies parkwide.

As technological advancements have helped invent "immersive and interactive environments . . . expanding our theoretical grasp of technologies of representation,"[8] park visitors have become participants, not just tourists, included within the screen story. As a result, theme park screen tourism invites and incorporates park guests into immersive, film-driven, authentic staged settings, that is, towards "imagination of reality,"[9] rather than just tourism.

With these developments, perspectives within theme parks have shifted away from traditional "behind-the-scenes" viewing and more towards participants themselves becoming a fully immersed participant in a movie. Themed environments within destinations such as Walt Disney World and Universal Studios have emerged as the leaders by setting the stage of an inclusive, live cinema experience. This has dramatically changed the guest expectation of theatrically screened story experiences within the theme park milieu. Interactive backlot park tours led to recreations of filmic scenes, finally evolving to fully immersive show viewings with characters, settings, and audience interaction.

The progression of popular themed cinema has expanded the immersive experience of screen tourism. An early example of guest interaction is the opening of E.T. Adventure at Universal Studios (1990). *Here, the* visitors' names were programmed into the system to *verbalize "good-bye"* to riders along with their names, providing interactivity and incorporation of guests into the storyworld. This interaction even elevated the experience to a place where the characters themselves became the viewers.

Today, themed participation in parks has expanded from the original E.T., Jaws, and Indiana Jones attractions *to entire lands within parks*, such as Jurassic Park, *Star Wars:* Galaxy's Edge, and the Wizarding World of Harry Potter (see Figure 12.3). These lands are leading themed experiences forward for fans and park visitors alike, allowing them to become saturated beyond "traditional limits or unknowns."[10] Furthermore, the evolution of this experience is made manifest as guests exchange seats in a theatre for live shared experiences, fully interacting with stories stemming from

FIGURE 12.3 Some architecture from The Wizarding World of Harry Potter.

Source: Photograph by the author.

the pages of Peter Benchley's *Jaws*, George Lucas's *Stars Wars*, Michael Creighton's *Jurassic Park*, and J.K. Rowling's epic *Harry Potter* novel series.

Universal's Jurassic Park iterations depict its screened cinema immersion using landscaped scenes of overgrown vegetation and with the help of audio depictions of dinosaur roars. The island escape, taken from the book and blockbuster film, brings immersion for park guests through visual recreations of living and breathing dinosaurs. And within Islands of Adventure's Raptor Encounter, guests can observe and interact with Blue, the carnivore in her habitat. This immerses the visitor beyond the typical selfie photo.

Disney's Pandora: The World of *Avatar* brings in moments for guests to play a role in the world with the rides Avatar Flight of Passage and Na'vi River Journey. Valley of Mo'ara encapsulates the environmental factors accompanied by visitor immersion through visual and interactive experiences. These include a 3D, simulated flight on the wings of a Mountain Banshee over the vast moon, a boat ride through a bioluminescent rainforest with a Na'vi Shaman character singing, and a hike under floating mountains surrounded by a drum circle, infusing elements of the film's visual voice.

Star Wars: Galaxy's Edge incorporates the fictional planet of Batuu within the gates of Disney's Hollywood Studios. This allows for interaction with familiar characters like Chewbacca and Stormtroopers, and activities such as piloting the Millennium Falcon or becoming recruits for the Resistance. Park visitors can immerse themselves in Savi's workshop to construct their own custom lightsaber, choosing the colour of the blade and design of the hilt, allowing the visitor to live their dreams from the character-driven cinema franchise.

Even though it is no longer the newest fully immersive themed land in a park, Universal's Wizarding World of Harry Potter is still considered by many to be the most authentic. A significant example of this authenticity is Ollivander's Wand Shop. Drawn from J.K. Rowling's *Harry Potter* series of books and films, choosing an interactive wand "allows players to take on the role of wizards casting spells in the Harry Potter theme parks lands."[11] Wands are an impactful symbol within the Wizarding World of Harry Potter, allowing wizards young and old to immerse themselves in a world of magic. The use of a transmitter in the wand's tip allows magic to come alive throughout the park with interactive experiences, engrossing park visitors as they take on the film characters' roles within themed environments. The guests' process of obtaining their wand intimately showcases the impetus of immersive interactivity within the film-themed shop.[12] Immersive participation at the individual level rather than the group level enhances the story experience towards the guest, setting the scene for the guest's experience.

Theme parks worldwide bring guests in to relive classic cinematic moments. Selecting a wand, building a lightsaber, or interacting with a "live" raptor invites visitors to participate with and become immersed in fantasy from themed film backdrops.[13] And though many books and films are not interactive, they become platforms for engagement within theme parks. While imaginative films are affecting, a guest's memory of notable film scenes is significantly enhanced when they are physically experienced within immersive themed environments. The next section will illustrate a final stage in theme park screen tourism: that created by the properties produced at theme parks.

Theme-Park-Created Screen Tourism

One of the most fascinating aspects of the rise of theme parks as a form of screen tourism is that they have become reciprocal sites of screen tourism activity. Not only do they frequently act as locales for fans of particular on-screen properties, but they also create screen properties and drive tourism back to these places. Transmedia storytelling, or when a story is "dispersed systematically across multiple delivery channels" (such as books, screen media, and video games), is a strategy common to theme park companies as well.[14] In conceptions of this approach,[15] properties are spread across several mediums, with businesses strategically employing nodes in a web of story links for corporate synergy and greater audience participation. The Lucasfilm Story Group, for instance, was assembled to create a robust and connected storyworld for the vast narrative of *Star Wars*.[16]

Nonetheless, the *Star Wars:* Galaxy's Edge area at Disney's Hollywood Studios and Disneyland was one of the later nodes in that franchise's web.[17] While both behind-the-scenes and immersive models of theme park screen tourism have yielded impressive experiences, theme parks are reflected on less in the literature related to transmedia webs. If an intellectual property is successful and it is owned by or can get a licensing agreement with a theme park operator, then a dimensional version of that film, television show, or video game may eventually be created in the physical space of the theme park. Occasionally, however, a theme park property either inspires its own transmedia development or is closely established alongside other forms of media. In these cases, it is the theme park's contribution that drives the success of the property and creates a new landscape for screen tourism.

Theme parks have informed other mediums in the past. The film *Star Wars: The Last Jedi* has scenes inspired by theme park rides (e.g. a scene on Crait, a nod to the Star Tours ride, which is now actually in the Star Tours ride, Stahler 2017), while *Jurassic Park* movies frequently play on theme park tropes. Andrew Nelson[18] sees film and theme parks as synergistic mediums that informed one another. Brian Bishop[19] considers theme parks to be the "evolution of franchise movies," and Director Steven Spielberg backed up this claim by noting that the E.T. Adventure ride at Universal (original 1991) was indeed the evolution of the 1982 film *E.T. The Extra-Terrestrial*, in fact the "only sequel to the movie."[20] A handful of productions were planned to connect in multiple mediums at once as seen with the Efteling ride Symbolica or its outdoor show Raveleijn, with television shows and books debuting alongside. In some cases, theme park companies produce popular series that then become integrated with the parks, as in the popular *Boonie Bears* series developed by the company Fantawild. Some attractions have even come to overshadow their screen properties in their markets, for instance Disney's Tron Lightcycle Power Run or the Waterworld stunt spectacular at many Universal theme parks.

Potentially the most relevant cases of reciprocal influence involve theme park attractions that have later become popular television shows or movies. This is evident with the Efteling's Sprookjesbos/Fairytale Forest area of the theme park. A land dating from the 1950s that is continually expanded, its success led to the creation of a 2006 children's television show called *Sprookjesboom*/Fairytale Tree, which generated a character that would later be made into a prominent animatronic figure in the park. The TV shows spawned a live stage production in the park and a merchandise line, so it is a synergistic vehicle for the theme park. Now audiences may be influenced by this property even before visitation since it became "accessible 24/7" and part of a person's daily life.[21]

Over the years, industry leader Disney has introduced television shows to explore aspects of theme parks including the recent *The Imagineering Story, Behind the Attraction*, and *The Magic of Disney's Animal Kingdom*. They have also released films associated with recent park projects (e.g. *Third Man on the Mountain*, 1959, *The Black Hole*, 1979, or *Tower of Terror*, 1997) and several that were created decades after a ride opened including *The Country Bears* (2002), *The Haunted Mansion* (2003), *Mission to Mars* (2000), *Tomorrowland* (2015), and *Jungle Cruise* (2021). Though several of these films had moderate success, none would usurp the popularity of the source material, with the Haunted Mansion and Jungle Cruise attractions especially having decades-old fan cultures surrounding them.

By far, the most successful film of Disney's theme park-related attempts is the *Pirates of the Caribbean* franchise. The Pirates of the Caribbean ride opened at Disneyland, California, in 1967 and spawned several additional theme park versions in Orlando (1973), Tokyo (1983), Paris (1992), and Shanghai (2016). The experience invites riders to board a boat and go into the shady realms of pirates as they ransack places, steal, and burn a city down. It uses environmental storytelling rather than overt narration, with the setting made obvious through architecture, props, music, sound, visual effects, lighting design, and more than 100 animatronic figures of people and animals. The ride places guests directly in the middle of the action, with scenes occurring on both sides of them at once and several vignettes displaying the antics of the pirates. In all versions, the famous "Yo Ho" song can be heard, one of the soundtracks that sparked visitor appreciation of Disney theme park music. The ride is highly acclaimed for setting a new standard with its immersive design and is often considered the "crown jewel" of the Disney park attractions.[22] The late, legendary Imagineer Marty Sklar[23] deemed it the "most valuable single property every created in the theme park business."

The film franchise of the same name would also meet with great success. *Pirates of the Caribbean: The Curse of the Black Pearl* (2003) was the first franchise film. Additional movies were released in 2006, 2007, 2011, and 2017, with another expected in the future. Altogether, these films earned more than $4.5 billion at the box office and have been especially strong performers in international markets.[24]

While there is certainly enthusiasm surrounding the Pirates rides, search engine results, and guest questions over the years indicate that many do not realize that the ride is the original. This confusion was compounded by the fact that in 2006 Disney began "retconning" the Jack Sparrow character, new dialogue, and music from the film into the rides, changing, and complicating the attraction narratives.[25] Regardless of one's knowledge of which property came first, the Pirates of the Caribbean ride has become an ironic bit of screen tourism, with guests visiting the ride that references the film series, which was inspired by and alluded to the theme park ride in the first place.

Disney's later entrance into the mainland Chinese market meant several important things for their Pirates attraction. Firstly, unlike in other regions, there was not a pre-existing emotional connection with the original ride. Thus, the Shanghai iteration, Pirates of the Caribbean: Battle for the Sunken Treasure (see Figure 12.4), is largely based on the movie franchise. This ride was created for a "brand new audience" according to Imagineer Luc Mayrand, one that recognized the movies and not the original attractions.[26] Though it would use a new backstory, contain new characters, and be set in a new place, a particular island in the Caribbean,[27] favourite characters like Jack Sparrow and Davy Jones are present, giving some familiarity to the experience. Mayrand grew up with the "mystique of the original attraction," so being able to break from that was "actually kind of freeing."[28] This use of invention makes the storyline the most different of all the versions, but its application of the characters and film music make it an even stronger instance of a

FIGURE 12.4 The entrance to Pirates of the Caribbean: Battle for the Sunken Treasure.

Source: Photograph by the author.

site for screen tourism. People "love the pirate life and the pirate world" according to Mayrand,[29] and those who appreciate the film might find that while not authentic in terms of being an actual site of production, it is a required fan space.

The newness of the Shanghai ride and conceptions of the designers also ensured that it is the most technologically advanced version. It was meant to "raise the bar" for the attractions industry and was modified in response to Chinese guests that desired a "more active" experience.[30] The Shanghai form includes a ride system created for this attraction, a new engine to program the ride, several endings, an even larger scale than the original, and an integration of practical effects with large media screens.[31] This version of the ride combines the immersive sensibilities of the current design paradigm and older ideas such as controlling rider direction but likewise channels the affordances of cinema by reproducing aspects of the film watching experience including the use of extra-large ride vehicles and giant screens. Mayrand explains that it is a "continuous movie" but "completely dimensional" in addition to changing the focus to be the story of the riders.[32]

Finally, its existence in the era of immersive worlds[33] and the popularity of the franchise meant that rather than a stand-alone attraction (as in California, Paris, or Tokyo) or one that has evolved into a sub-land (as in Orlando), Pirates of the Caribbean: Battle for the Sunken Treasure opened as part of Treasure Cove, a single intellectual property-based theme land. It is no longer a sub-land of Adventureland and is in fact separate from Shanghai Disneyland's Adventure Isle. Treasure Cove includes the Pirates ride, the Explorer Canoes, the stage show Eye of the Storm: Captain Jack's Stunt Spectacular, a walkthrough pirate galleon Siren's Revenge, kid's play area Shipwreck Shore, table service restaurant Barbossa's Bounty, snack stands, and two retail shops. All of these experiences combine into a powerful landscape where guests can play at being pirates and celebrate their fandom of the *Pirates of the Caribbean* film series.

The Theme Park as Participatory Playground

There is an evolution of theme park tourism from sites that re-enact filmmaking to immersing visitors in storyworlds and then finally spawning the narratives that reciprocally become media tourism sites. Parts of this shift may be attributed to design paradigms altering or other medium-specific instances. Nonetheless, a major part of the transition is due to changes that impacted the preferences of visitors including the ubiquity of immersion as a feature of entertainment, participatory culture, and convergence. Theme parks already represent a synthesis of other entertainment forms, a form of convergence, and transmedia webs came together to satiate fans' desire for many layers of engagement. The immersive design paradigm and the media penchant for interactivity collided, creating spaces conducive to role-playing. The elaborately realized land of the Wizarding World of Harry Potter at several Universal Studios parks is a site of fan participation[34] and according to Jenkins[35] is the "closest the community is going to come to an actual homeland." Despite the commercial nature of the theme park, a site like this becomes an affective landscape, an "emotional geograph[y] of community, meaning, and place."[36] As Kim[37] found, the more "emotional involvement" the audience develops with the film or television show, the more likely they are to visit film tourism locations.

Popular culture and media tourism are driven by this fervent fan culture and the representation of media properties in physical space brings actual tourism. Gyimóthy et al.[38] theorize a cycle wherein a pop culture representation sparks media convergence, which leads to identification, appropriation, and narrative transformation. Irimiás, Mitev, and Michalkó[39] studied TV screen tourism and connected it to the theory of narrative transportation, a perspective in several disciplines that narrative (often written, literary instances, though movies or video games can do this as well) has the power to transport a reader or viewer the way that travel does. They find that television consumption leads to a form of escapism, escapism to immersion, and immersion to travel intention. Strategic media convergence can contribute to the success of real-world spaces with synergistic practices, for instance film locations "alluding to the fictive narratives"[40] or advertising that sells place, as *The Hobbit* film generated,[41] a screen tourism instance that brought economic benefits to New Zealand. Destination development can be propelled by on-screen popular culture place-making,[42] with screen associations bringing tourists to real places. The success of media-based lands has illustrated that this happens with the simulated spaces of the theme park despite their inauthenticity as filming locations. For some screen tourists, however, it is not true authenticity that matters but simply association with favourite films.[43]

In screen tourism sites, mental transportation transforms into the physical transportation of travel. Dimensionality, a site's quality of existing in multi-sensory, physical space, is paramount to the tourism experience, and it is the foundation of the success of theme parks. According to Ateljevic,[44] tourism is "inseparable from the spaces and places in which it is created, imagined, perceived and experienced" (p. 369). It is, as Gyimóthy et al[45] asserted, a "negotiated reproduction of space." Theme parks are likewise a form of "spatial transmedia"[46] and imaginary *gateway* worlds materially realized in physical space. Theme parks became nodes in the transmedia storytelling environment, with their primary affordance being their dimensionality. The aforementioned Wizarding World of Harry Potter is "structured around places and not events," unlike the books or films.[47] While interactions with other people are possible through talking to family and friends or employees, being surrounded by the physical world of magic is a core part of the experience. Fan practices are conducted in these spaces, with everything from the individual wand quest or sporting a robe to the ritualized group activities at the former fan event A Celebration of Harry Potter. The pandemic particularly demonstrated how hard theme parks are to replicate due to their materiality and the core of the experience being exploring dimensional space. Theme parks go beyond just the physical space to be multi-sensory storyworlds and affective landscapes whose immersive properties powerfully connect with emotion and memory. It is no wonder that fans now embark on "media pilgrimages" of varying intensity[48] or that theme parks have had a long association with the concept of pilgrimage.[49]

Conclusion

From rides that allow participants to "peek behind the curtain" to see where films are made (e.g. backlot tours), to physically demonstrating the complex processes involved in filmmaking (e.g. The Indiana Jones Epic Stunt Spectacular), to locations that make the audience feel as though they are actually experiencing a tangible world that is based on a fictional story (e.g. The Wizarding World of Harry Potter), theme parks have provided years of joy for fans of film and television. This is true whether the fans enjoy a specific story or are simply enamoured with the filmmaking process itself. The power of this locative experience is even strong enough to flip the relationship, where theme parks have become source material for original, hugely popular film franchises like *Pirates of the Caribbean*.

The evolution of theme parks as sites for fan participation and pilgrimage started with simple tours and grew to encompass highly complex, thoroughly immersive experiences. When a guest can not only see a prop from a film, such as a wand from a Harry Potter film, but also actually hold that prop in their hands, and then wave it to "cast a spell," causing water to "magically" begin to spout from an umbrella, the experience becomes much more tangible, and thus much more emotional. The heightened enthusiasm created by these significant design innovations creates generates a synergy that film and television companies can leverage to drive sales in both the broadcast and the theme park mediums. And as in the case of *Pirates of the Caribbean*, the theme park experience itself can create enough enthusiasm to spawn a brand new, massively popular, worldwide film franchise. With new parks and attractions constantly under development, and with technological advancements always on the horizon, it seems likely that this relationship will continue to grow and evolve alongside the development of new tourist experiences, including screen tourism.

Notes

1 Rubin, "Global Attractions Attendance Report 2011."
2 Hudson, "Europe's Grand Old Theme Parks."
3 Cross and Walton, *The Playful Crowd: Pleasure Places in the Twentieth Century.*
4 Blackwood, "Dream Factory Tours: The Universal Pictures Movie Tour Attraction in the 1960s," 518.
5 Blackwood, 517.
6 "Universal Studios Hollywood – History Timeline."
7 Gennawey, *Universal versus Disney: The Unofficial Guide to American Theme Parks' Greatest Rivalry.*
8 Wardrip-Fruin and Montfort, *The New Media Reader.*
9 Baudrillard, *Simulacra and Simulation.*
10 Haraway, *Simians, Cyborgs, and Women.*
11 Eddy et al., "Hacking Droids and Casting Spells."
12 Firestone and Clark, *Harry Potter and Convergence Culture.*
13 Waysdorf and Reijnders, "Immersion, Authenticity and the Theme Park as Social Space."
14 Jenkins, "The Revenge of the Origami Unicorn: Seven Principles of Transmedia Storytelling."

15 Harvey, "A Taxonomy of Transmedia Storytelling"; Williams, *Theme Park Fandom: Spatial Transmedia, Materiality and Participatory Cultures.*
16 Baker-Whitelaw, "How the Lucasfilm Story Group Controls Every Corner of the 'Star Wars' Franchise"; Harrigan and Wardrip-Fruin, *Third Person: Authoring and Exploring Vast Narratives.*
17 Williams, "From Star Tours to Galaxy's Edge: Immersion, Transmediality, and 'Haptic Fandom' in Disney's Theme Parks."
18 "CINEMA FROM ATTRACTIONS: STORY AND SYNERGY IN DISNEY'S THEME PARK MOVIES."
19 "Theme Parks Are the Inevitable Evolution of Franchise Movies."
20 Gennawey, *Universal versus Disney: The Unofficial Guide to American Theme Parks' Greatest Rivalry.*
21 Hover and Vugts, "Efteling and the Fairytale Tree Storytelling between Tradition and Innovation."
22 Surrell, *Pirates of the Caribbean: From the Magic Kingdom to the Movies.*
23 *Dream It! Do It!: My Half-Century Creating Disney's Magic Kingdoms.*
24 Clark, "Disney Is Reportedly Trying to Reboot 'Pirates of the Caribbean' – and This Chart Shows Why."
25 Schweizer and Pearce, "Remediation on the High Seas: A Pirates of the Caribbean Odyssey."
26 Palicki, "Pirates, Treasure & WDI's Luc Mayrand."
27 Mayrand in *2017 TEA Summit Day 2 – Pirates of the Caribbean: Battle for the Sunken Treasure.*
28 Palicki, "Pirates, Treasure & WDI's Luc Mayrand."
29 Palicki, "Pirates, Treasure & WDI's Luc Mayrand."
30 Palicki, "Pirates, Treasure & WDI's Luc Mayrand."
31 Mayrand and Seruto in *2017 TEA Summit Day 2 – Pirates of the Caribbean: Battle for the Sunken Treasure.*
32 *2017 TEA Summit Day 2 – Pirates of the Caribbean: Battle for the Sunken Treasure.*
33 Lukas, *The Immersive Worlds Handbook: Designing Theme Parks and Consumer Spaces.*
34 Baker, "Universal's Wizarding World of Harry Potter: A Primer in Contemporary Media Concepts"; Godwin, "Theme Park as Interface to the Wizarding (Story) World of Harry Potter"; Waysdorf and Reijnders, "Immersion, Authenticity and the Theme Park as Social Space."
35 "The Night of a Thousand Wizards."
36 Gupta and Medappa, "Nostalgia as Affective Landscape: Negotiating Displacement in the 'World City'," 1683.
37 "Audience Involvement and Film Tourism Experiences: Emotional Places, Emotional Experiences."
38 "Popculture Tourism: A Research Manifesto."
39 "Narrative Transportation and Travel: The Mediating Role of Escapism and Immersion."
40 "Popculture Tourism: A Research Manifesto."
41 Li et al., "The Economic Impact of On-Screen Tourism: The Case of The Lord of the Rings and the Hobbit."
42 Lundberg, Ziakas, and Morgan, "Conceptualising On-Screen Tourism Destination Development."
43 Rittichainuwat et al., "Authenticity in Screen Tourism: Significance of Real and Substituted Screen Locations."
44 "Circuits of Tourism: Stepping beyond the 'Production/Consumption' Dichotomy."
45 "Popculture Tourism: A Research Manifesto."
46 Williams, *Theme Park Fandom: Spatial Transmedia, Materiality and Participatory Cultures.*
47 Jenkins, "The Night of a Thousand Wizards."
48 Couldry, "Pilgrimage in Mediaspace: Continuities and Transformations."
49 Fjellman, *Vinyl Leaves: Walt Disney World and America*; Kottak, "Disney Myth and Ritual"; Moore, "Walt Disney World: Bounded Ritual Space and the Playful Pilgrimage Center."

Bibliography

"2017 TEA Summit Day 2 – Pirates of the Caribbean: Battle for the Sunken Treasure, 2017. www.youtube.com/watch?v=P8717fVMfCU&ab_channel=ThemedEntertainmentAssociation.

Ateljevic, Irena. "Circuits of Tourism: Stepping Beyond the 'Production/Consumption' Dichotomy." *Tourism Geographies* 2, no. 4 (2000): 369–88.

Baker, Carissa. "Universal's Wizarding World of Harry Potter: A Primer in Contemporary Media Concepts." In *Harry Potter and Convergence Culture: Essays on Fandom and Exploring the Expanding Potterverse*. Jefferson: McFarland, 2018.

Baker-Whitelaw, Gavia. "How the Lucasfilm Story Group Controls Every Corner of the 'Star Wars' Franchise." *Daily Dot*, July 16, 2016. www.dailydot.com/parsec/star-wars-celebration-lucasfilm-story-group/.

Baudrillard, Jean. *Simulacra and Simulation*. Translated by Sheila Faria Glaser, 1st ed. Ann Arbor: University of Michigan Press, 1994.

Bishop, Bryan. "Theme Parks Are the Inevitable Evolution of Franchise Movies." *The Verge*, June 3, 2017. www.theverge.com/2017/6/3/15732968/disney-guardians-of-the-galaxy-mission-breakout-pandora-world-of-avatar-theme-parks.

Blackwood, Gemma. "Dream Factory Tours: The Universal Pictures Movie Tour Attraction in the 1960s." *Historical Journal of Film, Radio and Television* 38, no. 3 (2018): 516–35.

Clark, Travis. "Disney Is Reportedly Trying to Reboot 'Pirates of the Caribbean' – and This Chart Shows Why." *Business Insider Australia*, October 25, 2018. www.businessinsider.com.au/pirates-of-the-caribbean-rebooted-because-of-worldwide-box-office-2018-10.

Couldry, Nick. "Pilgrimage in Mediaspace: Continuities and Transformations." *Etnofor* 20, no. 1 (2007): 63–74.

Cross, Gary, and John Walton. *The Playful Crowd: Pleasure Places in the Twentieth Century*. New York: Columbia University Press, 2005.

Eddy, Ray, Carissa Baker, Robert Macy, John Murray, and Anastasia Salter. "Hacking Droids and Casting Spells: Locative Augmented Reality Games and the Reimagining of the Theme Park." In *Proceedings of the 31st ACM Conference on Hypertext and Social Media*, 37–46. New York: Association for Computing Machinery, 2020. https://doi.org/10.1145/3372923.3404801.

Firestone, Amanda, and Leisa A. Clark, eds. *Harry Potter and Convergence Culture: Essays on Fandom and the Expanding Potterverse*. Jefferson: McFarland & Company, 2018.

Fjellman, Stephen. *Vinyl Leaves: Walt Disney World and America*. London: Routledge, 1992.

Gennawey, Sam. *Universal Versus Disney: The Unofficial Guide to American Theme Parks' Greatest Rivalry*. Unofficial Guides, 2014. https://www.target.com/p/universal-versus-disney-the-unofficial-guide-to-american-theme-parks-greatest-rivalry-unofficial-guides-by-sam-gennawey-hardcover/-/A-87376402.

Godwin, Victoria. "Theme Park as Interface to the Wizarding (Story) World of Harry Potter." *Transformative Works and Cultures* 25 (2017).

Gupta, Hemangini, and Kaveri Medappa. "Nostalgia as Affective Landscape: Negotiating Displacement in the 'World City'." *Antipode* 52, no. 6 (2020): 1688–709.

Gyimóthy, Szilvia, Christine Lundberg, Christine Lindstrom, Maria Lexhagen, and Mia Larson. "Popculture Tourism: A Research Manifesto." In *Tourism Research Frontiers: Beyond the Boundaries of Knowledge*. Bingley: Emerald, 2015.

Haraway, Donna. *Simians, Cyborgs, and Women: The Reinvention of Nature*. London: Routledge, 2013.

Harrigan, Pat, and Noah Wardrip-Fruin. *Third Person: Authoring and Exploring Vast Narratives*. Cambridge: MIT Press, 2017.

Harvey, Colin. "A Taxonomy of Transmedia Storytelling." In *Storyworlds Across Media*, 278–94. Lincoln: University of Nebraska Press, 2014.

Hover, Moniek, and Olaf Vugts. "Efteling and the Fairytale Tree Storytelling Between Tradition and Innovation." *Trendrapport Toerisme, Recreatie En Vrijetijd* (2013): 272–75.

Hudson, Shaney. "Europe's Grand Old Theme Parks." *BBC Travel*, June 25, 2012. www.bbc.com/travel/article/20120626-europes-grand-old-theme-parks.

Irimías, Anna, Ariel Zoltán Mitev, and Gábor Michalkó. "Narrative Transportation and Travel: The Mediating Role of Escapism and Immersion." *Tourism Management Perspectives* 38 (2021).

Jenkins, Henry. "The Revenge of the Origami Unicorn: Seven Principles of Transmedia Storytelling." *Confessions of an Aca-Fan*, December 9, 2009 (blog). http://henryjenkins.org/blog/2009/12/the_revenge_of_the_origami_uni.html.

———. "The Night of a Thousand Wizards." *Confessions of an Aca-Fan* (blog), July 21, 2010. http://henryjenkins.org/blog/2010/07/fear_and_loathing_in_hogworts.html.

Kim, Sangkyun. "Audience Involvement and Film Tourism Experiences: Emotional Places, Emotional Experiences." *Tourism Management* 33 (2012): 387–96.

Kottak, Conrad. "Disney Myth and Ritual." In *Cultural Anthropology*. New York: McGraw-Hill, 1994.

Li, Shi Na, Hengyun Li, Haiyan Song, Christine Lundberg, and Shujie Shen. "The Economic Impact of On-Screen Tourism: The Case of The Lord of the Rings and the Hobbit." *Tourism Management* 60 (2017).

Lukas, Scott. *The Immersive Worlds Handbook: Designing Theme Parks and Consumer Spaces*. Waltham: Focal Press, 2012.

Lundberg, Christian, Vassilios Ziakas, and Nigel Morgan. "Conceptualising On-Screen Tourism Destination Development." *Tourist Studies* 18, no. 1 (2017): 83–104.

Moore, Alexander. "Walt Disney World: Bounded Ritual Space and the Playful Pilgrimage Center." *Anthropological Quarterly* 53, no. 4 (1980).

Nelson, Andrew. "Cinema from Attractions: Story and Synergy in Disney's Theme Park Movies." *Cinephile* 4 (2008).

Palicki, Martin. "Pirates, Treasure & WDI's Luc Mayrand." *InPark Magazine*, 2017.

Rittichainuwat, Bongkosh, Eric Laws, Noel Scott, and Suphaporn Rattanaphinanchai. "Authenticity in Screen Tourism: Significance of Real and Substituted Screen Locations." *Journal of Hospitality & Tourism Research* 42, no. 8 (2018): 1274–94.

Rubin, Judith. "Global Attractions Attendance Report 2011." *TEA/AECOM*, 2012. https://www.teaconnect.org/images/files/TEA_26_543179_140617.pdf.

Schweizer, Bobby, and Celia Pearce. "Remediation on the High Seas: A Pirates of the Caribbean Odyssey." In *A Reader in Themed and Immersive Spaces*. Pittsburgh: ETC, 2016.

Sklar, Marty. *Dream It! Do It!: My Half-Century Creating Disney's Magic Kingdoms*. New York: Disney Editions, 2013.

Stahler, Kelsea. "You May Have Missed This Awesome Disneyland Reference in 'The Last Jedi'." *Bustle*, 16, 2017. www.bustle.com/p/theres-a-star-tours-reference-in-the-last-jedi-according-to-rian-johnson-7617079.

The Studiotour.com – Universal Studios Hollywood – Chronology. "Universal Studios Hollywood – History Timeline." *The Studiotour.com*. Accessed September 16, 2021. www.thestudiotour.com/ush/chronology.php.

Surrell, Jason. *Pirates of the Caribbean: From the Magic Kingdom to the Movies*. New York: Disney Editions, 2005.

Wardrip-Fruin, Noah, and Nick Montfort, eds. *The New Media Reader*. Cambridge: MIT Press, 2003.

Waysdorf, Abby, and Stijn Reijnders. "Immersion, Authenticity and the Theme Park as Social Space: Experiencing the Wizarding World of Harry Potter." *International Journal of Cultural Studies* 21, no. 2 (March 2018): 173–88. https://doi.org/10.1177/1367877916674751.

Williams, Rebecca. "From Star Tours to Galaxy's Edge: Immersion, Transmediality, and 'Haptic Fandom' in Disney's Theme Parks." In *Disney's Star Wars: Forces of Production, Promotion, and Reception*. Iowa City: University of Iowa Press, 2019.

———. *Theme Park Fandom: Spatial Transmedia, Materiality and Participatory Cultures*. Amsterdam: Amsterdam University Press, 2020.

13

HOBBITON 2.0, 20 YEARS ON

Authenticity and Immersive Themed Space

Robert Moses Peaslee and Bobby Schweizer

> *"Human beings have produced whatever nature became accessible to them."*
>
> *Neil Smith*[1]

> *"You've travelled hundreds, perhaps thousands of miles to get here, but sitting in front of the fireplace, a rich ale in your hand, and a frumpy cat lazing on the chair next you, it's impossible not to feel as though it was worth it. The smell of fresh-baked pastries mingles with the raucous congress of your fellow patrons, new folks arriving through the heavy wooden doors, and a general atmosphere of comfort and good cheer. Welcome to the Green Dragon Pub, the penultimate stop in your journey through Hobbiton – but I'm afraid you can't linger too long: your tour guide is indicating it's time to exit, yield the fireside to the next lucky entrant, and visit the gift shop one more time."*

A heterotopic space, Hobbiton in Aotearoa New Zealand is many things simultaneously. It was originally constructed in 1999 as a movie set for director Peter Jackson's three *Lord of the Rings* films (2001–2003) and reconstructed for *The Hobbit* trilogy (2012–2014). While Hobbiton has been open to tourists since 2002, its complete restoration for *The Hobbit* has been left more or less intact since then with the long-tail interests of tourist appreciation in mind and with Jackson on board as a partner in the tourism enterprise.[2] Since the release of *The Fellowship of the Ring* in 2001, it has become among the most visited locations in Aotearoa New Zealand and serves as the crown jewel in the nation's substantial portfolio of screen tourism products.[3] It is also a village (of sorts) and thus surely an affective landscape, its iconography evoking traditional notions of a small agrarian community. This is bolstered by the fact that Hobbiton was built on a farm, sitting on roughly 12 acres of the Alexander family's sheep and cattle grazing land, which itself is located in the town of Matamata that has historically anchored a thriving farming community in the Waikato region. This situation contributes directly to Hobbiton's significant

DOI: 10.4324/9781003327585-13

power as a symbol for a bucolic, pre-industrial innocence that invites feelings of ecological longing and nostalgia. Finally, Hobbiton is an immersive construction, inviting visitors to walk through, touch, smell, and taste (through themed food and beverage available onsite) the atmosphere it establishes through careful and ongoing attention to detail. Here, we operationalize "immersion" as "encapsulation," a process described by André Jansson as providing tourist sites with a kind of "separatedness and liminality" that seems always at risk of disruption.[4]

Hobbiton is encapsulated in the traditional sense because unlike most film locations that are temporarily transformed into a movie set and then are returned for normal use, the actual shooting location persisted and the site developed a stable aura of object authenticity; that is, Hobbiton as a place is, for those who go seeking it, an authentic object and therefore provides value to those who possess it, however fleetingly.[5] For nearly two decades, it has functioned as a destination for media pilgrimage that reflects Tolkien's (and Jackson's) stories' shared cultural significance, particularly for intrepid Western audiences in Europe and North America.[6] Encapsulation is not absolute. As Jansson points out, there is always threat of decapsulation, where "the magic is broken."[7] Decapsulation can crop up in myriad ways, some straightforward (such as inclement weather), others more complex or nuanced. The point is that, like all tourism sites, Hobbiton and the encapsulation that it promises are products. And, as Hobbiton very much did during the Covid-19 pandemic, sites such as these can diminish or even temporarily vanish. And yet in other ways, the imagined bubble of encapsulation is decidedly resilient. The dynamics of media-induced tourism are constantly in flux and what qualifies as immersive or liminal changes along with the media landscape. In fact, Jansson has recently expanded on how "encapsulation" itself has ballooned as the processes that create the "tourist imagination" have been influenced by prevalence of transmedia as a "dominant mode of cultural circulation."[8] Jansson suggests that in addition to the place itself and representations of the place, the "overall picture" must include the logistical realities of daydreaming about trips, watching others tour on social media, booking flights, downloading apps, stopping to find free Wi-Fi in a foreign country to check email, and constantly comparing the current trip to past vacations.[9] This is all a part of the "intertwining of media content and processes of production and consumption" that "has had and continues to have a profound effect on tourisms, the tourism system, and its related industries."[10] What media tourism *is* and what media tourism *can be* are in inextricably linked, and Hobbiton offers a noteworthy example of a media tourism destination that has seen the massive changes in mediatization and communication technologies during the opening decades of the 21st century.

We highlight the quote by Neil Smith in our epigraph not to overemphasize the materialist influence on this chapter, but rather to suggest, with the author, that place and nature are always-already constructs. As Smith writes, "a geographical space which is abstracted from society is a philosophical amputee:" Hobbiton is obviously constructed in the sense that it is built, but it is also constructed and continuously reconstructed in terms of its symbolic, encapsulating value.[11] This chapter thus describes not only a particular geographical space but also a particular set of characteristics, practices, and discourses which formulate in Hobbiton a nature that is at once deeply specific and confoundingly ambivalent. The core questions we pose are exactly what kind of product *is* Hobbiton today, and how does understanding its evolution provide clues regarding the future development of affective spaces? Considering the many things Hobbiton *is*, it is also worth considering what Hobbiton *could be* as a way of contrasting it with other similar forms of entertainment tourism. Tolkien's own writings on subcreation and world-building – whose prescience underpins transmedia storytelling through themed spaces – provides an avenue for imagining a version of Hobbiton that offers role-playing alongside costumed characters in a commercialized theme park "land."

Presenting a longitudinal case study derived from more than 30 participant observations onsite from 2006 to 2019, we engage with Hobbiton's two iterations as an affective tourist commodity. Firstly, Hobbiton 1.0, the location that was originally constructed in 1999, opened to tourists three years later, and by 2004 was established as a site of significant visitor interest despite being half-demolished and only partially recognizable. Hobbiton 2.0, secondly, exists today as fully interactive simulacrum of the fictional village portrayed in the films, complete with a functional Green Dragon Pub. Among other features, we explore the guided nature of the site tour which creates tension in Hobbiton's immersive appeal and raises questions about the typology of affective themed space, which we compare to the various forms of tourism related to the Harry Potter franchise. Additionally, we reflect upon Hobbiton's future as a media tourism site within the current mediatized tourism landscape. Hobbiton's longevity and trajectory provide a critical case study, and our analysis charts the evolution of a space that could be characterized simultaneously as, among other things, film location, museum, theme park, and landscape.

Hobbiton's (Status as a) Character

Tolkien's Shire, a region within which Hobbiton was one of the more provincial settlements, was meant to evoke the pre-industrial English countryside that was crucial to the author's storytelling process.[12] As Tolkien outlines in his seminal essay "On Fairy-Stories," the creation of a believable "secondary world" is central to the suspension of disbelief necessary in fantasy narratives.[13] If his stories had actually taken place in England, they would be beholden to the constraints of our world, or what Tolkien refers to as the "primary world." Conversely, Middle-earth serves as a "secondary world," a *subcreation* that "involves new combinations of existing concepts [that] become the inventions that replace or reset" the "defaults" of the Primary World.[14] For example, the *LOTR* novels are replete with references to herbs, produce, and other foodstuffs that are based in the reality of our primary world, even as they can be associated with fantastical scenarios rooted in the secondary world of Middle-earth (as with the Hobbits' love of tobacco and mushrooms).

The Hobbit and the *LOTR* trilogies, then, were imagined less as a fantasy than as an ancient, forgotten pre-history of our "primary world" allowing the reader to identify analogous concepts and places not explicitly referred to in the story. Part and parcel of the stories' verisimilitude (at least for Tolkien's main readership in the West) were his descriptions of Hobbiton grounded in the "primary world," including familiar landscapes with attendant cultural associations: the semiotics of the rolling hills of the landscape, the agrarian lifestyle, and the importance of community that is expressed most often through the media of song, food, and drink. As subcreation is an act that establishes a spectrum of difference from the primary world, the Hobbits serve an important purpose: the audience enters into Middle-earth from the perspective of characters who are both familiar and exotic as compared, for example, to the immortal Elves.[15] According to Tolkien's philosophies of subcreation, the Shire itself supported the secondary world in significant ways. It was a specific "fictional *place*" with a "distinct border partitioning it from the Primary World" whose remoteness meant that the audience could imagine it as a location that is too distant to conveniently visit themselves, but whose stories are plausible mythologies of a real place.[16]

Following the release of *The Fellowship of the Ring* in 2001, then, Hobbiton's existence for nearly 70 years only on paper and within audiences' imaginations suddenly became quite tangible. Hobbiton's capacity to provide visitors "a playful attempt to touch something untouchable" is greatly enhanced by the sparsely mediated quality of Tolkien's secondary worlds in the novels.[17] Tolkien wrote critically of literary Fantasy performed by actors on the stage, for example, because "drama is naturally hostile to Fantasy. . . . Fantastic forms are not to be counterfeited. . . . Men dressed up as talking animals may achieve buffoonery or mimicry, but they do not achieve Fantasy."[18] As Mark J. P. Wolf points out, however, these comments were "not directed at film or animation, nor could he (Tolkien) have envisioned the degree to which cinematic special effects would be able to appear photorealistic in their appearance or behaviour."[19] Middle-earth had been visualized in Tolkien's maps and drawings in the novels, and later by artists and illustrators, but decades went by before a photo-realistic moving image of Middle-earth was projected on a screen. Arguably, the island nation of New Zealand was a natural choice to visualize the landscape of Middle-earth precisely, because Tolkien did not significantly deviate from primary world defaults. Though fantastic, Middle-earth's many locations could be constructed using materials from Earth, thereby making Hobbiton, as the only place in the world where a constructed Middle-earth location set still stands, both more exceptional and more plausible as a place one might actually live.[20] Tolkien's approach to Middle-earth permits, or indeed encourages, reflective readings in which the audience might dwell simultaneously in the imaginary and real worlds.[21]

Hobbiton therefore presents a complex and capacious iconography symbolizing the Garden of Eden, pre-Industrial Age Europe, and ideals of proper community all at once. The goings-on of the outside world are regarded by hobbits as objects of disinterest, fear, or mistrust. Family ties are strong, war is geographically, and temporally distant, and foreign ideologies are absent. Tolkien himself viewed the Shire as a conservative, nationalistic representation of Englishness. As Mike Foster posits, one of the factors that contributed to the resurgent popularity of *The Lord of the Rings* in mid-1960s America was the way its story resonated with "the tenor of the times" for white, middle-class opponents of the Vietnam War who longed for "a homely ideal of benevolent, natural, ungoverned, mellow freedom like that of the Shire."[22] Living as a hobbit is synonymous with a particular variety of living well; it is naïve innocence made manifest.[23]

Hobbiton's value as a symbol in this regard is widely appreciated beyond Tolkien himself, as indicated not only by Jackson's framing of the village but also by a general idealization of pre-industrial life in films as diverse as *Koyaanisqatsi* (1982), *Dances With Wolves* (1990), and, indeed, *Avatar* (2009). César Graña (1971) has pointed out the value of the past and the traditional for both tourists and tourism providers, a value that we suggest is shared by producers and consumers of narrative:

> *The destruction of local traditions and the assault upon the "the past" perpetuated by industrialization and world-wide modernization seem to make large numbers of people susceptible to an appetite for relics of pre-industrial life. This appetite is so intense that it accounts in part for one of the major and most characteristically modern industries: tourism.*[24]

Hobbiton therefore provides access to a variety of explicit and subtle pleasures as a literary, screen, and tourism product, and perhaps it is in the last of these iterations, where its reliquary, tactile character can be packaged, sold, and remarked upon, that its aura is most pronounced.

Hobbiton and the Evolution of (Trans)Media Tourism

Hobbiton emerged at a time of significant ferment within themed entertainment tourism, which has evolved considerably alongside changing expectations and affordances around media attractions' relationship to their source text(s). Media or screen tourism – particularly that which is induced by film and television productions, many of which are adapted from pre-existing sources – can involve, among other delights, visits to filming locations, tours of production sites, and film-related theme parks.[25] According to Sue Beeton, Hobbiton can be classified according to a number of characteristics of on-location tourism: it is a primary travel motivator or a secondary part of a vacation, it attracts fan pilgrimages, and it is a commodity that has been constructed specifically to serve tourists who can participate in guided tours of the location.[26] Accordingly, while Hobbiton's evolution is singular in many ways, it certainly hasn't occurred in a vacuum. Media tourism increasingly complicates its destinations' character such that "the general tourist experience is constituted by the combination of its constructed image, the individual perception of this image, and its subsequent on-site consumption and experience."[27] Hobbiton is but one example from the diverse and changing landscape in which fans take tours, go to filming locations, and explore recreations in themed attractions.

A comparable example is provided by the Harry Potter franchise whereby fans-as-tourists find value in sojourns such as walking or bus tours of London[28] as well as the official destination attraction of the Warner Bros. Studio Tour London – The Making of Harry Potter in Watford, UK.[29] In unofficial tours, hosts or guidebooks share filming locations to an audience that ranges from casual tourists on holiday to super-fans, ready to re-create movie moments while dressed in robes. At the official Making of Harry Potter venue, on the other hand, visitors enter a studio-sized warehouse full of authentic sets, props, and memorabilia from the movies. *Object authenticity* is not always a motivating factor for Potter fans, however. Media tourists have also come to value what Matt Hills calls the *enchantment authenticity* within branded places that have been designed to encourage fan engagement.[30] This has made way for transmedia attractions and transmedia tourism[31] to serve as extensions of the stories that, in addition to being new sources of revenue, provide audiences new ways to engage. As Rachel Marie Gilbert identifies, the Wizarding World of Harry Potter in the Universal Studios theme park (WWoHP) is interpreted as authentic because it was built with detail and care under the purview of author J. K. Rowling, providing the space with a kind of authorial authenticity (Hobbiton can make similar claims based on its ongoing partnership with Jackson).[32]

While Hobbiton has become more commercialized and polished in its nearly 20 years of existence, it has so far resisted elements of immersive entertainment that have enveloped popular franchises such as *Harry Potter* and *Star Wars*. Tucked geographically away from the major sites of entertainment tourism in cities across the globe, Hobbiton troubles the boundary between reality and fantasy through its spatial isolation, its concatenation of primary and secondary worlds, and its artefactual authenticity as a movie set. As evidenced in an early-aughts New Zealand tourism board slogan, Tolkien's previously inaccessible secondary world had suddenly been connected to our primary world through the medium of Jackson's filming location choices: "The movie is fictional. The location isn't. Middle Earth is New Zealand."[33] The movies provided shared images that simultaneously withstood scrutiny by fans of Tolkien's books and established a canon for film audiences encountering Middle-earth for the first time. Of the many *LOTR* and *Hobbit* tourism sites in New Zealand, Hobbiton takes on symbolic and commercial importance precisely because it offers perhaps the best threshold into a particular physical location where reality and imagination are connected through an "object authenticity of place"[34]; in other words, Hobbiton as an object both confirms and reifies its own value through authoritative claims to resemblance and indexicality, and it has done so since visitors first starting arriving uninvited in 2001. Since being reconstructed in 2009, however, the modes by which Hobbiton achieves its object authenticity or pursues other authenticity claims in the minds of visitors have changed considerably.

Hobbiton 1.0 (2002–2009)

The character of the Hobbiton 1.0 tourist visit was unique in its complexity and ambivalence. While Hobbiton's somewhat unfinished character at this time led in some cases to a sense of disappointment on the part of visitors, in other cases it provided productive openings for affect that were only possible because the site allowed – indeed, required – the imaginative participation of those who encountered it. Brandon Zhao, an American visitor in 2007, speaks to this character of the site, evoking at the same time the richness of its transmedia associations. Compared to his previous journeys to theme parks or famous landmarks, for example,

> *[T]he experience [at Hobbiton] was more associative, it evoked my own memories – of the movies, yes – but also of the books, and maybe even who I was, how I felt when I first read the books. That sounds suspiciously like nostalgia. So, rather than take you to a site that has "real" historical significance, this kind of tourism trades on your memory of experiencing a work of fiction. I suppose it works best if that experience was personally historic.*

Hobbiton 1.0 was situated on a continuum, placed somewhere between the poles of unrecognizable and unmistakable, and how visitors encountered it had a great deal to do with what expectations, memories, or associations they brought with them.

Hobbiton's character at this time was less a matter of intent than happenstance. The property's initial development as a filming location started in 1998.[35] Following the completion of principal photography in December 2000, the site was left by New Line Cinema to the Alexander family's devices with the understanding that no visitors would be allowed to tour the location. Visitors to New Zealand, however, began arriving in Matamata and, in some cases, to the Alexander property uninvited, and soon the family and local authorities began efforts to systematize the tour product in consultation with New Line Cinema, the intellectual property holders.

The scene that greeted visitors from roughly 2002–2009 in what we are here calling "Hobbiton 1.0" proceeded quite modestly, with the Alexanders and their small team of locals utilizing little more than handheld set photos and a rich guide script to enhance the tour of the partially demolished location. By the point of initial field work onsite by Peaslee in 2007, some pieces had been restored, but even then, the site was made up solely of a series of plywood frontispieces painted white.[36] While many visitors found the site to be everything for which they had hoped, not all reviews were favourable. As Carl, Kindon, and Smith found in their survey of New Zealand tourists:

> *Respondents with a low satisfaction level named as the main reason the poor condition of the film set remains (particularly in the case of the former Hobbiton film set on the*

Rings Scenic Tours) or just the lack of film set remains, as in the case of the Safari of the Rings tours. . . . It seems that the closer the "real" landscapes approximate the hyper-real landscapes of the films, the more expectations are met or exceeded and the greater the satisfaction of the tourists with their tour experience.[37]

In Hobbiton's case, the affective value of its rudimentary landmarks was greatly enhanced by a series of onsite production stills (negotiated with and provided by New Line shortly after the commencement of tours onsite) that reconciled the visitor's position vis-à-vis both the current and former iterations of the space. Comparison between past and present states, between the stills placed onsite and the physical landscape, mediated by tour guides, underpinned Hobbiton 1.0's authenticity claims, leading visitors to the emotional reactions (joy, nostalgia, surprise, and so on) for which they had presumably paid and securing the product's value (see, e.g. Fig. 13.1). Robert Farrell, a 2007 visitor to Hobbiton from the United Kingdom, is an exemplar here: "the idea of visiting this 'most secret' of movie sets both fascinated and intrigued me. (And also, there was the 'thrill' of walking into a 'real-life Hobbit-Hole' to look forward to)!" Similarly, Marjorie Fenton, who also visited from the UK around the same time, characterized his time at Hobbiton 1.0 as "quite magical."

In addition to its physical appearance, another distinguishing feature of Hobbiton 1.0 was its relative solitude, particularly in comparison to what confronts a visitor in 2021. Prior to 2009, the Hobbiton tour began at the Matamata i-SITE Visitor Information Centre where participants purchased tickets before boarding one of several coaches, each named for an LOTR character, and began the 17km scenic ride through the Hinuera Valley to the village itself.[38] This extended liminal moment provided more than just transportation; it served as an increasingly meaningful journey towards and, ultimately, into a centre of cultural significance.[39] As visitors left their personal vehicles behind in Matamata, boarding the bus could effectively activate a feeling of whimsy, the first in a series of threshold moments that marked stages of closeness to and recognition of the Hobbiton so familiar to many visitors through the films. Each of these boundary moments, compounded by the narrative provided by the accompanying tour guide, leveraged a dynamic interplay of anticipation and revelation, such that tour participants found themselves "discovering" the site even as they were delivered to it. As André Jansson has suggested, "navigation" is an increasingly fundamental part of tourism, reflecting "the need to appropriate space in a cartographic manner."[40] While in 2021 visitors to Hobbiton do indeed approach the site using their personal vehicles and, presumably, some form of GPS, in Hobbiton 1.0, this navigational agency was denied them – all tourists could do was get themselves to the visitor centre, at which point they could no longer control their progress in any direct way. Elizabeth Abrehart, an English visitor to Hobbiton 1.0, touched on the affective value of this pleasurable disorientation:

I also remember wondering how Peter Jackson managed to shoot Lord of the Rings without getting any of "our world" in the frames. As we got onto the farm I saw how.

> *There was definitely a moment on the bus . . . that everything looked so familiar, and I felt the same peace that I feel when I watch the Shire in the films.*

The boundaries that marked the approach to the village included a stop at the Buckland Road entrance to the farmland, at which point the guide would exit the bus to open the first in a series of livestock gates. Marta Quint, a 2007 visitor from Austria, remarked that she "felt a bit excited when arriving at the house of farmer Alexander," and that each of the stops along the way out to the site, though provoking some impatience, also charmed with their "arcane character." From here, buses proceeded approximately 2.5 km into the property, with guides offering commentary on the Alexander family's history and farming practices, one of several places in the guide script where Hobbits and New Zealanders were concatenated on the basis of their shared provinciality, resourcefulness, and stolid common sense, a seemingly unconscious rhetorical manoeuvre that had the added effect of further destabilizing the boundary between primary and secondary worlds onsite. This narrative then dissolved into another outlining the "discovery" of the location and its eventual conversion into Hobbiton for the *LOTR* films. Arriving at an overlook to the northwest of the village, participants exited the bus for their first glimpse of the Party Tree and the lake, anchored by a nearby sign welcoming visitors to the Hobbiton Movie Set. Here the location's *bona fides* was established and visitors were provided the first in a series of photo opportunities actively prescribed by tour guides, each of which played an important role in establishing Hobbiton's authentic reputation when shared with others, as they could only have been captured at the "true" Hobbiton location.

The isolation provided by Hobbiton 1.0 underpinned a productive value discourse of exclusivity, one that celebrated participants for their intrepid character and assured them of the novelty of their experience. While tourism in New Zealand has been a cornerstone of the economy for decades, first surpassing dairy products as the leading export in 2004 and having maintained that position more or less to the present day,[41] the Hobbiton's surrounding Matamata/Piako District has historically been a tourism afterthought. According to Stats NZ, who track international visitors and related tourism metrics, Matamata/Piako saw only 2,211 international visitors in 1997. At the time we first visited Hobbiton in 2005, we were among 15,660 external travellers to the region. While this increase of more than 600 percent is substantial from a local perspective, it represents a very small portion of the total number of international visitors in 2005 to New Zealand (more than 8 million) and pales in comparison to the hundreds of thousands of international tourists to be found each year in places like Auckland, Rotorua, and Wellington.[42]

A typical day in Hobbiton during this time involved a battery of 5–7 tours onsite, each group ranging in size from only a few people to as many as 30–40.[43] In the vast majority of cases observed during our fieldwork pre-2009, only one group at a time was present in the Hobbiton village itself, with occasional and fleeting overlaps at the beginnings and conclusions of tours. This relative isolation allowed visitors to suspend disbelief, "have the place to themselves" for the better part of an hour (albeit with the oversight of a guide), and revel in the overlap of primary and secondary worlds. Guides, meanwhile, were allowed the time to share sometimes quite lengthy and detailed anecdotes. Bonnie Palmer, a Canadian visitor who lived in Hong Kong when she visited Hobbiton in 2007, captures the value of this approach:

> *I found that each spot around the hobbit holes became more alive as [the guide] explained which tree was which and how the distances between certain events was really quite short instead of days of travelling as was portrayed in the movie.*

Groups' movement through the space was controlled but relaxed, and participants were encouraged to linger and reflect, particularly in the moments after reaching Bag End and its commanding view of the location. There were several unmistakably emotional reactions to such denouements: quietly tearful handholding, extended periods of seated contemplation, and joyful recreations of Tolkien narrative scenarios. There was clear value in allowing visitors some time and autonomy to soak in the atmosphere of the site, one which was largely rural, serene, and visually stunning. As Brandon put it,

> *I found the [Hobbiton movie set] quite peaceful, not at all like other tours, which often have a hustle and bustle feel, or like theme parks, where the whole point is to hit as many attractions as possible to maximize the fun factor, to "make" happy memories.*

Three Engagements With Bag End and Hobbiton 2.0 (2010–Current)

In Figures 13.1–13.3, we see three different encounters with what might reasonably be called Hobbiton's most recognizable feature, Bag End. Each of these images constructs, reacts to, and negotiates a boundary. Boundaries are, of course, part and parcel of Couldry's theorizing of the mediated centre, made manifest in places where "key media-related categories and boundaries" reinforce, "indeed [help] legitimate, the underlying 'value' expressed in the idea that the media is our access point to our social centre."[44] Boundaries are also necessary to the kind of "theming" that Jansson characterizes as one form of encapsulation, at once defining the contours a particular theme, drawing the theme together around itself and tracing a circle in the sand, while also calling into being all things that, because they belong outside the theme, might serve to break the magic and decapsulate the themed space.[45] In these images, we see boundaries in both senses that provide a kind of heuristic for seeing how Hobbiton has evolved since 2010, Bag End operating in this case as microcosm.

In Figure 13.1, the boundary is symbolized by the glass p(l)ane that separates the man and the photograph. This boundary is primarily one between present and past, overlapping two iterations in time of Bag-End, the equivalent of a strip of

FIGURE 13.1 Visitors to Hobbiton 1.0 engage with Bag End and the nearby photo-board circa 2006.

film bent to overlap two frames from either end. Here, Bag End (physically present, indexically represented in photography, and hyper-realized in its production state by that same photographic image) is utilized as a background for a visitor photo taken in early 2007, two years prior to the reconstruction process undertaken for the *Hobbit* films. Because at this point Bag End was decapsulated by the removal of its set dressing, it is supplemented by the presence of the large set photo installed just in front of the structure itself.

While the photo demonstrates an attempt by Rings Scenic Tours to re-encapsulate Bag End by reinscribing the structure with object authenticity through visual comparison, the actions of the tourists in the picture suggest that the encapsulation remains incomplete. The man situates himself in such a way as to create the illusion, in his partner's photo, that he's standing before the "real" Bag End as it appeared in the films. The tension between encapsulation and decapsulation permeates this image, since the onsite photo simultaneously establishes its subject's fullness and emptiness on the register of authenticity. The photo suggests – and the visitor behaves in accordance with – the actual structure's representational poverty. But the complexity does not stop there, since the very act of taking the photo of the visitor "in" the photo of Bag End suggests the magical and therefore encapsulated nature of the structure (and, by extension, of Hobbiton) – where else on Earth could one make this photo?

In Figure 13.2, the boundary is the round frame of the door, and its character is not temporal but spatial, and not just spatial but also dimensional, a boundary between real and imagined, between the inside and outside of the fantasy. In this image, the visitor is captured from inside Bag End, which as a set for the *LOTR* films was constructed with enough room "inside" the structure for a small camera crew to film the surrounding landscape from the same point of view. Of all the photos available to visitors to Hobbiton 1.0, this is the money shot – the most iconic and most explicitly encouraged photo, one snapped by tour guides for every visitor to the site during our initial fieldwork. Such a photo suggests a greater degree of encapsulation in Hobbiton 1.0 than is suggested in Figure 13.1, since the tactility and depth of the space is verified by the notion of an inhabitable "inside" to Bag End. Moreover, it suggests that Bag End is occupied, that the tourist in the photo is not just visiting Hobbiton but also visiting *someone*. Finally, aside from a few sheep, the subject of the photo is utterly alone in Hobbiton, a feat that as we will discuss is effectively impossible in Hobbiton 2.0 without intervention. As an iconic image, this one and others like it, composed according to the explicit instructions of the guides ("stand here, scoot over . . . yes, perfect"), circulated the site's central value claim informally among vast networks of potential visitors for years. The image indicates two interconnected subjectivities for the tourist-viewer simultaneously: that of the encapsulated inhabitant of the secondary world (inside) and the intrepid but decapsulated traveller (outside), who simultaneously casts and breaks the spell that underpins the location's value. The tourism image often destabilizes the tourist self, since sharing such an image says, "I was *here*" and "I was a *tourist* here" at exactly the same time; Figure 13.2 is remarkable not for illustrating this tension but for its capacity to illustrate it so clearly. The lack of other people in the background removes a source of immediate decapsulation, which is the sense that "other tourists are there too," a negative value proposition that undermines our own individual good taste.

Figure 13.3, finally, captures tourist activity at Bag End as part of Hobbiton 2.0, circa 2017. The tensions remain in place, but they are configured in quite different ways. Here the fence – our third and final boundary – performs double-duty as both signifier and signified, a product of both the primary and the secondary world. It says "real" at the same time it says "no," signalling the boundary's predominantly spatial nature. Despite the fence's complicity in establishing the visitor's desire to enter, "you shall not pass." Overall, in Figure 13.3 Bag End is more encapsulating than its predecessors in the sense that it more fully replicates the secondary world of Hobbiton as represented in the books and movies.

In addition to featuring the artisanal wood and brick façade, Bag End 2.0's representational richness is supplemented by surrounding elements such as the bench, wheelbarrow, and verdant wildflower garden (evoking the care taken by Samwise Gamgee, the gardener). The fence in front of the visitors appears handmade according to the technologies of the Shire, and although it is obscured in

FIGURE 13.2 Hobbiton 1.0 visitor Natalie Brechtel poses for a photo taken by her tour guide, framed by the iconic round doorway of Bag End.

Source: Photo published with permission of the subject.

FIGURE 13.3 Visitors' crowd around Bag End in a July 2019 tour of Hobbiton.

this photo, the sign attached to the fence ("No admittance except on party business") firmly situates and encapsulates the site not only in the space but also in the timeline of the secondary world – the visitor finds themselves at Bag End hours or even minutes before Gandalf's arrival in *The Fellowship of the Ring*. And yet, the fence and the sign are more than mere props – they enact very real boundaries and direct visitors' movement. While the bench invites one to sit, the stairs invite one to climb, and the open round door (with the tantalizing visual suggestion of Bag End's interior) invite entry, none of these invitations is in good faith. A tactile engagement with these elements remains just out of reach, and one is permitted only to look and photograph, which of course people do. This creates significant congestion on typical days, when groups are deployed into Hobbiton approximately every 15 minutes. Both the photography, much of it undertaken with smartphones and intended for placement on social media profiles, and the congestion engender a decapsulation made all the more jarring by the sight's carefully encapsulated presentation.

Together, the aforementioned comparisons of Bag End, help us understand how the separatedness and liminality that Hobbiton 1.0 provided through its relative separation from the world around it is somewhat diminished today, even as it presents what Jansson describes as a more "encapsulated," themed space.[46] Hobbiton welcomed its one millionth tourist in 2015 and, following this milestone in the years pre-Covid, had increased to nearly 650,000 visitors a year.[47] While the increase in raw numbers is part of Hobbiton's changed character, there are several other factors in play that have been equally, if not more significant to its transformation. This transformation is not at all straightforward, since the very act of more explicitly theming the location invites, paradoxically, the forces of decapsulation.

For example, the village remains accessible only via buses that travel in and out of the location via a gated farm road, but the Hobbiton tour now begins just outside that gate at the Hobbiton Visitor Centre, a full-service facility complete with café, gift shop, toilets, and sophisticated crowd control mechanisms such as waiting areas, group collection points, and continuously updated departure times. As a result, the adventure of visiting Hobbiton in its current iteration unfolds in a much more circumscribed fashion compared to Hobbiton 1.0, with its long and languid approach to and movement through the space. The contemporary visit is compacted. On a typical day, the Visitor Centre is a hive of activity that is characterized by the cacophonic comings and goings of dozens of tour buses, hundreds of rental vehicles, and the constant movement of Hobbiton employees and guests from around the world.[48] The topological difference between Hobbiton 1.0 and 2.0 limits visitors to a more circumscribed geographical space and their movements unfold within a more compressed, more policed amount of time.

In Hobbiton 1.0, the typical visitor trajectory was essentially linear ("there and back again"), moving purposefully from an outermost threshold of Matamata's visitor centre to the innermost centre of value, Bag End. Returning to the visitor centre was not part of the scripted tour taken part in by the guides – most said nothing after this point, aside from thanking participants as the bus approached the disembarkation point. It is safe to characterize Bag End as the symbolic destination of the Hobbiton 1.0 journey.

In Hobbiton 2.0, Bag End is visited about two-thirds of the way through the tour. Since it is no longer permits visitors entry inside, the time standing in front of Bag End is effectively de-accentuated by a combination of guide anecdotes and jostling for pictures, a congestion point along the tourist circuit of egress through the space rather than its climax. One could posit that in Hobbiton 2.0 the Green Dragon Pub, a feature that was invoked only by onsite photoboards in Hobbiton 1.0, is now the destination; it is a large and recognizable element of the films' mise en scene positioned as the end of visitors' path around the property. Beckoning entry, the Green Dragon offers the comfortable ambience of a fully operational, provincial public house, complete with real beer and a cat stretched out, as if on retainer, before the cosy fireplace.

In the evolution from 1.0 to 2.0, Hobbiton's signature moment is thus a good deal more difficult to discern. Prior to the site's renovation, Bag End's primacy onsite derived in part from its position at the narrative centre of the secondary world, but also from its capacity to deliver navigational and tactile pleasures superior to those offered by any surrounding elements. In Hobbiton 2.0, however, the haptics of Bag End (entering, looking out from, and performing within) are essentially gone, transplanted across the lake to the Green Dragon, which also happens to be adjacent to the onsite gift shop.

A final relevant change effected by the magnified hyperreality of Hobbiton 2.0 concerns the role of the tour guides.[49] In Hobbiton 1.0, many guides we interviewed were locals converted to the tourism industry by opportunity and necessity, which invested their delivery with certain kinds of provincial authenticity. As mentioned earlier, Hobbiton 1.0 relied heavily on the successful delivery of intertextual package comprised in equal parts of the space, the "documentary" evidence provided by production stills, and the stories told by tour guides. The latter provided context for many of the objects and settings encountered onsite and were key to establishing and maintaining a sense of value around the tour.

Furthermore, the guides' stories and insider status also made each tourist engagement an interpersonal, collaborative act of reconstruction. Hobbiton 1.0, in a way, could not exist independent of someone there telling its story and someone else actively listening. Hobbiton 2.0, meanwhile, speaks its object authenticity all by itself. As a result, the Hobbiton tour guide's function has atrophied from a completely essential, active co-creator of the space to a position of informational and ambulatory dispatch: "let's keep moving, folks." It is not criticism but rather diagnosis to suggest that guides are less essential for finding Hobbiton 2.0 valuable: if one only imagines having a tour guide lead one through WWoHP or Disneyland, the awkwardness, redundancy, or inappropriateness of such a scenario is immediately appreciated.

Hobbiton has therefore evolved substantially in its second decade beyond the central fact of its reconstruction for both filming and tourism purposes. Gone is the somewhat unclear, tenuous relationship with New Line Cinema and all its intellectual property restrictions. Hobbiton 2.0 is a fully commercialized space steadfastly anchored by the semiotic markers of Peter Jackson's *LOTR/Hobbit* product empire. Hobbiton 2.0's appeals are more immersive in nature, offering previously unimagined four-dimensional tactility anchored by the kinds of food and drink for which hobbits have become known. Hobbiton 2.0 therefore invites visitors to engage the space in way that is at once more playful but also more passive – to move through a fictional space as if it actually existed, but to do so with less effort. All that is missing to make Hobbiton 2.0 a "ride" – a term that suggests a thrilling, immersive passivity – is a conveyance. For now, visitors do still walk through the village, but guides, now focused largely on crowd control at the expense of visitors' leisurely indulgences, make sure the walking is brisk, orderly, and resolutely in one direction.

Hobbiton 3.0?

> *What might be coming? Nothing has been announced. But one can only imagine where Universal Creative will take us. Perhaps we will be able to take a carriage ride through the Shire and visit the home of Bilbo Baggins. Better still, we could soon be wandering the pristine stone and water passageways of Rivendell for a visit with Elrond. Can't you just imagine a spectacular Mordor coaster-style attraction?*
>
> *Eric Davis, The Orlando Parkhopper*[50]

Here, Eric Davis offers a glimpse of a parallel development in themed entertainment, one that allows us to imagine a hypothetical alternative to Hobbiton 2.0 in the 2020s – a *LOTR*-themed "land" replete with rides and attractions. In fact, rumours resurface every few years about Universal Studios possibly adding the franchise to one of its parks.[51] Looking back at the changing approach Universal Studios has taken in its theme parks over half a century, with its slow transition from the primacy of film-industry tourism into story-world tourism, the rumour is credible.

We have already alluded to the fact that commercial activity in Hobbiton 2.0 is greatly enhanced in comparison to its predecessor. While in Hobbiton 1.0 there was a small gift shop at the Visitor Centre (marking the beginning and end of the tour), it was for much of its early history quite limited in terms of its size and availability of merchandise. Also, the implicit or explicit appeals to purchase souvenirs were somewhat modest in Hobbiton 1.0. In 2021, enormous, well-stocked, and relatively expensive gift shops are present at the new Visitor Centre and onsite (inside the physical and psychical boundaries of the Hobbiton village). Visitors can shop in either location, or perhaps in both locations, prior to and after the tour. This also allows groups to shop after the tour and then be shuttled way, rather than being brought back to the Visitor Centre for additional shopping and increasing the potential for unpleasant congestion.

Alongside its commercial imperatives, Hobbiton 2.0 could also be characterized effectively as a "legacy project," a term used recently to describe the *Game of Thrones* Studio Tour at Linen Mill Studios outside Belfast, Northern Ireland.[52] Implicit in the word "legacy" is that there is something historically and culturally significant about the fact that a particular production unfolded in a particular place, suggesting that what is left behind is both a tourist space and a museum of sorts.

This kind of complexity has been parsed out by Sheela Agarwal and Gareth Shaw in their study of heritage, screen, and literary tourism, wherein they note that each of these forms of tourism "involve the portrayal of the past and, in doing so, promulgate nostalgia and backward-looking thinking."[53] Hobbiton 1.0 certainly exhibited some archival qualities – particularly in terms of its use of relics (things left behind by the production, such as a piece of set dressing) and the photo boards, each of which provided a kind of historical documentation that spoke to the site's authenticity or veracity – but its incomplete character complicated matters. In Hobbiton 2.0, and in other "legacy projects" which attempt to archive sets for touristic consumption, the nature of the product (and perhaps the audience) may beg analysis in comparison to the product and audiences of a more traditional theme park.

Is Hobbiton's character today, while more robust in its textual verisimilitude (and therefore a more exact copy of the secondary world), actually in aggregate poorer because, in its hyperreality, it discourages the playful, implicit bleeding together of the primary/secondary worlds on the part of those who enter it?[54] Or is it Hobbiton's tenuous resistance to the whims of synergistic, conglomerated entertainment that give it its authentic aura? As of this writing, Hobbiton has re-opened for tours after its pandemic-induced closure. How it and other, comparable products will change as the result of decreased global air travel, hardening of national boundaries, and other pandemic-related outcomes has yet to be seen, but Hobbiton finds itself in a remarkable position to define its identity, its legacy, and perhaps themed, transmedia tourism writ large, going forward.

Notes

1 Neil Smith, *Uneven Development: Nature, Capital, and the Production of Space,* 3rd ed. (Athens, GA: University of Georgia Press, 2008), 81.
2 Tom Hunt, "Jackson's Hobbiton set to Remain Intact for Tours," *The Dominion Post* (NZ), February 11, 2011, www.pressreader.com/new-zealand/the-dominion-post/20110210/281681136354764.
3 Many of these screen tourism products derive their value from the Jackson films.
4 André Jansson, "A Sense of Tourism: New Media and the Dialectic of Encapsulation/Decapsulation," *Tourist Studies* 7, no. 1 (2007): 9.
5 Robert Moses Peaslee, "One Ring, Many Circles: The Hobbiton Tour Experience and a Spatial Approach to Media Power," *Tourist Studies* 11, no. 1 (2011): 37–53.
6 Regarding media pilgrimages, see Nick Couldry, *Media Rituals: A Critical Approach* (New York: Routledge, 2003) and Roger C. Aden, *Popular Stories and Promised Lands: Fan Cultures and Symbolic Pilgrimages* (Tuscaloosa, AL: Alabama UP, 1999).
7 Jansson, "A Sense of Tourism," 9.
8 André Jansson, "Introduction: In the Juncture of Media Convergence and Tourism – Towards a Research Agenda," in *The Routledge Companion to Media and Tourism*, ed. Maria Månsson et al. (New York, NY: Taylor & Francis Group, 2020), 14.
9 Jansson, "Introduction," 15–17.

10 Maria Månsson et al., "Invited Contribution – the Janus Face of Transmedia Tourism: Towards a Logistical Turn in Media and Tourism Studies," in *The Routledge Companion to Media and Tourism* (New York, NY: Taylor & Francis Group, 2020), 1.
11 Smith, *Uneven Development*, 107.
12 Rumeana Jahangir, "The Hobbit: How England Inspired Tolkien's Middle Earth," *BBC News*, December 7, 2014, www.bbc.com/news/uk-england-29787528.
13 J. R. R. Tolkien, "On Fairy-Stories," in *Tolkien on Fairy-Stories*, ed. Verlyn Flieger and Douglas A. Anderson (London: Harper Collins, 2014), 52.
14 Mark J. P. Wolf, *Building Imaginary Worlds: The Theory and History of Subcreation* (New York: Routledge, 2012), 24.
15 Wolf, *Building Imaginary Worlds*, 32.
16 Wolf, 25–26.
17 Peaslee, "One Ring, Many Circles," 42.
18 Wolf, *Building Imaginary Worlds*, 256.
19 Wolf, 256.
20 Unlike, for instance, the secondary world of Pandora featured in James Cameron's *Avatar* films where there is a clear separation between the fantasy realm and non-diegetic world, there is a closeness between Hobbiton and our primary world. Following the global success of *Avatar* at the box office in 2009, two *Avatar* sequels are currently being produced in Aotearoa New Zealand while two others are in development.
21 Michael Saler, *As If: Modern Enchantment and the Literary Prehistory of Virtual Reality* (New York: Oxford University Press, 2012), 160–161.
22 Mike Foster, "America in the 1960s: Reception of Tolkien," in *J.R.R. Tolkien Encyclopedia: Scholarship and Critical Assessment*, ed. Michael D. C. Drout (New York: Routledge, 2006), 14–15.
23 Tolkien understood Sauron – "the Enemy" – to derive in part from what he thought of as "the Machine," defined in opposition to nature as "all use of external plans or devices (apparatus) instead of development of the inherent inner properties or talents – or even the use of these talents with the corrupted motive of dominating: bulldozing the real world or coercing other wills. J.R.R. Tolkien, Humphrey Carpenter, and Christopher Tolkien, *The Letters of J.R.R. Tolkien* (Boston: Houghton Mifflin, 1981), 145. For further discussion, see Joshua Hren, "Tolkien and the Technocratic Paradigm," *New Blackfriars* 99, no. 1079 (2018): 97–107.
24 César Graña, *Fact and Symbol* (New York: Oxford University Press, 1971), 98.
25 Sue Beeton, *Film-induced Tourism* (Clevedon, UK: Channel View Publications, 2005), 11.
26 Beeton, *Film-induced Tourism*, 10.
27 Stefan Roesch, *The Experiences of Film Location Tourists* (Bristol, UK: Channel View Publications, 2009), 104.
28 Christina Lee, "'Have Magic, Will Travel': Tourism and Harry Potter's United (Magical) Kingdom," *Tourist Studies* 12, no. 1 (April 1, 2012): 52–69.
29 Katherine Larsen, "(Re)Claiming Harry Potter Fan Pilgrimage Sites," in *Playing Harry Potter: Essays and Interviews on Fandom and Performance*, ed. Lisa S. Brenner (Jefferson, NC: McFarland, 2015), 38–54; Matthew Freeman, "Transmedia Attractions: The Case of Warner Bros. Studio Tour – The Making of Harry Potter," in *The Routledge Companion to Transmedia Studies*, ed. Matthew Freeman and Renira Rampazzo Gambarato (New York: Routledge, 2019), 124–130.
30 Matt Hills, "The Enchantment of Visiting Imaginary Worlds and 'Being There': Brand Fandom and the Tertiary World of Media Tourism," in *Revisiting Imaginary Worlds: A Subcreation Studies Anthology*, 1st ed., ed. Mark Wolf (New York: Routledge, 2016), 245.
31 Matthew Freeman, "Transmedia Attractions,' 124–30; Ross Garner, "Finding Nemo's Spaces: Defining and Exploring Transmedia Tourism," *JOMEC Journal* 14 (November 12, 2019): 11–32.
32 Rachel Marie Gilbert, "A Potterhead's Progress: A Quest for Authenticity at the Wizarding World of Harry Potter,' in *Playing Harry Potter: Essays and Interviews on Fandom and Performance*, ed. Lisa S. Brenner (Jefferson, NC: McFarland, 2015), 24–37.

33 Beeton, *Film-induced Tourism*, 88.
34 Peaslee, "One Ring, Many Circles," 44.
35 Robert Moses Peaslee, "'The Man From New Line Knocked on the Door': Tourism, Media Power, and Hobbiton/Matamata as Boundaried Space," *Tourist Studies* 10, no. 1 (April 2010): 57–73.
36 New Line Cinema prohibited any further reconfiguration of their filmed state.
37 Daniela Carl, Sara Kindon, and Karen Smith, "Tourists' Experiences of Film Locations: New Zealand as 'Middle-Earth'," *Tourism Geographies* 9, no. 1 (January 1, 2007): 56–59.
38 While the Matamata i-SITE was the primary embarkation point, Rings Scenic Tours, Ltd. also operated a tour out of the visitor hub of Rotorua, roughly one hour's drive to the southeast. Private tour groups also planned to visit the location using their own buses.
39 Peaslee, "One Ring, Many Circles," 40.
40 Jansson, "A Sense of Tourism," 12.
41 Stats NZ (Tatauranga Aotearoa), "Tourism Satellite Account: The Contribution Made by Tourism to the New Zealand Economy" (Wellington, NZ: Stats NZ Information Center, 2019), 15–16.
42 International visitor data culled from Stats NZ, "International Visitor Survey: Place Visited (TLA), 2014', accessed June 4, 2020, http://nzdotstat.stats.govt.nz/wbos/Index.aspx?DataSetCode=TABLECODE7574.
43 Homogenous private groups were often larger, but at this time were relatively infrequent.
44 Nick Couldry, *Media Rituals: A Critical Approach* (New York: Routledge, 2003), 2.
45 Jansson, "A Sense of Tourism," 9.
46 Jansson, 8.
47 Travel PR News Editors, "Hobbiton Movie Set in New Zealand Welcomed Its One Millionth Visitor," *Travel PR News*, March 26, 2015, https://travelprnews.com/hotbbiton-movie-set-in-new-zealand-welcomed-its-one-millionth-visitor-876787671234567890/travel-press-release/2015/03/26/; Sharnae Hope and Te Aorewa Rolleston, "Tourists shall Now Pass: Waikato's Biggest Players Welcome Border Reopening Dates," *Stuff* (NZ), March 16, 2022, www.stuff.co.nz/waikato-times/news/300542210/tourists-shall-now-pass-waikatos-biggest-players-welcome-border-reopening-dates.
48 Hobbiton's director Russell Alexander requested a cap of 3,500 visitors per day to the site in May of 2019, which could result in an increase from roughly 300,000 to one million visitors per year. This has caused some consternation among local residents and authorities, particularly regarding safety and infrastructure. Hope Sharnae, "Hobbiton Boss Drops Ball," *Waikato Times* (NZ), May 30, 2019, 5.
49 Jean Baudrillard, *Simulacra and Simulation*, trans. Sheila Faria Glaser (Ann Arbor, MI: University of Michigan Press, 1994), 1. Baudrillard here defines the hyperreal as "the generation of models of a real without origin or reality."
50 Eric Davis, "Universal Acquiring Rights to Lord of the Rings Attractions," *MiceChat* (blog), January 3, 2013, www.micechat.com/19049-universal-lord-of-the-rings-universal-park/.
51 Scott Edelman, "Is Universal Building a Lord of the Rings Theme Park?," *SYFY WIRE* (blog), December 14, 2012, www.syfy.com/syfywire/is-universal-building-a-lord-of-the-rings-theme-park; Josh Young, "Universal Orlando Is Quietly Working on A 'Lord of The Rings' Land?," *Theme Park University*, January 2, 2020, https://themeparkuniversity.com/universal/universal-orlando-is-quietly-working-on-a-lord-of-the-rings-land/; Ken Storey, "Oh Lordy, the Universal Orlando 'Lord of the Rings' Rumor Is Back," *Orlando Weekly*, January 14, 2020, www.orlandoweekly.com/Blogs/archives/2020/01/14/oh-lordy-the-universal-orlando-lord-of-the-rings-rumor-is-back.
52 Belfast Telegraph, "Northern Ireland *Game of Thrones* Studio Tour to Open Westeros to the Public in 2020," *Belfast Telegraph*, April 3, 2019, www.belfasttelegraph.co.uk/entertainment/news/northern-ireland-game-of-thrones-studio-tour-to-open-westeros-to-the-public-in-2020-37980026.html.
53 Sheela Agarwal and Gareth Shaw, *Heritage, Screen, and Literary Tourism* (Bristol, UK: Channel View Publications, 2017), 61.
54 Baudrillard, *Simulacra and Simulation*, 1.

Bibliography

Aden, Roger C. *Popular Stories and Promised Lands: Fan Cultures and Symbolic Pilgrimages*. Tuscaloosa, AL: Alabama University Press, 1999.

Agarwal, Sheela, and Gareth Shaw. *Heritage, Screen, and Literary Tourism*. Bristol, UK: Channel View Publications, 2017.

Baudrillard, Jean. *Simulacra and Simulation*. Translated by Sheila Faria Glaser. Ann Arbor, MI: University of Michigan Press, 1994.

Beeton, Sue. *Film-Induced Tourism*. Clevedon, UK: Channel View Publications, 2005.

Belfast Telegraph. "Northern Ireland Game of Thrones Studio Tour to Open Westeros to the Public in 2020." *Belfast Telegraph*, April 3, 2019. www.belfasttelegraph.co.uk/entertainment/news/northern-ireland-game-of-thrones-studio-tour-to-open-westeros-to-the-public-in-2020–37980026.html.

Carl, Daniela, Sara Kindon, and Karen Smith. "Tourists' Experiences of Film Locations: New Zealand as 'Middle-Earth'." *Tourism Geographies* 9, no. 1 (January 2007): 49–63.

Couldry, Nick. *Media Rituals: A Critical Approach*. New York: Routledge, 2003.

Davis, Eric. "Universal Acquiring Rights to Lord of the Rings Attractions." *MiceChat*, January 3, 2013 (blog). www.micechat.com/19049-universal-lord-of-the-rings-universal-park/.

Edelman, Scott. "Is Universal Building a Lord of the Rings Theme Park?" *SYFY Wire* (blog), December 14, 2012. www.syfy.com/syfywire/is-universal-building-a-lord-of-the-rings-theme-park.

Foster, Mike. "America in the 1960s: Reception of Tolkien." In *J. R. R. Tolkien Encyclopedia: Scholarship and Critical Assessment*, edited by Michael D. C. Drout. New York: Routledge, 2006.

Freeman, Matthew. "Transmedia Attractions: The Case of Warner Bros. Studio Tour – The Making of Harry Potter." In *The Routledge Companion to Transmedia Studies*, edited by Matthew Freeman and Renira Rampazzo Gambarato. New York: Routledge, 2019.

Garner, Ross. "Finding Nemo's Spaces: Defining and Exploring Transmedia Tourism." *JOMEC Journal* 14 (November 12, 2019): 11–32.

Gilbert, Rachel Marie. "A Potterhead's Progress: A Quest for Authenticity at the Wizarding World of Harry Potter." In *Playing Harry Potter: Essays and Interviews on Fandom and Performance*, edited by Lisa S. Brenner. Jefferson, NC: McFarland, 2015.

Graña, César. *Fact and Symbol*. New York: Oxford University Press, 1971.

Hills, Matt. "The Enchantment of Visiting Imaginary Worlds and 'Being There': Brand Fandom and the Tertiary World of Media Tourism." In *Revisiting Imaginary Worlds: A Subcreation Studies Anthology*, edited by Mark J. P. Wolf, 1st ed. New York: Routledge, 2016.

Hope, Sharnae, and Te Aorewa Rolleston. "Tourists Shall Now Pass: Waikato's Biggest Players Welcome Border Reopening Dates." *Waikato Times (NZ)*, March 16, 2022. www.stuff.co.nz/waikato-times/news/300542210/tourists-shall-now-pass-waikatos-biggest-players-welcome-border-reopening-dates.

Hren, Joshua. "Tolkien and the Technocratic Paradigm." *New Blackfriars* 99, no. 1079 (2018): 97–107.

Hunt, Tom. "Jackson's Hobbiton Set to Remain Intact for Tours." *The Dominion Post (NZ)*, February 11, 2011. www.pressreader.com/new-zealand/the-dominion-post/20110210/281681136354764.

Jahangir, Rumeana. "The Hobbit: How England Inspired Tolkien's Middle Earth." *BBC News*, December 7, 2014. www.bbc.com/news/uk-england-29787528.

Jansson, André. "A Sense of Tourism: New Media and the Dialectic of Encapsulation/Decapsulation." *Tourist Studies* 7, no. 1 (2007): 5–24.

———. "Introduction: In the Juncture of Media Convergence and Tourism – Towards a Research Agenda." In *The Routledge Companion to Media and Tourism*, edited by Maria Månsson et al. New York, NY: Taylor & Francis Group, 2020.

Larsen, Katherine. "(Re)Claiming Harry Potter Fan Pilgrimage Sites." In *Playing Harry Potter: Essays and Interviews on Fandom and Performance*, edited by Lisa S. Brenner. Jefferson, NC: McFarland, 2015.

Lee, Christina. "'Have Magic, Will Travel': Tourism and Harry Potter's (United) Magical Kingdom." *Tourist Studies* 12, no. 1 (April 1, 2012): 52–69.

Månsson, Maria et al. "Invited Contribution – the Janus Face of Transmedia Tourism: Towards a Logistical Turn in Media and Tourism Studies." In *The Routledge Companion to Media and Tourism*. New York, NY: Taylor & Francis Group, 2020.

Peaslee, Robert Moses. "'The Man from New Line Knocked on the Door': Tourism, Media Power, and Hobbiton/Matamata as Boundaried Space." *Tourist Studies* 10, no. 1 (April 2010): 57–73.

———. "One Ring, Many Circles: The Hobbiton Tour Experience and a Spatial Approach to Media Power." *Tourist Studies* 11, no. 1 (2011): 37–53.

Roesch, Stefan. *The Experiences of Film Location Tourists*. Bristol: Channel View Publications, 2009.

Saler, Michael. *As If: Modern Enchantment and the Literary Prehistory of Virtual Reality*. New York: Oxford University Press, 2012.

Sharnae, Hope. "Hobbiton Owner Admits 'Dropping Ball' on Community Engagement." *Waikato Times (NZ)*, May 29, 2019. www.stuff.co.nz/waikato-times/news/113021976/hobbiton-owner-admits-dropping-ball-on-community-engagement.

Smith, Neil. *Uneven Development: Nature, Capital, and the Production of Space*, 3rd ed. Athens, GA: University of Georgia Press, 2008.

Stats NZ. "International Visitor Survey: Place Visited." *TLA*, 2014. Accessed June 4, 2020. http://nzdotstat.stats.govt.nz/wbos/Index.aspx?DataSetCode=TABLECODE7574.

———. *Tourism Satellite Account: The Contribution Made by Tourism to the New Zealand Economy*, 15–16. Wellington, NZ: Stats NZ Information Center, 2019.

Storey, Ken. "Oh Lordy, the Universal Orlando 'Lord of the Rings' Rumor Is Back." *Orlando Weekly*, January 14, 2020. www.orlandoweekly.com/Blogs/archives/2020/01/14/oh-lordy-the-universal-orlando-lord-of-the-rings-rumor-is-back.

Tolkien, J. R. R. "On Fairy-Stories." In *Tolkien on Fairy-Stories*, edited by Verlyn Flieger and Douglas A. Anderson. London: Harper Collins, 2014.

Tolkien, J. R. R., Humphrey Carpenter, and Christopher Tolkien. *The Letters of J.R.R. Tolkien*. Boston: Houghton Mifflin, 1981.

Travel PR News Editors. "Hobbiton Movie Set in New Zealand Welcomed Its One Millionth Visitor." *Travel PR News*, March 26, 2015. https://travelprnews.com/hotbbiton-movie-set-in-new-zealand-welcomed-its-one-millionth-visitor-876787671234567890/travel-press-release/2015/03/26/.

Wolf, Mark J. P. *Building Imaginary Worlds: The Theory and History of Subcreation*. New York: Routledge, 2012.

Young, Josh. "Universal Orlando Is Quietly Working on A 'Lord of The Rings' Land?" *Theme Park University* (blog), January 2, 2020. https://themeparkuniversity.com/universal/universal-orlando-is-quietly-working-on-a-lord-of-the-rings-land/.

14

SWORDS, SANDALS, AND SELFIES

Videogame-Induced Tourism

Erik Champion

Introduction: Videogame Tourism

Many tourists to real places have already been there, virtually and imaginatively, in video games.[1] Videogame tourism, also known as game-induced tourism, may appear to be a recent development, academic researchers as recently as 2018 declared their paper to be the first on the subject,[2] but the phenomenon has been around since at least 1999.[3]

It is difficult to say exactly why videogame tourism has only been so recently explored as a serious subject of study. Perhaps videogames have been considered too light and insubstantial to be worthy of serious study. Perhaps tourism organizers did not consider videogames could influence people to undertake meaningful tourism experiences, or that they are too insular and don't interact well with other media. Film critics and directors have suggested that compared to film, videogames will never have the artistic depth of films,[4] are too reductive, and lack empathy.[5] Such criticisms would also call into question the value of video games to promote tourism on any deep and meaningful level.

I argue that the aforementioned premises are out of date or have been overplayed. Videogame tourism is not an isolated bubble of limited significance, undertaken by a few teenagers. The phenomenon ranges across generations, countries, and genres. No longer restricted to a skimpy, low-resolution vignette of a caricatured fantastical setting, their power, capacity, and cross-platform capacity to depict and respond to players' decisions, as well as their increasing ability to include personal content have transformed them into international transmedia platforms. Their game engines are used to develop and publish shorts, machinima, and feature films, as well as augmented, mixed, and virtual environments. Game companies are also forming alliances with carmakers now that assisted driving technologies improve,

DOI: 10.4324/9781003327585-14

seating becomes larger (thanks to electric platforms), and information system is increasingly an entertainment medium.

Movie-related game franchises carry the legacy and memory of films, so they influence far beyond the release date and into the hands of newer (and older) generations and games are even influencing the power structure of Hollywood.[6] As an example of their increasing strategic value, Microsoft spent US$2.5 billion for Minecraft in 2014, but nearly US$69 billion in 2022 for the game company Activision-Blizzard.[7] And global movie industry profits exceed those of the film industry.[8]

Polish tourism has been helped by *the Witcher* game series.[9] Italy is reaping the benefits of videogame tourism thanks to the *Assassin's Creed* Series (especially Assassin's Creed: Renaissance).[10] This is not only a standard Western phenomenon, but videogames also influence tourism to Bogotá and La Habana.[11] Gamers in China can buy historically correct fashion for their real self, first worn by their avatar: Tencent's game *Honor of Kings* also features contemporary fashion from famous fashion house Burbury.[12]

One can tour videogames such as kingdoms and planets; they range from phone-based applications to huge, spectacular, and memorable cities and landscapes, stretching across simulated and fictional cultures and eras. They can be personalized, modified, and shared, offering peaceful and less peaceful modes, integrating with social media, played offline, online, in museums, galleries, and as escape rooms.

Videogame tourism is increasingly hybrid, in format and technology, in game genre and audience. The major game engines Unity and Epic Games' Unreal Engine can now publish games by corporations and individuals beyond Windows Apple and Unix (Linux) to webpages, social media platforms, smartphones, and game consoles, to virtual reality headsets or VR rooms (CAVES), and mixed reality systems. Any defining boundaries between games, virtual worlds, and real places have been irretrievably blurred. Some VR games feature real-world recreations,[13] such as New Orleans, with zombies (an extended later version is called the *Tourist Edition*). There are even physical location VR games.[14] You can "tour" medieval fantasy worlds via *Skyrim VR*,[15] or explore alien worlds (*No Man's Sky* on PlayStation VR).[16]

Why Videogames Are Seldom Linked to Tourism

Given the apparent success already of videogame tourism, why is there still resistance and a lack of evaluation data? I suggest at least three misconceptions are holding back the development and research of videogame tourism.

Misconception 1: Games Are Purely Fantastical Forms of Light-Hearted Escape

Many misconceptions have become embedded in the public psyche, they don't need to be fully proven, they only need to sound "right." Unfortunately, and perhaps due to the popular examples of space invaders, hungry ghosts, and moustached plumbers of amazing physical athleticism (but questionable driving skills), video games have become associated with fantastical light entertainment themes and figures capable of being played in their entirety in one sitting (or standing, if an arcade game).

Games have been famously described as having a magic circle; the player must complicitly agree to the "logic" of that magic circle; normal physics and consequences in the outside world are separated from the consequences and "logic" of the game. However, the continual importance of fantasy (thematic imagination), and the early video games' association with teenagers and children, has obscured for non-players the serious and powerful messages and themes that computer games, as participatory interactive media, can convey.

While it is true that gameplay is a form of escapism, escapism lies at the heart of cultural activity itself, the creation of another, better world. The American cultural geographer Yi Fu Tuan has famously declared culture itself is a form of escape, "Seeing what is not there lies at the foundation of all human culture."[17] Games are both cultural artefacts and cultural processes; their underlying mechanics are learnt through trial and error; the player learns to see "what is not there." They are scaffolds rather than scenery, as the player has to learn to navigate, explore, determine, and act through the very act of gameplay.

Games can also run the full gamut of entertainment's affective spectrum, from action through comedy to tragedy.[18] They are not only light-hearted cultural escape machines, but they are also now rich and atmospheric, aesthetic, and affective realms, because increasingly sophisticated players with their increasingly powerful platforms and devices demand these new levels of attraction and challenges.

Designed to challenge, games are not normal software. They are not designed to be easy to learn and master, but easy to learn and *difficult* to master. An unchallenged player will leave the game, so designers must work out how to continually entice, challenge, and reward, but also thwart the increasingly experienced player to keep them as the audience. Computer games are still imaginative, imagination is required to enter the "magic circle" of games, and fantasy can provide a broad explanation of gameplay and goals, but without having to overload the player (and the computer) with unnecessary detail.

That is why video games provide experiential realism (not to be confused with the philosophical position), the experience is not photo-realistically accurate but contains *conceptual fidelity*.[19] Experiential realism does not mean capturing the realism of a real-world scene (with its extra demands on lighting, geometry, and physics), but capturing the experiential quality of the moment so that it is retained and recalled as a memory, a feeling of being in another place. Another example of experiential realism is the Notre Dame model of *Assassin's Creed: Unity*. The designers were not allowed to directly copy some of the interior due to copyright issues, but they used the opportunity to induce a sense of realism. Their aim was to create a feeling of realism and authenticity.

> *Had the cathedral's roof been historically accurate in Unity – i.e., spireless – players would have missed out on the experience of seeing and climbing a monument they recognize.*[20]

As the game designer Jeff Minter writes, open world (sandbox) games such as Bethesda's *Skyrim* and *Oblivion* provide enjoyable landscapes to tour, as intrinsically aesthetically valuable experiences.

> *I was playing Oblivion on the Xbox 360, riding along on my horse on a path overlooking the Imperial City, and it just looked so lovely I felt compelled just to stop, get off my horse and stand there for a while taking in the view. . . . Transforming the environment from just being a space where the action took place to . . . well, the kind of place that you actually want to just stop and gaze out at for a while because it's beautiful. Somehow this new appreciation of the beauty of the game world made it feel more like you'd been there.*[21]

Games can also be dark and toxic, simulating and inspiring dark, dangerous, and difficult tourism While games can be seen as light tourism, there is at least one computer game which is clearly not inspiring tourism in the form of light-hearted escapism. On 26 April 1986, roughly 80 km from Kyev, Ukraine's capital, Chernobyl Nuclear Plant's Reactor No. 4 exploded. How many died and will die from the radiation is still debated, but most articles agree the explosion unleashed 400 times as much radiation as the bomb dropped on Hiroshima. Radiation spread quickly through Ukraine, western Russia, and Belarus, and may take centuries, millennia, or even 20,000 years to return to normal levels. And the giant concrete shell encasing the worst of the radiation will need to be replaced within a century.

Despite these dangers, Chernobyl has been visited by players of the atmospheric and eerie *S.T.A.L.K.E.R.* game since the first game of the series in 2007, even though the game designers explicitly warned players against visiting the radioactive site. B&D argue that S.T.A.L.K.E.R.[22] induced considerable tourism despite dangers[23] such as eating, drinking, wandering into the forest, or disturbing the dust. Others have called visits inspired by the game as dark and toxic tourism.[24]

S.T.A.L.K.E.R.: Shadow of Chernobyl[25] was a highly popular game – one of three in the series, with a new game about to be published (but now postponed due to the war in Ukraine) – in the radioactive wasteland of Chernobyl, with attempts to atmospherically capture the experience of visiting the radioactive site.

Unusually for a game, it was based less on the Tarkovsky film and more on the Russian book *Roadside Picnic*, published in 1972.[26] The game borrowed from both the 1972 book and the 1979 film adaption of that book, but set the result inside Chernobyl's Exclusion Zone. Although labelled a role-playing game, the character survives and shoots rather than develops, but the original game was so difficult the players were forced to observe the environment before fighting.

Given the apocalyptical setting of the original game and its intention to inform rather than inspire visitation, and ongoing revelations as to the long-term dangers of Chernobyl, many might conclude that the Exclusion Zone is free of tourists. But the Exclusion Zone has instead turned out to be a tourist attraction. Far from it, there are bus tours, three-day "illegal tourism" hikes and spontaneous raids by organized subcultures, even the regulated tourism is creating new forms of pollution. Multiple tourists behave badly, not wearing protective clothing, stealing objects, but most dangerously of all, kicking up radioactive dust.[27]

The game tourists have an advantage over the others – "One can venture to say that the game creates new must-see places that are obvious only to players."[28] S.T.A.L.K.E.R's success shows games and game mods can convey the affective power of lethal, monstrous, and forbidden landscapes but it also reveals their indirect educational value. Another feature of games (and virtual reality applications) is that they are very suitable for explaining processes as they are themselves procedural.

Misconception 2: Videogames Are Crude and Limited

The research that goes into video games is often under-appreciated. *Max Payne*[29] featured a highly accurate 3D environmental simile of the New York underground, (and, ironically, tourists featured in *Max Payne 3*). *LA Noire* was a dedicated simulation of LA in the 1930s, based on careful research of the urban planning. Many articles were led to believe that game company Ubisoft, creators of *Assassin's Creed: Unity*, offered to supply their extensive and highly detailed 3D model from of the Notre Dame to engineers and heritage architects so the fire-ravaged cathedral could be accurately rebuilt.[30] Although this appears to have been an exaggeration,[31] the example reveals the impressive detail of modern 3D game assets.

The size, scale, and detail of their landscapes are also evolving at an impressive rate. The 2019 game *Assassin's Creed: Odyssey* has approximately 90 square miles of land and sea, 100 quests, and 28 cities and islands that the player can traverse.[32] This pales in comparison with the science fiction-themed, procedurally driven open world game *No Man's Sky*, which contains 255 galaxies and "18 quintillion planets."[33]

Games can also contain well-researched cultural knowledge. For example, many tourists to Poland have been inspired by the video game *The Witcher*[34] (and the recent Netflix series). As Jaworowicz-Zimny writes, the game contained

> [N]arrative set in a fantasy world resembling medieval Eastern Europe which is deeply rooted in Slavic traditions, customs and mythology. The Polish origins of the franchise caused most dedicated fans to develop a broader interest in Poland, with some even being inspired to visit the country.[35]

Misconception 3: Games Don't Educate or Inform

Recent research has suggested that games are affective,[36] and players can form attachment with the characters, story line, and specific scenes.[37] Following in Varutti's footsteps,[38] just as museum experiences can and perhaps should be affective, games can be as well, their paths are often curated, the player's senses are tempted.

In 2007, a new computer (and console) game hit the shelves. It was an adventure, first person shooter 3D game, with a twist. The protagonist was an assassin, a traveller, and also a mythical figure from the past, controlled by a player from the present. *Assassin's Creed* is a highly successful game series available on computers and consoles, featuring time travel and almost mythical or religious powers granted to a central character in each game version who appears to be primarily an assassin in a past time and culture.

However, the series has evolved and expanded since the very first game, landscape screenshots and videos can now be staged, taken, filtered, and shared and the game can now be explored without violence, in a "Discovery Mode." According to the Ubisoft employees that reporter Keiza MacDonald interviewed,[39] *Assassin's Creed* is a type of museum, providing engaging access to lost places. "We give access to a world that was lost" said Jean Guesdon. "Discovery Tour will allow a lot of our players to revisit this world with their kids, or even their parents."

Touring With Assassin's Creed

The 2017 game, *Assassin's Creed: Origins'* Discovery Tour mode "promises" educational enlightenment, visitors can explore points of interest with an optional audiovisual commentary (Thier, 2018; Walker, 2018); while the next in the series, the 2019 game *Assassin's Creed: Odyssey* (Figure 14.1), has an additional Story Creator Mode so players can add personalized quests.

Both games also feature advanced camera editing facilities, you can change the settings on the camera, move it independently of the avatar, and take videos as well during gameplay which can then be shared with other players. Media can also be shared via the map with other players.

Beyond the polaroid fun of sharing landscape selfies with other players and ancient history voyeurs across the Internet,[40] there is also the prospect of real-world

FIGURE 14.1 Assassin's Creed: Odyssey.

Source: Screenshot by author.

physical and hybrid games. Disregarding the critically panned movie adaption of the game series, *Assassin's Creed* has become a transmedia phenomenon. Game company Ubisoft created escape game VR and virtual tours inside physical exhibitions such as *Assassin's Creed VR – Temple of Anubis*).[41] The potential to link with cultural tourism and heritage organizations is fascinating,[42] given the large variety of scenes, settings, and potential quests from the both the games and player-generated content.

Assassin's Creed has of course historical scene visualizations (that could also be ahistorical, conjectural,[43] or even counterfactual). But *Assassin's Creed* is also *counterfictional*, inspired by fiction, mechanics, embedding in this fantasy game actual historical (or historically inspired and researched) elements. It also has an interesting simulation of time. In its representation of current and past places has hybridity of place where the modern character is caught in an interstitial hybrid place where they have to escape from shadowy antagonists from the past era. *Assassin's Creed* is *chronotopical*: the player travels across time. [44] Van Nuenen argues this is a form of anti-tourism, "the series provides a remarkable amalgam of historical tourism, action-adventure gaming tropes, and a Dan Brown-esque narrative."

We could thus view the touristic potential of the popular game *Assassin's Creed: Odyssey*[45] in at least two major ways, as a previsitation visualization tool in discovery tour mode, and as a metaphor to explore in more general and speculative terms how questing and historical dilemmas and conflicts could be incorporated into both fan tourism and cultural/historical tourism.[46]

World-sprawling action-centric role-playing games (RPGs) such as *Assassin's Creed* have particularly useful features for virtual tourism, in their development of the player's character and significance, they often project or proclaim a *uniqueness of place*, and feature imagined or historically inspired distinctive characters.

Adding Features to Assassin's Creed

In Mafkereseb Bekele's doctoral project, we have developed mixed reality (MR) maps.[47] The player, wearing, say a Microsoft HoloLens, can walk around a museum or gallery or outdoor space, and objects virtually appear (in 2D, 3D, or animated 3D), appearing to be placed in the real world. The player can move or edit them. Using a GIS-formatted data exported into a game engine, we are able to export map data into mixed reality, creating maps that appear in front of or below players, and 3D objects on them that animated and responsive.

I can easily see mixed reality levels or artefacts taken from *Assassin's Creed* and viewed in the real world (Figure 14.2). We also developed collaborative mixed-reality scenarios where two players, each wearing Microsoft HoloLens mixed reality headgear, were given two different views and MR items of the same historical journey. They must communicate with each other to learn what each other sees. Different cultural perspectives of the same basic reality could also be developed.

FIGURE 14.2 Walkable 3D/MR maps with a HoloLens wearer inside a reconstructed 16th-century Dutch ship.

Source: Photo courtesy of Dr Hafizur Rahaman.

FIGURE 14.3 Agathe Limouzy, Toulouse, 2018 student intern project.

Source: Photo by author.

Secondly, intern Agathe Limouzy developed a virtual reality (VR) scenario of a past civilization for two players using a game engine (Figure 14.3). One player wore a typical HTC Vive VR headset, and controllers. The other player looked at a screen and a LEAP camera hidden in their bandana tracked their hands. The game can require two players to control the same virtual body, but they independently control different limbs, or can see different views or objects. Again, this could be developed for games like Assassin Creed.

Thirdly, with Andrew Dekker, we examined whether biofeedback (galvanic skin response and heartbeat) could indirectly control the atmosphere of a horror game. The player's heartbeat and GSR affected the music, the hostility of the non-playing characters (NPCs), their ability to perceive, and also the shaders of the game. It was in effect cinematic biofeedback.

The last project, with Hafizur Rahaman, developed phone-based audio augmented reality (AR). If you point the camera on your phone at say, local flora, the phone will recognize the plant and say its local indigenous name back to you. This is particularly relevant in Australia, where there are over 250 Aboriginal languages, but extrapolating further, phone-based AR could augment the landscape one travels in with suitable scenes, music, or audio, created or spoken in various languages or by various characters.

Conclusion

I hope to have shown that videogames induce tourism that it is a special kind of tourism, and can include affective landscapes (although it is an interesting question as to how associated with a game is the newly re-experienced landscape).

I have suggested that video games can be powerful, richly dramatic, and atmospheric, and they can provide for a level of experiential realism suitable for screen tourists. Mixed and augmented reality, various deployments of VR, and biofeedback could all add to the atmospheric and intangible aspects of landscape, both aesthetically, and as an extra information layer.

For example, in future iterations, *Assassin's Creed* discovery tour may be extended to include more interaction and personalization, perhaps with features from the Story Creator tool. There is also great potential in using the NPC bystanders more effectively to convey information about the place, and for the landscape to be more interactive, or even be augmented with real-time data from the real world. The NPCs in the games could also be used more dramatically in reaction to the gamer's avatar committing good or bad deeds, or to convey more situated aspects of certain places or behaviours.

Videogames have been with us for at least 70 years. Videogame-induced tourism has been around a few decades, so is relatively young and immature. Just as ethnographic study has been undertaken on virtual world visitors, there also needs to be more research on videogame-induced tourism. And research is still unclear on the *specific* affective power of games:

> *[R]esearch has yet to explain how video games actually induce individuals' emotions or actions towards a given destination at the pre-travel stage.*[48]

Notes

1 In this chapter I will treat computer games (Mac, PC, Unix etc.), and video games (arcade games and console games) as interchangeable even though that is not strictly accurate.
2 Louis-Etienne Dubois and Chris Gibbs, "Video Game – Induced Tourism: A New Frontier for Destination Marketers," *Tourism Review* 73, no. 2 (2018), https://doi.org/10.1108/TR-07-2017-0115, https://doi.org/10.1108/TR-07-2017-0115.
3 Carlos Ramírez-Moreno and Dale Leorke, "Promoting Yokosuka through Videogame Tourism: The Shenmue Sacred Spot Guide Map," (2020).
4 Jeremy Reimer, "That time Roger Ebert Said Games Will Never be as Worthy as Movies," *Ars Technica*, April 26, 2020, https://arstechnica.com/gaming/2020/11/5657-2/. NB Ebert walked back a little on his position here: www.rogerebert.com/roger-ebert/okay-kids-play-on-my-lawn.
5 Ravi Sinha, "Steven Spielberg and George Lucas Criticize Current State of Gaming, Praise Kinect-Filled Future," *Gaming Bolt*, April 26, 2022, 2013.
6 Janko Roettgers and Nick Statt, "Game Engines Are Changing Hollywood's Power Structure," ed. Goli Sheikholeslami, *Protocol, Axel Springer SE*, March 24, 2022, www.protocol.com/newsletters/entertainment/game-engines-hollywood.
7 Anshool Deshmukh, "Visualizing the Biggest Gaming Company Acquisitions of All-Time," ed. Jeff Desjardins, *Visualcapitalist.com* (Online news), April 24, 2022, Visualizing the Biggest Gaming Company Acquisitions of All-Time.
8 Wallace Witkowski, "Videogames Are a Bigger Industry Than Movies and North American Sports Combined, Thanks to the Pandemic," *MarketWatch* (Webpage), *MarketWatch*, April 26, 2021, www.marketwatch.com/story/videogames-are-a-bigger-industry-than-sports-and-movies-combined-thanks-to-the-pandemic-11608654990.
9 Aleksandra Jaworowicz-Zimny, "3. The Witcher Novels and Games-inspired Tourism in Poland," in *Contents Tourism and Pop Culture Fandom: Transnational Tourist Experiences*, ed. Yamamura Takayoshi and Seaton Philip (Channel View Publications, 2020).

10 Bruno H. M. Carvalho, Gabriella G. C. Bertozzi, and Cynthia Correa, "Video Games Generating Tourist Demand: Italy and the Assassin's Creed Series," in *Sustainable Tourism Development Futuristic Approaches*, ed. Anukrati Sharma (Oakville, Ontario, Canada; Palm Bay, Florida, USA: Apple Academic Press, 2019).
11 Osvaldo Cleger, "Virtual Flânerie: Gaming and Video Tourism in Bogotá and La Habana," in *New Directions in Flânerie: Global Perspectives for the Twenty-First Century*, ed. Kelly Comfort and Marylaura Papalas (New York, USA: Routledge, 2021).
12 Christina Criddle, "Burberry Designs Skins for Honor of Kings Characters," *BBC, BBC*, April 6, 2021, www.bbc.com/news/technology-56511343.
13 *The Walking Dead: Saints & Sinners – Standard Edition,* (Skydance Interactive LLC, 2020), PS4.
14 David Jagneaux, "E3 2018 Hands-On: Assassin's Creed VR Experience Is a Wireless Maze Featuring a Bow and Arrow," ed. David Jagneaux, *UPLOAD, Online*, April 25, 2022, https://uploadvr.com/preview-e3-2018-assassins-creed-vr-wireless/.
15 *The Elder Scrolls V: Skyrim VR*, v. 3 April, 2018 (Bethesda Game Studios, 2018).
16 *No Man's Sky,* 2020 ed. (SIE/Hello Games, 2020).
17 Yi-Fu Tuan, *Escapism* (Baltimore, MD: Johns Hopkins University Press, 1998), 6.
18 Erik Champion, "Entertaining the Similarities and Distinctions Between Serious Games and Virtual Heritage Projects," *Entertainment Computing* 14 (2016), https://doi.org/10.1016/j.entcom.2015.11.003.
19 "I might believe something exists and is 'real' even if it does not look 'realistically' like anything I know; as long as I believe it may affect me" Erik Champion, "Experiential Realism and Digital Place-Making," *Metaverse Creativity: Building, Performing, Learning and Authorship in Online 3D Worlds* (2016), www.intellectbooks.co.uk/journals/view-journal,id=179/.
20 Simone de Rochefort, "Assassin's Creed Unity can't help rebuild Notre-Dame, and that's OK," *Polygon*, November 24, 2021, www.polygon.com/features/22790314/assassins-creed-unity-notre-dame-restoration-accuracy.
21 Jeff Minter, "No Man's Sky, eh. What a Load of Old Smeg," *No Man's Sky, eh. What a load of old smeg*, http://minotaurproject.co.uk/blog/?p=380.
22 S.T.A.L.K.E.R.: Shadow of Chernobyl.
23 "Chernobyl Tourism: Exclusion Zone Becomes Official Tourist Site Amid Health Warnings," The Washington Post, 2019, accessed April 7, 2022, www.washingtonpost.com/travel/2019/07/12/ukraine-wants-chernobyl-be-tourist-trap-scientists-warn-dont-kick-up-dust/.
24 Ganna Yankovska and Kevin Hannam, "Dark and Toxic Tourism in the Chernobyl Exclusion Zone," *Current issues in Tourism* 17, no. 10 (2014).
25 Acronym from https://stalker.fandom.com/wiki/Stalker.
26 Arkady Strugatsky and Boris Strugatsky, *Roadside Picnic* (Chicago Review Press, 2012).
27 Katie Mettler, "Ukraine Wants Chernobyl to Be a Tourist Trap. But Scientists Warn: Don't Kick Up Dust," *The Washington Post,* July 12, 2019.
28 Magdalena Banaszkiewicz and Anna Duda, "To be a S.T.A.L.K.E.R. On Architecture, Computer Games and Tourist Experience in the Chernobyl Exclusion Zone," in *Tourism Fictions, Simulacra and Virtualities*, ed. Maria Gravari-Barbas, Nelson Graburn, and Jean-Francois Staszak (Routledge Taylor & Francis Group, 2020).
29 *Max Payne* (Remedy Entertainment Oyj, 2001).
30 Keza MacDonald, "Assassin's Creed Creators Pledge €500,000 to Notre Dame," *The Guardian*, no. Games, April 18, 2019, www.theguardian.com/games/2019/apr/17/assassins-creed-creators-pledge-500000-notre-dame-restoration.
31 de Rochefort Assassin's Creed Unity can't help rebuild Notre-Dame, and that's OK.
32 Ian Phillpott-Kenny, "Comparing Skyrim's Map Size to Assassin's Creed Odyssey," *GAMERANT*, March 30, 2020, https://gamerant.com/skyrim-map-size-assassins-creed-odyssey/.

33 Samit Sarkar, "A Brief Tour of a Tiny Corner of No Man's Sky," *Polygon*, April 1, 2016, www.polygon.com/2016/4/1/11340322/no-mans-sky-preview-ps4.
34 *The Witcher* (CD Projekt, 2007).
35 Jaworowicz-Zimny, "3. The Witcher Novels and Games-inspired Tourism in Poland."
36 Federica Pallavicini, Ambra Ferrari, and Fabrizia Mantovani, "Video Games for Well-Being: A Systematic Review on the Application of Computer Games for Cognitive and Emotional Training in the Adult Population," Systematic Review, *Frontiers in Psychology* 9 (7 November 2018, 2018), https://doi.org/10.3389/fpsyg.2018.02127, www.frontiersin.org/article/10.3389/fpsyg.2018.02127.
37 M. Jawad Sajid et al., "Video Gaming a New Face of Inducement Tourism: Main Attractors for Juvenile Gamers," *International Journal for Social Studies* 4 (2018).
38 Marzia Varutti, "Affective Encounters in Museums," in *Heritage Ecologies*, ed. Torgeir Rinke Bangstad and Þóra Pétursdóttir (London, UK; New York, USA: Routledge, 2021).
39 Keiza MacDonald, "We Give Access to a Lost World: Assassin's Creed's New Life as a Virtual Museum," *The Guardian*, March 27, 2018, Education, 2018, www.theguardian.com/games/2018/mar/27/assassins-creeds-origins-discovery-tour-virtual-museum-ancient-egypt-ubisoft.
40 A. Politopoulos et al., "History Is Our Playground": Action and Authenticity in Assassin's Creed: Odyssey," *Advances in Archaeological Practice* 7, no. 3 (2019), https://doi.org/10.1017/aap.2019.30.
41 "How Ubisoft designed and refined its Assassin's Creed VR escape rooms," (Blog), updated October 14, 2019, www.gamasutra.com/view/news/352175/How_Ubisoft_designed_and_refined_its_Assassins_Creed_VR_escape_rooms.php.
42 Ismail Shaheer, "The Nexus of Video Games and Heritage Attractions," *Current Issues in Tourism* 25, no. 9 (May 3, 2022, 2022), https://doi.org/10.1080/13683500.2021.1915254.
43 Michał Mochocki, "Heritage Sites and Video Games: Questions of Authenticity and Immersion," *Games and Culture* 16, no. 8 (2021).
44 Tom Van Nuenen, "Touring the Animus: Assassin's Creed and Chronotopical Movement," *Loading . . .* 10, no. 17 (2017): 23. https://journals.sfu.ca/loading/index.php/loading/article/view/195
45 D. Thier, "Assassin's Creed Origins Discovery Tour May Be The Best Assassin's Creed Ever," *Forbes*, 2018, accessed February 21, 2018, www.forbes.com/sites/davidthier/2018/02/21/assassins-creed-origins-discovery-tour-may-be-the-best-assassins-creed-ever/#4a279ac15d32.
46 Lai Ma, "Video Game as an Immersive Interactive Virtual Interpretation of Historic Heritages, Taking Assassin's Creed as an Example" (Masters diss. Columbia University, 2021), https://doi.org/10.7916/d8-4vnq-d355.
47 Bekele et al., "The Influence of Collaborative and Multi-Modal Mixed Reality: Cultural Learning in Virtual Heritage," *Multimodal Technologies and Interaction* 5, no. 12 (2021).
48 Jianhua Yolanda Dong et al., "How Do Video Games Induce Us to Travel?: Exploring the Drivers, Mechanisms, and Limits of Vide o Game-Induced Tourism," in *Audiovisual Tourism Promotion*, ed. Diego Bonelli and Alfio Leotta (Singapore: Palgrave Macmillan 2021), 153.

Bibliography

Associated Press. "Chernobyl Tourists Bring a New Kind of Pollution to Ground Zero." *Euronews*, March 24, 2019 (YouTube). www.youtube.com/watch?v=jEYQEs2ltOU.

Banaszkiewicz, Magdalena, and Anna Duda. "To Be a S.T.A.L.K.E.R. On Architecture, Computer Games and Tourist Experience in the Chernobyl Exclusion Zone." Chapter 11 In *Tourism Fictions, Simulacra and Virtualities*, edited by Maria Gravari-Barbas, Nelson Graburn, and Jean-Francois Staszak, 197–210. London: Routledge Taylor & Francis Group, 2020.

Bekele, Mafkereseb Kassahun, Erik Champion, David A. McMeekin, and Hafizur Rahaman. "The Influence of Collaborative and Multi-Modal Mixed Reality: Cultural Learning in Virtual Heritage." *Multimodal Technologies and Interaction* 5, no. 12 (2021).

Carvalho, Bruno H. M., Gabriella G. C. Bertozzi, and Cynthia Correa. "Video Games Generating Tourist Demand: Italy and the Assassin's Creed Series." In *Sustainable Tourism Development Futuristic Approaches*, edited by Anukrati Sharma, 305–26. Oakville, Ontario, Canada; Palm Bay, Florida: Apple Academic Press, 2019.

CD Projekt. *The Witcher*. Projekt, CD, 2007.

Champion, Erik. "Entertaining the Similarities and Distinctions Between Serious Games and Virtual Heritage Projects." *Entertainment Computing* 14 (2016): 67–74. https://doi.org/10.1016/j.entcom.2015.11.003.

———. "Experiential Realism and Digital Place-Making." *Metaverse Creativity: Building, Performing, Learning and Authorship in Online 3D Worlds*, 2016. www.intellectbooks.co.uk/journals/view-journal,id=179/.

"Chernobyl Tourism: Exclusion Zone Becomes Official Tourist Site Amid Health Warnings." *The Washington Post*, 2019. Accessed 7 April 2022. www.washingtonpost.com/travel/2019/07/12/ukraine-wants-chernobyl-be-tourist-trap-scientists-warn-dont-kick-up-dust/.

Cleger, Osvaldo. "Virtual Flânerie: Gaming and Video Tourism in Bogotá and La Habana." In *New Directions in Flânerie: Global Perspectives for the Twenty-First Century*, edited by Kelly Comfort and Marylaura Papalas, 196–215. New York: Routledge, 2021.

Criddle, Christina. "Burberry Designs Skins for Honor of Kings Characters." *BBC*, April 6, 2022. www.bbc.com/news/technology-56511343.

de Rochefort, Simone. "Assassin's Creed Unity Can't Help Rebuild Notre-Dame, and That's Ok." *Polygon*, November 24, 2021. www.polygon.com/features/22790314/assassins-creed-unity-notre-dame-restoration-accuracy.

Deshmukh, Anshool. "Visualizing the Biggest Gaming Company Acquisitions of All-Time Jeff Desjardins ed." *Visualcapitalist.com*, April 24, 2022 (Online news).

Dong, Jianhua Yolanda, Louis-Etienne Dubois, Marion Joppe, and Lianne Foti. "How Do Video Games Induce Us to Travel?: Exploring the Drivers, Mechanisms, and Limits of Vide O Game-Induced Tourism." In *Audiovisual Tourism Promotion*, edited by Diego Bonelli and Alfio Leotta, 153–72. Singapore: Palgrave Macmillan, 2021.

Dubois, Louis-Etienne, and Chris Gibbs. "Video Game – Induced Tourism: A New Frontier for Destination Marketers." *Tourism Review* 73, no. 2 (2018): 186–98. https://doi.org/10.1108/TR-07-2017-0115.

The Elder Scrolls V: Skyrim Vr. v. 3. Bethesda Game Studios, April 2018.

Harrell, Ethan, and James Marson. "Apocalypse Today: Visiting Chernobyl, 25 Years Later." *TIME*, April 26, 2011. http://content.time.com/time/health/article/0,8599,2067562,00.html.

"How Ubisoft Designed and Refined Its Assassin's Creed Vr Escape Rooms." October 14, 2019. www.gamasutra.com/view/news/352175/How_Ubisoft_designed_and_refined_its_Assassins_Creed_VR_escape_rooms.php.

Jagneaux, David. "E3 2018 Hands-On: Assassin's Creed Vr Experience Is a Wireless Maze Featuring a Bow and Arrow, David Jagneaux ed." *Upload Online*, April 25, 2018. https://uploadvr.com/preview-e3-2018-assassins-creed-vr-wireless/.

Jaworowicz-Zimny, Aleksandra. "3. The Witcher Novels and Games-Inspired Tourism in Poland." In *Contents Tourism and Pop Culture Fandom: Transnational Tourist Experiences*, edited by Yamamura Takayoshi and Seaton Philip, 46–61. Bristol: Channel View Publications, 2020.

Khan, Gulnaz. "See Photos Taken on Illegal Visits to Chernobyl's Dead Zone," *National Geographic*, October 10, 2017. www.nationalgeographic.com/travel/destinations/europe/ukraine/exclusion-zone-chernobyl-ukraine/.

Ma, Lai. "Video Game as an Immersive Interactive Virtual Interpretation of Historic Heritages, Taking Assassin's Creed as an Example." Masters diss., Columbia University, 2021. https://doi.org/10.7916/d8-4vnq-d355.

MacDonald, Keiza. "We Give Access to a Lost World: Assassin's Creed's New Life as a Virtual Museum." *The Guardian*, March 27, 2018. Accessed June 16, 2020. www.theguardian.com/games/2018/mar/27/assassins-creeds-origins-discovery-tour-virtual-museum-ancient-egypt-ubisoft.

———. "Assassin's Creed Creators Pledge €500,000 to Notre Dame." *The Guardian*, April 18, 2019. Accessed November 5, 2019. www.theguardian.com/games/2019/apr/17/assassins-creed-creators-pledge-500000-notre-dame-restoration.

Max Payne. Oyj, Remedy Entertainment, 2001.

Maza, Cristina. "Are Tourists Safe for Chernobyl? As Danger Drops, Problems Rise." *Newsweek*, May 6, 2019 (Online). Tech & Science. www.newsweek.com/chernobyl-tours-safe-visit-rave-risk-dangerous-1414562.

Mettler, Katie. "Ukraine Wants Chernobyl to Be a Tourist Trap: But Scientists Warn: Don't Kick up Dust." *The Washington Post*, July 12, 2019.

Minter, Jeff. "No Man's Sky, Eh. What a Load of Old Smeg." http://minotaurproject.co.uk/blog/?p=380.

Mochocki, Michał. "Heritage Sites and Video Games: Questions of Authenticity and Immersion." *Games and Culture* 16, no. 8 (2021): 951–77.

Morris, Holly. "The Stalkers: Inside the Bizarre Subculture That Lives to Explore Chernobyl's Dead Zone," *Slate: News & Politics*, September 26, 2014. https://slate.com/news-and-politics/2014/09/the-stalkers-inside-the-youth-subculture-that-explores-chernobyls-dead-zone.html.

No Man's Sky. Games, SIE/Hello, 2020.

Pallavicini, Federica, Ambra Ferrari, and Fabrizia Mantovani. "Video Games for Well-Being: A Systematic Review on the Application of Computer Games for Cognitive and Emotional Training in the Adult Population, [In English]. Systematic Review." *Frontiers in Psychology* 9 (November 7, 2018). https://doi.org/10.3389/fpsyg.2018.02127. www.frontiersin.org/article/10.3389/fpsyg.2018.02127.

Phillpott-Kenny, Ian. "Comparing Skyrim's Map Size to Assassin's Creed Odyssey." *Gamerant*, March 30, 2020. https://gamerant.com/skyrim-map-size-assassins-creed-odyssey/.

Politopoulos, A., A. A. A. Mol, K. H. J. Boom, and C. E. Ariese. "History Is Our Playground: Action and Authenticity in Assassin's Creed: Odyssey." *Advances in Archaeological Practice* 7, no. 3 (2019): 317–23. https://doi.org/10.1017/aap.2019.30.

Ramírez-Moreno, Carlos, and Dale Leorke. "Promoting Yokosuka Through Videogame Tourism: The Shenmue Sacred Spot Guide Map." 38–63, 2020. https://www.researchgate.net/publication/362270144_Promoting_Yokosuka_through_videogame_tourism_The_Shenmue_Sacred_Spot_Guide_Map.

Reimer, Jeremy. "That Time Roger Ebert Said Games Will Never Be as Worthy as Movies." *Ars Technica*, April 26, 2022. https://arstechnica.com/gaming/2020/11/5657-2/.

Roettgers, Janko and Nick Statt, "Game Engines Are Changing Hollywood's Power Structure, Goli Sheikholeslami ed." *Protocol. Axel Springer SE*, March 24, 2022. www.protocol.com/newsletters/entertainment/game-engines-hollywood.

Sajid, M. Jawad, Qingren Cao, Li Xinchun, Manzoor Ali Brohi, and M. Fahad Sajid. "Video Gaming a New Face of Inducement Tourism: Main Attractors for Juvenile Gamers." *International Journal for Social Studies* 4 (2018): 52–56.

Sarkar, Samit. "A Brief Tour of a Tiny Corner of No Man's Sky." *Polygon*, April 1, 2016. www.polygon.com/2016/4/1/11340322/no-mans-sky-preview-ps4.

Shaheer, Ismail. "The Nexus of Video Games and Heritage Attractions." *Current Issues in Tourism* 25, no. 9 (May 5, 2022): 1356–60. https://doi.org/10.1080/13683500.2021.1915254.

Sinha, Ravi. "Steven Spielberg and George Lucas Criticize Current State of Gaming, Praise Kinect-Filled Future." *Gaming Bolt*, April 26, 2022.

Strugatsky, Arkady, and Boris Strugatsky. *Roadside Picnic*. Chicago: Chicago Review Press, 2012.

Tsymbal, Evgenii, and Muireann Maguire. "Tarkovsky and the Strugatskii Brothers: The Prehistory of Stalker." *Science Fiction Film & Television* 8, no. 2 (2015): 255–77. https://doi.org/10.3828/sfftv.2015.17.

Tuan, Yi-Fu. *Escapism*. Baltimore, MD: Johns Hopkins University Press, 1998.

Van Nuenen, Tom. "Touring the Animus: Assassin's Creed and Chronotopical Movement." *Loading . . .* 10, no. 17 (2017). https://journals.sfu.ca/loading/index.php/loading/article/view/195

Varutti, Marzia. "Affective Encounters in Museums." In *Heritage Ecologies*, edited by Torgeir Rinke Bangstad and Þóra Pétursdóttir, 129–44. London and New York: Routledge, 2021.

The Walking Dead: Saints & Sinners – Standard Edition. LLC, Skydance Interactive, 2020. PS4.

Witkowski, Wallace. "Videogames Are a Bigger Industry Than Movies and North American Sports Combined, Thanks to the Pandemic." *MarketWatch*, April 26, 2022 (Webpage). www.marketwatch.com/story/videogames-are-a-bigger-industry-than-sports-and-movies-combined-thanks-to-the-pandemic-11608654990.

Wojdon, Joanna. "14 Gaming the Polish Past." *Public History in Poland* (2021): 92.

Yankovska, Ganna, and Kevin Hannam. "Dark and Toxic Tourism in the Chernobyl Exclusion Zone." *Current issues in Tourism* 17, no. 10 (2014): 929–39.

INDEX